SURVIVAL
IN THE
DOLDRUMS

SURVIVAL IN THE DOLDRUMS

The American Women's Rights Movement, 1945 to the 1960s

Leila J. Rupp and Verta Taylor

New York   Oxford
OXFORD UNIVERSITY PRESS
1987

Oxford University Press

Oxford   New York   Toronto
Delhi   Bombay   Calcutta   Madras   Karachi
Petaling Jaya   Singapore   Hong Kong   Tokyo
Nairobi   Dar es Salaam   Cape Town
Melbourne   Auckland

and associated companies in
Beirut   Berlin   Ibadan   Nicosia

Published by Oxford University Press, Inc.,
200 Madison Avenue, New York, New York 10016

Oxford is a registered trademark of Oxford University Press

Library of Congress Cataloging-in-Publication Data
Rupp, Leila J., 1950–
Survival in the Doldrums.
Bibliography: p.   Includes index.
1. Feminism—United States—History—20th century.
2. Women—United States—History—20th century.
3. Women's rights—United States—History—20th century.
4. United States—Social conditions—1945–
I. Taylor, Verta A.   II. Title.
HQ1420.R86   1987   305.4′2′0973   86–23486
ISBN 0-19-504938-1 (alk. paper)

2 4 6 8 9 7 5 3 1

Printed in the United States of America
on acid-free paper

To the families we come from
and to those who nurture us now

Preface

Feminism in the fifties? The decade following World War II has been, for those interested in the history of the American women's movement, more important for what did not happen than for what did. Presumably these were years of domesticity and conformity for American women, not years of discontent and protest. We began our research wondering what had happened to the American women who had fought for equal rights and full citizenship for women during World War II when they confronted the "feminine mystique," the glorification of traditional femininity, of the 1950s. Knowing what the record had to say about the strong opposition feminists faced in the postwar decades, we did not expect to discover many women carrying on the struggle for women's rights. As the work progressed, however, we found that a great deal was going on. Eventually we came to see the period from 1945 to the mid-1960s as a particular stage of the women's rights movement that survived from the suffrage struggle of the early decades of the century to connect with the resurgent movement of the 1960s. Our research provides evidence, then, of the continuity of the American women's movement.

In talking about our research on feminism in the 1940s, 1950s, and early 1960s to groups of both academic and community feminists, we found they were not surprised there were women who saw themselves as part of a women's movement in these years. Accustomed to media pronouncements of the death of feminism in the 1980s, our audiences could readily understand how the widespread conviction that the women's movement had died in 1920 could come to dominate the public and scholarly view of those decades. For they, no more than we, did not accept the view that the women's movement was by the 1980s a thing of the past. We also began to see that the causes and consequences of feminist survival in the fifties might offer lessons, some of them cautionary, for the women's movement in the 1980s.

For we undertook this work as feminists involved in the women's movement. Quite frankly, if we had not been accustomed to having feminist existence challenged, we might never have questioned the scholarly record and undertaken this work. Nevertheless, throughout the process of developing our research objectives, examining documentary sources, interviewing women, and analyzing the data, we tried—to the extent that it was possible—not to approach our topic with preconceived conclusions that grew out of either a particular theoretical orientation or political perspective. Instead, it was our objective to try to let the sources themselves speak to us about the character of feminist activism in this period, even when what we found did not conform to our expectations as scholars or feminists. We set out to find what was missing from the historical record, but we were also interested in what our research might tell us about the women's movement in different periods of its history.

One of us is a historian and the other is a sociologist. Despite the difficulties of combining historical narrative and a sociological framework, we found our cross-disciplinary collaboration satisfying. Although stages of research, conceptualization, and writing sometimes relied on our different disciplinary skills and experience, we are full coauthors and have, in that spirit, listed our names in alphabetical order.

First and foremost, we would like to thank the women whom we interviewed. Their willingness to take the time to talk with us about their activities in the 1940s and 1950s is greatly appreciated. We are also grateful to the staffs of the various archives that made this work possible, especially the Schlesinger Library at Radcliffe College.

Further, we are grateful to the following institutions for financial support during the course of our research: the National Endowment for the Humanities for a two-year Basic Research Grant (1981–1983); the Radcliffe Research Scholars Program for a fellowship for research at the Schlesinger Library (Rupp, 1979–1980); the Graduate School of Ohio State University for two research grants (Rupp, 1978–1979; Taylor, 1979–1980); the College of Humanities of Ohio State University for a one-quarter research leave (Rupp, 1979–1980); and the College of Humanities and the College of Social and Behavioral Sciences of Ohio State University for a grant-in-aid to support our local interviewing project (1982–1983).

Portions of chapters 2, 3, 5, and 8 appeared in somewhat different form in "The Women's Community in the National Woman's Party, 1945 to the 1960's," published in Signs: Journal of Women in Culture and Society, copyright 1985, the University of Chicago Press; "The Survival of American Feminism: The Women's Movement in the Postwar Period," which was printed in Reshaping America: Society and Institutions, edited by Robert H. Bremner and Gary W. Reichard, copyright 1982, the Ohio State University Press; and "The Women's Movement Since 1960: Structure, Strategies, and New Directions," published in American Choices: Social Dilemmas and Public Policy Since 1960, edited by Robert H. Bremner, Gary W. Reichard, and Richard J. Hopkins, copyright 1986, the Ohio State University Press. We appreciate permission to quote from the Bancroft Library oral histories and from collections housed at the Schlesinger Library, Radcliffe College, the Western Historical Manuscript Collection, and the Princeton University Library. Detlev F. Vagts was kind enough to grant permission to quote from the Mary Beard papers, as well as Elisabeth Burger to quote from the Frieda Miller papers, and we also thank Elizabeth Chittick for permission to quote from the National Woman's Party papers.

We are greatly indebted to a number of scholars who read and commented on the entire manuscript: Cynthia Harrison, Susan M. Hartmann, Joan Huber, Marlene Longenecker, John D. McCarthy, E. L. Quarantelli, Gary Reichard, and Laurel Richardson. In addition, we are grateful to those who read parts of the manuscript: Carl Brauer, William Form, Rebecca Klatch, Frances Kolb, Lauren Krivo, Tahi Mottl, Toby Parcel, and Mayer N. Zald. We would also like to thank a number of scholars who shared their then-unpublished work with us, including Carl Brauer; D'Ann Campbell; Nancy Cott; Madeline Davis, Liz Kennedy, and Avra Michelson; John D'Emilio; Myra Marx Ferree and Frederick Miller; Amelia Fry; Nancy Gabin; J. Craig Jenkins and Craig M. Eckert; Ethel Klein; Cynthia Patterson; and Ruth Meyerowitz.

We were fortunate to have a great deal of able assistance from students and staff members of the Center for Women's Studies, the Department of Sociology, and the Department of History at Ohio State. We would particularly like to thank Judy DiIorio, Sigrid Ehrenburg, Grace Moran, Mary Rhodes, and Susann Rivera, who assisted us in a variety of ways in the early stages of our work; Andrea Friedman, Phyllis Gorman, and Mary Sullivan, who helped in such thankless tasks as footnote checking; Mary Irene Moffitt and Mary Rhodes, who conducted some interviews; Suzanne Hyers, who administered the financial side of our grant from the National Endowment for the Humanities; and Susan Moseley and Cindy Brown, who typed a long manuscript and our endless footnotes quickly and beautifully.

Susan Rabiner has been a most supportive editor. Her enthusiasm for this project, from the moment we met, has meant a great deal to us. We are also appreciative of the work of our copy editor, Linda Grossman. Finally, we would like to thank those to whom we dedicate this book. They have on occasion provided refuge in Arkansas, Michigan, New Jersey, North Carolina, and Washington, D.C., as well as at home in Columbus, and they have given love and support over the years.

Columbus, Ohio   L. J. R.
July 1986   V. T.

A Note on Language

In referring to the offices of organizations we describe, we have retained the terminology used at the time despite our preference for gender-free titles. Thus we refer to women as "chairmen" when that was the term used. It seemed to us historically inaccurate to change this usage.

Contents

SURVIVAL
IN THE
DOLDRUMS

1

Introduction

Well, I'm sure there are people that would just love to think that
when Betty Friedan wrote *The Feminine Mystique* that started the
whole thing except for some ineffectual old ladies that sat in that
vine-covered building and made repeated phone calls to congressional leaders.

CARUTHERS BERGER, 1982

With these words, a National Woman's Party member who had been active
in the fight to include sex discrimination in Title VII of the 1964 Civil Rights
Act expressed her impatience with those who overlook the connections
between the resurgence of the women's movement in the 1960s and the
activities of the "ineffectual old ladies" of the National Woman's Party in
the preceding years. Like other women working for women's rights in the
late 1940s and 1950s, Caruthers Berger saw herself as part of a movement.

In looking at how these women described their efforts during the 1950s
to improve the status of women, we found that many used the term "women's
movement" or its nineteenth-century equivalent, "woman movement," to
refer to their activities. Members of the National Woman's Party were particularly likely to use the term "movement." For example, the chairman of
the Woman's Party wrote to historian Miriam Holden in 1947 to explain
that Holden had been included on the board of the organization because
"your ideas about the Woman Movement and regard for women makes you
worthy."[1] In the same vein, a former Woman's Party staff member, testifying
about the ability of a coworker, indicated that the woman had been "a
distinguished member of the feminist movement for many years."[2] In appealing for help on a project, two Woman's Party members asserted in 1949
that "[a]s this is a vital time in the history of the world and especially in the
history of the Woman Movement, it is important that all understanding
women in this country should help now."[3]

Women in other organizations also used the term "movement" to describe their activities. In 1953, the guiding force behind the Connecticut
Committee for the Equal Rights Amendment called for unified action on
the proposed amendment, since "[a]lready several million women have a
vested interest in the movement."[4] With similar optimism, in 1955 a woman

3

4 SURVIVAL IN THE DOLDRUMS

dedicated to winning recognition for the pioneers of the women's movement ruminated in her diary on the anniversary of the first women's rights convention, which she called "the first group protest against the injustices toward women amounting to tyranny and slavery." It resulted, she wrote, "in an *organized* effort known as the Woman Movement, which has continued to this day to secure a status of equality for women established in 'the law of the land.' "[5]

Women such as these did not confine their use of the term "movement" to correspondence within their organizations or networks. For example, the chairman of the Woman's Party wrote to the American ambassador to the United Nations in 1948 to ask him to support the appointment of one or two Woman's Party members as official advisors on women to the U.S. delegation, insisting that "the Woman Movement is sufficiently important to have its representative in the delegation."[6] And the woman quoted above on the first women's rights convention suggested to Cecil B. DeMille in 1957 that he make a film on the Woman Movement along the lines of "The Ten Commandments."[7]

Examples of such usage are common in the literature and correspondence of women's organizations interested in women's rights in the late 1940s and 1950s. Yet scholars by and large have failed to take note of the existence of an organized women's rights movement in this period. As Charles Tilly suggests in a discussion of group mobilization, it "takes confidence, even arrogance, to override a group's own vision of its interests in life."[8]

What, then, might it mean to consider these women as having been part of a social movement? For these women saw themselves not as creating a movement but as maintaining one. Women's rights supporters of the 1950s linked their activism to suffrage activities in the early years of the century. Many had participated in the last victorious years of the suffrage struggle, and they often recalled the glory of suffrage days. Florence Kitchelt, for example, the leader of the Connecticut Committee for the ERA, asked a fellow suffragist in 1950 whether she ever felt, "as I do, that the 'modern' woman is missing something very thrilling, uplifting as well as unifying, in not being able to take part in a suffrage campaign? Those were the days!"[9] National Woman's Party member Katharine Hepburn, mother of the actress, compared the ERA struggle to the fight for suffrage in the 1940s. "That whole period in my life I remember with the greatest delight," she said. "We had no doubts. Life was a great thrill from morning until night."[10]

Indeed, for those former militant suffragists who had picketed the White House and endured imprisonment and the brutality of forced feeding, the day in 1920 when Alice Paul unfurled the National Woman's Party banner from the balcony of their headquarters to celebrate the long-awaited ratification of the suffrage amendment would have no equal. The euphoria of that day was captured in a poem by feminist writer Charlotte Perkins Gilman, published in the Woman's Party *Suffragist*:

"See now the marching millions, row on row,
With steady eyes and faces all aglow,
They come! they come! a glad triumphant band,—
Roses and laurels in their pathway strow—
Women are free at last in all the land!"[11]

The Woman's Party had hoped to film for the public record a ceremonious signing into law of the Nineteenth Amendment, but the Secretary of State issued the proclamation from his home, unobserved by any of the triumphant suffragists. The *New York Times* reported that differences between the militant suffragists and the mainstream suffragists of the National American Woman Suffrage Association as to who should be present at the signing could not be ironed out, so the Secretary of State had decided to issue the proclamation privately in order to avoid a clash at his office.[12] It did not bode well for the post-suffrage years.

Yet the split between the militants and the mainstream suffragists, although it continued to plague the women's movement, did not lead to its death. Alice Paul, the mastermind of the militant suffrage struggle, turned her thoughts to the next step in the fight for women's rights, throwing her weight behind a proposal for a constitutional amendment guaranteeing equality, which she pushed through a conference called by the National Woman's Party in 1921. Some of her contemporaries denounced her dictatorial control of the conference, and the decision to organize around the Equal Rights Amendment (ERA) alienated a wide range of former suffragists, from members of the newly formed League of Women Voters to socialists like Crystal Eastman. Yet it is significant that the Woman's Party invited a range of women's organizations to the 1921 conference and charted its program in the context of an existing, if rapidly fragmenting, women's movement.[13]

In anticipation of the suffrage victory, the National American Woman Suffrage Association had transformed itself into the nonpartisan League of Women Voters, dedicated to educating women for their new citizenship responsibilities and advocating a broad range of reforms. Another new organization, the National Federation of Business and Professional Women's Clubs, had grown out of war work among business and professional women. It, like the League of Women Voters, supported a range of goals designed to appeal to the newly enfranchised electorate. These two large groups, joined by eight other women's organizations, established the Women's Joint Congressional Committee in 1920 to serve as a clearinghouse for lobbying on such issues as independent citizenship for women, maternity legislation, and a child labor amendment. This coalition, in conjunction with the Women's Bureau of the Department of Labor (also a wartime creation) and the National Women's Trade Union League, targeted for defeat the newly proposed Equal Rights Amendment because it feared it would eliminate the protective labor legislation it favored.

Despite the opposition of these reform-oriented women's organizations, the National Woman's Party worked persistently on behalf of the ERA in the years after 1921. In the face of increasing hostility between the two camps, cooperation developed on only a few issues: jury service for women, equal nationality rights for women, and the struggle against legislation forbidding the employment of married women during the Great Depression. In addition to work for the ERA, the Woman's Party fought to repeal discriminatory legislation at the state level, supported women candidates for Congress and women appointees to federal positions, and worked to build an international women's movement. Alice Paul, in fact, directed a great deal of her attention to international affairs from the mid-twenties until the outbreak of World War II. Her work culminated in the founding of the World Woman's Party in 1938. But the war forced her attention back to the United States, and in 1942 she took over the national chairmanship of the Woman's Party for the first time since suffrage days. Our story begins shortly thereafter.

We focus here on the years between 1945 and the resurgence of the women's movement in the 1960s because this period seemed to us crucial in understanding the fate of the women's movement. The traditional view of the women's movement holds that it died in 1920 and was resurrected in the 1960s. Yet we knew that feminist activity had continued throughout the interwar years, and we were curious about the fate of women who had, during World War II, worked for what they termed "full citizenship" for women. Although we suspected that these women had not been converted or overwhelmed by "the feminine mystique" in the 1950s, we began our research by asking what had happened to supporters of women's rights in the postwar period, an era that followed on the heels of a war that had brought women into areas previously reserved for men and preceded the emergence of a large visible women's movement, yet is traditionally characterized as a period of extreme domesticity for American women. Feminism and feminist activism have little place in either the popular image or the scholarly record of women's lives in the years after World War II.[14] To the contrary, in terms of women's movement activity, this period has been characterized as "the bleak and lonely years."[15]

Compared with the 1960s and early 1970s, when social movements and social protest flourished in American society, the 1950s appear to be a period of conformity and contentment in which there was little significant social movement activity of any kind, feminist or otherwise. Yet social movement scholars have recently begun to trace the roots of both the civil rights and gay rights movements to the early 1950s.[16] This new research provides evidence that movements as broad-based, well-organized, and sustained in their efforts as those of the 1960s and 1970s did not spring fullblown out of nowhere anymore than did the injustices which they attempted to correct; instead, they had complex and inconspicuous origins. We set out to ask whether women were really as satisfied and contented with the limited roles society prescribed in the 1950s as we have been led to believe or whether there were some, even if not large numbers, who challenged the cultural prescriptions.

We did not, of course, discover a vibrant and massive women's rights movement in the fifties. On the contrary, the women active on behalf of women's rights were relatively few in number, mostly survivors of the suffrage struggle who maintained their commitment in a period inhospitable to feminism. In 1955, Betty Friedan, author of *The Feminine Mystique* and first president of the National Organization for Women, was thirty-four and on the verge of discovering women's inequality in American society for herself. Gloria Steinem, media-proclaimed feminist spokesperson of the 1970s and editor of *Ms*. magazine, was twenty-one years old and no doubt unaware of the persistent efforts of women's rights activists. Neither was part of the women's rights movement of the fifties. Yet both played leading roles in the liberal branch of the women's movement in the 1960s and 1970s. Why? How did it come to pass that the women active in the resurgent women's movement were such a different group from those feminists of the 1940s and 1950s?

The answer emerges as the story of the women's rights movement in the fifties unfolds. Despite the lack of contact between most women who came to represent the liberal branch of the women's movement in the seventies and the women's rights activists of the fifties, we believe that this period can best be understood as a link in the chain stretching from the early women's rights movement of the 1840s to the women's movement of the 1980s. There are two overlapping groups of readers—historians and feminists—who may find their assumptions about the women's movement challenged by this argument. Both groups, for different reasons, may question the boundaries we have drawn in exploring the women's rights movement.

Historians, on the one hand, may assume that there is nothing about the groups and networks of women we describe in the 1950s that sufficiently differentiates them from other groups of women to warrant considering them a women's rights movement. This research focuses on the organizations and individuals who saw themselves as the heirs of the suffrage movement, worked on activities to promote what traditionally had been defined as "women's rights," and identified with feminism. By focusing on these groups and individuals, we by no means intend to suggest that these were the only women involved in social movement activity or even the only women working to improve the lives of women. In the 1940s and 1950s, there were women involved in the emergent civil rights movement, the peace movement, the Communist Party, and the birth control movement, to name only a few. But these other activists distinguished themselves from, and often worked against, women's rights activists.

Feminists, on the other hand, may be reluctant to claim the groups and individuals we describe here because the women's rights movement in this period was not a broadly based, grass-roots, progressive movement. In order to emphasize the unique composition and style of the women's rights movement in this period, we coin the term "elite-sustained" to describe it.[17] Both feminist activists and scholars may be uncomfortable with the label "elite-sustained," for it admits outright what scholars of social movements have long known but been reluctant to state—that many movements start and continue as minorities, even though they can act as harbingers of significant social and political change. In the contemporary American women's movement, women of color, working-class women, and Jewish women have criticized the racial, class, and cultural biases associated with the limited goals that have traditionally been defined by the interests of white, middle-class women as inappropriate for a movement focusing on gender inequality. We are entirely in sympathy with this critique, but this has not led us to broaden our definition of the women's rights movement in the 1950s. We believe that it is essential to portray as accurately as possible the history of a relatively encapsulated stage of the movement in order to understand better what could become of the women's movement in the future. It is no coincidence that the movement in the doldrums defined women's rights in such a limited way. Suffice it to say that the portrait of the women's rights movement in the fifties that has emerged from our research is not one that conforms to contemporary feminist expectations.

The Approach

Scholars of social movements have, in recent years, shifted from an emphasis on studying discontents, grievances, and psychological dispositions as explanations for the rise and decline of social movements to studying the structure of social movement organizations, their position in the political system, and the factors that influence their success or failure.[18] This newer work has been loosely grouped together under the term "resource mobilization theory" because of the emphasis it places on the role of resources—money, expertise, access to publicity, and the support of influential groups and individuals outside the movement—in determining the nature, course, and outcome of a movement.

Among the major contributions of the resource mobilization approach is its emphasis on the dynamic nature of social movements. For a long time, the most interesting question for sociologists of social movements has been why social movements emerge. When new groups come into being that challenge the status quo by attempting to alter accepted relations of power, the suddenness and the seeming unexpectedness of their claims compel us to take note. Intent on understanding how it is that such departures from the ordinary come about, sociologists have tended to be more interested in studying movements in periods of emergence and growth than in cycles of continuance and decline. By focusing on beginnings, sociologists have overlooked the historical continuity of certain social movements based on long-standing and fundamental social cleavages in American society. A more dynamic view of social movements helps to illuminate social movement continuity by recognizing that the same general social movement might adopt different structural forms and strategies at various periods in its history, depending upon whether it is in a stage of formation, success, continuance and survival, or decline.[19] Scholars have, in fact, found this to be the case by applying the concepts of resource mobilization theory to understand the evolution of the American civil rights movement.[20]

This book follows the latter tradition in exploring feminist activism in the doldrums. We hope to contribute to an understanding of the historical continuity of the American women's movement by analyzing the activities of groups and individuals that promoted women's rights in a period of continuance and decline. The women's rights movement that survived in 1945 was largely, although not exclusively, a remnant of the early twentieth-century suffrage movement. In conjunction with other preexisting and newly formed groups, the National Woman's Party fashioned a kind of movement able to perpetuate feminist ideals and aims in the unfavorable atmosphere of the postwar years. Although feminists in the fifties did little to expand the base of support for women's rights, they did create an environment in which committed feminists pursued their goals and found a community of like-minded women. At the same time, feminist survival had its costs. To characterize this cycle of the women's rights movement, we introduce the concept of an elite-sustained movement. We intend this book as a contribution, then, to both the history of the women's movement and to the sociology of social movements.

Data Sources

This book is based primarily on research in archival collections and interviews with women active in the period. We approached our archival research by exploring both the beginning and end of this stage of activity. We began with the National Woman's Party papers and the League of Women Voters papers because these were the successor organizations to the two major factions of the suffrage movement. Next we analyzed the records of the President's Commission on the Status of Women (1961–1963) because feminist scholars had identified the activities of this group as significant in precipitating the rise of the feminist movement of the late 1960s.

We then turned to the papers of other organizations and the personal collections of individuals active on behalf of women's rights. Most of these were housed either at the Schlesinger Library at Radcliffe College or the Library of Congress. The diaries, autobiographies, personal correspondence, and memoranda of individual women were important sources, and sometimes the only sources, of information about the activities of women's groups other than the National Woman's Party.

As is so often the case with historical work, the extent to which we have been able to reconstruct the contours and activities of the women's rights movement has been affected by the availability of written records of organizations and individuals. The National Woman's Party played a central role in women's rights activism, but we also know a great deal more about the Woman's Party than other groups because the leadership maintained a complete record of correspondence and official documents that is now available on microfilm. In contrast, the archives of the National Federation of Business and Professional Women's Clubs for this period were not accessible to researchers. Although there is inevitably more detail about the Woman's Party than other organizations in our account, we were able to gather a great deal of information about a variety of groups and networks in the papers of individual participants.

Our second major source of data was open-ended semistructured, tape-recorded interviews with twelve leaders and key members of the most central groups, and with selected informants who provided a general overview of feminist activism in the period. Where available, we also made use of transcribed interviews with important individuals conducted by other researchers. In addition, we conducted thirty-three interviews with women involved in a variety of women's organizations in Columbus, Ohio, in order to determine the extent of feminist activity at the local level and to see whether or not the efforts of local activists were linked with those of national leaders. We undertook all of these interviews not only to obtain information missing in the documentary sources, but also to check the validity of our sources and our interpretations. Without the participation of the women we interviewed, much of what went on would have remained unrecorded.

Outline of Chapters

This book analyzes the survival of an American women's rights movement from the end of World War II to the resurgence of the women's movement in the mid-1960s. We begin Chapter 2 by setting out the social context of women's rights activism. We contrast major social trends affecting women's lives—increasing labor force participation and the Baby Boom—with the cultural ideal of the Happy Housewife, and consider expressions of women's discontent in the context of the social climate of antifeminism. In Chapter 3, we focus on the National Woman's Party, the major group that survived the suffrage struggle and spearheaded the women's rights movement in the post-1945 period. We expand our coverage in Chapter 4 to a range of organizations, individuals, and networks that joined the National Woman's Party to promote equal rights. Chapter 5 considers commitment and community, two phenomena that were vital to feminist survival in this period. In Chapter 6 we explore the lives of the individuals who sustained the struggle for women's rights by sketching portraits of seven feminists.

The relationship of the women's rights movement to other major social movements of the time—McCarthyism, the labor movement, and the civil rights movement—is the focus of Chapter 7. Chapter 8 details the connections that this stage of feminist activism made with the early manifestations of the larger and more visible women's movement that began to grow in the mid-1960s. Finally, Chapter 9 concludes by considering the movement's impact, applying insights from social movement theory to analyze the concept of an elite-sustained movement, and exploring what it means to suggest that the American women's movement has a continuous existence.

2

The Social Context of
Women's Rights Activism

Women active on behalf of women's rights in the post-1945 period perceived the social environment as an inhospitable one. In November 1945, for example, one National Woman's Party member wrote to the group's leader about the forces directed at pushing women back into the home at the end of the war. She herself thought there was nothing wrong with women war workers going back to their kitchens if it was what they wanted and was not the result of "an underground movement put forward by the N[ational] A[ssociation of] M[anufacturers], etc."[1] Another woman in 1946 wondered if "those who were living at the beginning of the last Dark Ages . . . knew the darkness had descended!"[2] Yet another remarked the same year that the "position of women is certainly deteriorating."[3] These feminists feared that the postwar years would be repressive for women and inhospitable to women's rights activism.

As it turned out, they were right. Although scholars such as Jo Freeman, Joan Huber and Glenna Spitze, and Ethel Klein suggest that large-scale social trends in the 1950s were affecting women's lives in fundamental ways that prepared the ground for the resurgence of the women's movement in the 1960s, feminists were unable to mobilize large numbers of women to their cause.[4] To understand feminist activism in the fifties, we must have a sense of the social environment in which they struggled and the changes in women's lives.

Increased labor force participation is one of the factors cited in all explanations of the resurgence of the women's movement in the 1960s. After a dramatic drop at the end of World War II, the rate of women's participation in the labor force continued to grow in the postwar years. This was largely the result of certain changes in the occupational structure, specifically increases in organizational size and a rise in the service occupations, that brought about greater job opportunities and massive labor shortages in areas that had come to be defined as "female occupations," such as clerical and secretarial work, teaching, and sales.[5]

TABLE 2.1. Women in the Labor Force, 1940–1965

Year	Total (1,000)	Labor Force Participation Rate	
		Total	Married
1940	13,840	27.4%	16.7%
1944	18,449	35.0	25.6
1947	16,323	29.8	21.4
1950	17,795	31.4	24.8
1955	20,154	33.5	29.4
1960	22,516	34.8	31.7
1965	25,952	36.7	35.7

Source: U.S. Bureau of the Census 1983b, p. 383. Adopted from Ferree and Hess, *Controversy and Coalition*, p. 4.

Furthermore, the female labor force was increasingly composed of older and married women. Young single women were in short supply because of the low birthrate of the 1930s, the increase in numbers of eighteen- to nineteen-year-old women enrolled in school, the decline in the age of marriage, and a rise in the birthrate immediately after the war.[6] By the 1950s, employers, however reluctant, were more likely to hire married and older women. Thus not only did the labor force participation rate for women increase from 27 percent in 1940 to 31 percent in 1950 to 35 percent by 1960, but the comparable rate for married women rose from 17 percent in 1940 to 25 percent in 1950 to 32 percent by 1960.[7]

Other factors cited in explanations of the resurgence of the women's movement include declining fertility (the birthrate dropped after 1957, when the postwar Baby Boom leveled out), rising divorce rates, and rising levels of education.[8] What we see in the post-1945 period is a social environment in which the preconditions for the development of a large and broad-based women's movement were beginning to fall into place. Yet the public image of women's lives denied the reality of these changes. Eugenia Kaledin captures the essence of this contradiction when she states, in her history of American women in the fifties, that this became the decade "when many women first confronted the reality that American society made no concessions to their problems as wives or mothers or divorcees or underpaid workers."[9]

The Paradox of Woman's Place

Nothing makes clearer the role strain and inconsistency women faced in the post-1945 years than the contradictions between the reality of their lives and the public discourse that framed only one picture of the perfect woman, the wife who put home and family first. World War II had brought millions of women into the labor force for the first time, moved millions from their old jobs into ones previously reserved for men, and opened up the military to women. The end of the war brought a concerted effort to push women out of their new jobs into lower paying, less secure ones in order to make room for returning veterans.[10] One of the weapons in the battle to return to prewar ways—less direct than the immediate postwar layoffs of women in industry—was the intensification of propaganda designed to sell women their "place." The "Happy Housewife" of the 1950s is an image so familiar it need not be belabored: the smiling, pretty, suburban matron, devoted mother of three, loyal wife, good housekeeper, excellent cook. Like all American social imagery, especially in the conformist 1950s, it is an image determinedly white and middle class, in total disregard of the diversity of American women and thus a typical product of the "snowblindness" and middle-class myopia of American society.[11]

Betty Friedan presented the feminine mystique as a new phenomenon created in the postwar period when in fact it was simply a postwar version of the traditional ideal, the successor to nineteenth-century "True Womanhood."[12] The assumptions behind the traditional image of women are fairly simple—women and men differ in fundamental ways, since biological differences have profound social and psychological consequences; therefore "woman's place" in society centers on the home and family and the complementary roles of wife and mother. Women's magazines, along with television, movies, and advertising, sold the ideology of "woman's place" and thus provide a clear picture of the cultural ideal set up for women.[13] A *Good Housekeeping* article, "Most Likely to Succeed," summed up the message in typical fashion: success for women meant marriage and family, not brilliance, so the college woman with a C average was in fact "more likely to succeed" than the extraordinarily intelligent woman so voted by her classmates.[14] Put another way in *Ladies' Home Journal*, a woman loved and needed by a man would become "the fragile, feminine, dependent, but priceless creature every man wants his wife to be." Using revealing commercial metaphors, the author noted that there is "actual cash value" in being kind to women, since a happy wife is an "asset" and the proper actions will pay "immediate and lasting dividends."[15] Numerous articles lauded traditional domesticity in an attempt to glorify the housewife's role. From "I'd Hate to be a Man" to "I'm Lucky! Lucky!," the magazines told women that, despite what their own experiences might tell them about life in a patriarchal society, it was better to be a woman.[16]

The concept of "woman's place" found legitimacy not only through its long tradition, but also, in the 1950s, in the developing intellectual school of functionalism, which gave added, scholarly weight to the traditional division of labor. The functionalist perspective, in attempting to apply the objectivity of the natural sciences to the social sciences, seeks to analyze social behavior or institutions in terms of the consequences or functions they have for maintaining the larger society or social system in which they exist. But functionalism, as it came to be applied to an analysis of women's roles in this period, was used by some conservative scholars to pronounce the status quo the best of all possible worlds. In the popular application, functionalism came to mean that society required the traditional division of labor between women and men; society functioned best with women in the home caring for the children while men earned the living.[17]

The feminine mystique, then, was simply the traditional ideal dressed in fifties garb. The irony was that conformance to the ideal, which had never been possible for poor women or, in fact, for any women in the labor force, was out of the question for a growing number of white middle-class women as the decade progressed.

The post-1945 period saw a continuation and acceleration of the long-term trend of increased employment of women outside the home. Recognizing the irreversibility of this change on the very eve of the decade of the 1950s, the director of the Women's Bureau speculated that the United States "was approaching a period when for women to work is an act of conformity."[18] What rising labor force participation and the changing composition of the female labor force meant, translated into the lives of individual women, was that more and more women returned to work after or remained at work while raising families, even during the 1950s when the twin roles of homemaking and motherhood represented the core of women's social identity. Even married middle-class women worked outside the home. In fact, the improved standard of living for the middle class depended on the employment of women, and many families only attained middle-class status through the earnings of the woman.[19] As employment outside the home increasingly became more commonplace for white middle-class women, the justification tended to be financial need—not just for bread alone, but for a second car, a house in the suburbs, a college education for the children.

But because more and more women went to work outside the home and contributed their earnings to the family income did not mean that public attitudes had changed. Although the public had rather readily accepted the need for female labor during the war, postwar opinion continued to disapprove of the employment of married women. In 1945, only 18 percent of a Gallup Poll sample approved of a married woman's working if she had a husband capable of supporting her. Attitudes did begin to change—by 1967, 44 percent approved when asked the same question—but it is clear that no wholesale reversal of traditional notions had occurred as a result of the war.[20]

The increased participation of middle-class women in the labor force came at a time when these women's reproductive roles also expanded. The postwar Baby Boom resulted from women marrying younger and bearing more children. Fewer married women remained childless—only 6.8 percent in the 1950s compared to 14 percent in the 1900s—and the number of women bearing second and third children increased enough to raise the birthrate from 86 offspring per 1,000 women of childbearing age in 1945 to 123 in 1957.[21] Women began their childbearing earlier and lived an increasing number of years after their children had grown. Many women returned to work after completing their childrearing responsibilities, but even mothers of young children were moving into the labor force. In 1950, 12 percent of women with children under six were employed, and this figure increased to 19 percent by 1960. In 1955, 35 percent of women with children six years and older worked.[22] Employed mothers of young children still tended to come more frequently from lower-income families, but clearly the domestic ideal fit with the daily reality of ever fewer women. While the traditional picture of middle-class women isolated in the home while their children were young is not wholly inaccurate, more and more women began to work during the years of their children's growth. Perhaps as a result, a 1947 *Ladies' Home Journal* survey showed that twice as many people believed that women's lives were harder than men's.[23]

Accompanying the Baby Boom was a renewed emphasis on family life. Suburbanization underscored the existence of the nuclear family as a discrete unit by eliminating old extended family and neighborhood ties, often leading to isolation for women who did not work outside the home.[24] Women's magazines pushed togetherness for the suburban family, and a "do-it-yourself" craze encouraged leisure-time pursuits within the four walls of home. Critics denounced suburbia as a matriarchy but, despite a daytime world of women and children, the family, husband in command, remained the essential unit. In 1957, 80 percent of the respondents to a national poll believed that people who chose not to marry were sick, immoral, and neurotic.[25] Of course, the ideal of the happy suburban family was a far cry from the reality of life for the millions of Americans living in poverty, out of reach of the suburban dream. In 1962, the publication of Michael Harrington's *The Other America* cracked the image of the affluent society by documenting the existence of poverty in the midst of plenty. But in the 1950s, as Harrington pointed out, the spatial development of the American city removed poverty from the experience of middle-class suburbanites. "In the 1950's," Harrington wrote, "America worried about itself, yet even its anxieties were products of abundance. . . . There was introspection about Madison Avenue and tail fins; there was discussion of the emotional suffering taking place in suburbs."[26]

The Baby Boom meant an increase in childrearing responsibilities—as always, assigned to the mother's sphere. Employed women juggled their roles with little or no help from the government, which had quickly scrapped the meager day-care programs that had been set up during the war. In addition, the 1950s brought a new emphasis on the quality of childrearing, including a popularized Freudian notion of the crucial importance of a child's first years, the encouragement of breast feeding, and the emergence of a corps of childrearing experts, such as Dr. Spock, who warned of the dire consequences of anything less than full-time attention from a mother for her children's well-being.[27] All of this came at least partly in response to factors that decreased the time spent in child care, such as the development of commercially prepared infant food, the increase in kindergarten attendance, and, despite the barrage of literature recommending breast feeding, the massive shift to bottle feeding (from 35 percent of mothers in 1946 to 63 percent in 1956).[28] Even the sex object of the 1950s, the full-breasted Marilyn Monroe, emphasized the physical attributes of motherhood. Although the 1953 Kinsey report on female sexuality, in some ways a harbinger of change, accepted the importance of women's sexual fulfillment and activated a flurry of publicity in the women's magazines, the popular imagery of the 1950s continued to subordinate women's sexuality to reproduction.[29]

The contradictions between the ideal and the reality for even white middle-class women, as well as the conflicts between work and family roles created by increasing labor force participation and the rising birthrate, hinted of an atmosphere of dissatisfaction among some American women. Marilyn French's 1977 novel *The Women's Room* explores the discontent of middle-class suburban women in the 1950s; judging by the popularity of the book and its rapid transformation into a television movie, it struck a responsive chord among many.[30] Even at the time, the media sometimes assumed that American women had grievances. In 1947, *Life* defined the American woman's dilemma in terms of her decision whether or not to work, assuming that staying at home would not help a wife to keep up with her husband's interests, although admitting that full-time work combined with motherhood could prove exhausting.[31] In 1949, *Life* proclaimed that "[s]uddenly and for no plain reason the women of the U.S. were seized with an eerie restlessness" and went on to picture women filing for divorces, beating their husbands, appearing in public in scanty clothing, and swimming the English Channel.[32] A special issue of *Life* in 1956 focused on the problems of American women. The introduction suggested that those problems might be the consequence of a preoccupation with rights, since women no longer seemed to cherish their "privileges," which *Life* defined as femininity, childbearing, and devotion to beauty. "Historians of the future may speak of the 20th Century as 'the era of the feminist revolution,'" *Life* proclaimed.[33]

The culmination of the scattered reports of dissatisfaction came when Betty Friedan discovered in 1956 and 1957 that her Smith College alumnae classmates were often discontented, troubled, perhaps even wretched. These highly educated and privileged women found housework and childrearing insufficient outlets for their energy. Fearing just such an outcome from higher education for women, Mills College president Lynn White, Jr., in his 1950 book *Educating Our Daughters*, had urged a college education for women that would prepare them for marriage and motherhood.[34] Advertisers—the "captains of consciousness" who sought to expand markets by manipulating consumers—urged women to find fulfillment through buying. Television, the powerful new medium of the fifties, depended on advertising revenues, and thus the advertisers' attempt to sell products not only reached an unprecedented mass audience, but also affected the programming itself.[35] Not coincidentally, television images of women reinforced the traditional stereotypes. Consumerism assigned the American woman the weighty task of buying and eased the crisis of overproduction that beset the American economic system, but did not solve women's problems.[36]

The social trends affecting women's lives in this period gave rise, then, to grievances among even the most privileged women. In her examination of American women in the 1950s, Eugenia Kaledin goes far enough to state that "in the broadest sense the 1950's may finally be seen as the most active period of consciousness raising for modern American women; women pushing baby carriages still may have time to think."[37] But women's grievances did not receive a sympathetic hearing in the fifties. Instead discontented women confronted a social climate of antifeminism.

The Social Climate of Antifeminism

In any consideration of the antifeminism that pervaded the period after World War II, it is important to take into account the larger social and political climate of the Cold War era, a period that was generally unfavorable for social protest and reform. The mood of conformity described and analyzed by sociologists in the 1950s found its counterpart in the consensus school of history, which set forth an interpretation of American history as free of class conflict.[38] Academic theories, however, fade to insignificance beside the political climate embodied in McCarthyism. The times were certainly not auspicious for mass protest and movements interested in social change.

The media played a major role in perpetuating the social climate of antifeminism and thwarting the possibility of mobilizing discontented women. As we have already seen, the media, for the most part, portrayed women as happy housewives whose lives centered around their homes and families. This dominant image served to deny that women as a group had grievances.

In addition, the mass media began to criticize American women. Throughout World War II, the government and media had praised women for their contributions to the war effort and urged ever greater involvement. But as the war drew to a close, a vicious attack on women began to take shape. Even before the fighting ended, a *Life* editorial attacked the wartime performance of American women and accused them of being helpless and hopeless, lazy, apathetic, and ill-informed.[39] Returning GI's, in a manner reminiscent of the post-World War I period, lamented their return to American women after their experiences with their more "womanly" European counterparts who sought to please their men.[40] In the tradition of Philip Wylie—author of the venomous attack on American mothers, *Generation of Vipers*—a psychiatric consultant to the Secretary of War blamed mothers for the shockingly high number of men rejected for military service on psychological grounds.[41] Both feminists and antifeminists seized upon the poor mental health of American boys to prove their points. While the feminists argued that women frustrated by the limitations of their traditional roles produced overprotected and neurotic sons, the antifeminists countered that the problem lay precisely in women's rejection of traditional roles. If only women would renounce feminist goals and devote themselves to wife-hood and motherhood, all would be well in American society. In this vein, Agnes Meyer, a prominent journalist whose husband owned the Washington *Post*, surveyed the American home front during the war and blamed women who cared more for their rights than for their children for the problems she identified.[42]

The most influential attack on feminism came from Ferdinand Lundberg and Marynia Farnham, authors of the vicious diatribe, *Modern Woman: The Lost Sex*.[43] From a Freudian perspective, Lundberg and Farnham analyzed feminists as severe neurotics responsible for the problems of American society and urged federally subsidized psychoanalysis for feminists, cash subsidies for motherhood, and other measures to restore American women, and thus the American family, to health. The impact of *Modern Woman* was far-reaching. One contemporary study of the feminist movement described Marynia Farnham as "possibly the most frequently quoted writer on the modern woman."[44] A *Ladies' Home Journal* article noted the increasing attacks on women and cited Wylie's theory of "Momism" and *Modern Woman*.[45] Feminists viewed the Farnham book with alarm. The National Woman's Party called a special meeting to decide on strategy for countering the threat, and one Woman's Party member reported in an interview that the book set the movement back a decade.[45] Mary Beard, author of a classic work in women's history, *Woman as Force in History*, wrote to the president of Radcliffe College out of concern for the impact of the book on Harvard men. She had little doubt that the "psychiatrists and sexologists will be the victors" and that women would become more confused than ever.[47] The president responded with an expression of concern "about what seems to be a rising wave of 'anti-feminism' of which *Modern Woman: The Lost Sex* is symptomatic."[48]

Lundberg and Farnham were by no means alone in their denunciations of feminism. A "noted figure in criminal psychopathology" explained in *Collier's* why women's progress toward emancipation was dangerous and why women did not want full equality under the law.[49] *Ladies' Home Journal* published an article entitled "Should Women Vote?," thus questioning the only major advance in women's political status in the twentieth century.[50] *This Week*, a Sunday newspaper supplement, carried an article by a Barnard College sociologist who argued the need for women to give up their jobs, bear children for the good of the race, and submit to the personal ascendancy of men. He advised husbands to tell their wives plainly that they were going to be the bosses, for then the women would be very angry, threaten to leave, and love their husbands to distraction. "And that is why the men must win the postwar battle of the sexes," he concluded. "They cannot afford defeat; the women cannot afford victory."[51]

Such collective attacks on women served to discredit those who had continued, in this period, to work toward women's rights in American society. Another way that the media affected the movement was through the tendency to focus on individual women rather than on the issues. For example, the media carried attacks on particular women leaders. An angry letter to the editor of the Washington *Star*, for example, complained about a feature on Anna Lord Strauss, president of the League of Women Voters: "Whenever I see these smug pictures of women who have abdicated their normal functions and entered the field of politics and the like I instinctively say failure and slacker. Such women have flunked at their own jobs and yet pretend to tell men what they should do in their normal field."[52] In the same vein, a syndicated newspaper article attacked Sally Butler, president of the National Federation of Business and Professional Women's Clubs, as the "boss-lady of the National Federation," noting that "[t]his rebellion has been growing since the war.... One of these days the ladies are going to have to decide between the kitchen or the council chamber."[53]

The media also used the testimonials of ex-feminists to discredit feminism. In a manner reminiscent of the confessions of ex-feminists that had appeared after World War I and the suffrage victory, women who had once identified as feminists declared that feminism was dead. Dorothy Thompson, journalist and radio commentator, proclaimed her belief in the superiority of men.[54] Writer Cornelia Otis Skinner lashed out at "those 'what-women-are-doing' enthusiasts who still go under the outdated term of feminists."[55] Doris Fleischman, a successful public relations counsel who had kept her own name while married for twenty-six years, explained in "Notes of a Retiring Feminist" why she decided to give it up.[56] In 1955, Fleischman—now Bernays—published a book called *A Wife Is Many Women* that explicitly responded to the antifeminist climate by glorifying the many roles that women filled in the modern family.[57] Such testimonials from women who had once identified as feminists contributed to the impression that feminism was old-fashioned and irrelevant to women in the postwar world. Like attacks on specific women, these confessions focused attention on individuals at the expense of the issues they supported.

Finally, the media hampered feminist activism by failing to give it much publicity. Seldom if ever were the activities of feminists portrayed as part of a larger women's rights movement, which contributed to the widespread belief that there was no such thing in the 1940s and 1950s. A leader of the Lucy Stone League, as a result, insisted that bad publicity was better than none at all.[58] Women active on behalf of the ERA complained when newspapers did not report on the amendment's progress in Congress. One woman realized "how unimportant men thought" the ERA was "when in all the lists of pending important legislation, I never saw or heard it mentioned."[59] Even news of the defeat of the ERA was used for publicity by its supporters, who took advantage of anti-ERA articles and editorials to demand equal time for rebuttal. When Alice Hamilton, who changed her mind on the ERA seven years later, wrote an article for *Ladies' Home Journal* entitled "Why I am Against the ERA," the head of the Industrial Women's League for Equality submitted a response. When the magazine rejected it, she sought help from the National Woman's Party leadership, which, after much correspondence, failed to gain satisfaction, and dismissed the magazine as conservative.[60] Such incidents illustrate the media's reluctance to provide publicity for feminist causes.

Despite the different ways in which the media attacked or ignored the women's rights movement, some serious feminist literature was published during this period. Occasional feminist articles appeared even in the women's magazines. Margaret Hickey, a major figure in the National Federation of Business and Professional Women's Clubs throughout the period, served as an editor for the *Ladies' Home Journal*, which she perceived to be a magazine "devoted to the encouragement and recognition of the power of women."[61] Some women's magazine articles supported such an optimistic view. For example, Susan B. Anthony II, grandniece of the famous suffragist, in 1945 predicted a rosy future for American women based on their increased participation in the labor force. *Good Housekeeping* in 1948 printed an article entitled "Do Women Have Equal Rights?" and answered "no." Anthropologist Ashley Montagu, in a preview of his 1953 book, *The Natural Superiority of Women*, argued in *Ladies' Home Journal* that women are genetically superior.[62] Women's rights activists applauded such articles and attacked manifestations of antifeminism. When a *Life* editorial in January 1945 denounced women for failing to earn the respect paid them during the war years, women's rights advocates mobilized to respond. Jane Grant, a leader of the Lucy Stone League, called for a deluge of letters that would "give them an idea of what solidarity can be."[63]

Because of minimal and mostly negative coverage of the women's rights movement by the media, supporters seized eagerly on any feminist literature that came their way. A number of important feminist works appeared during these years, and women in the movement read them avidly. Mary Beard's *Woman as Force in History* (1946) won praise for its recognition of women's achievements but engendered a great deal of controversy because of its attack on the pioneer feminists for their emphasis on the subjugation of women. Ashley Montagu's *The Natural Superiority of Women* (1953), too, sparked debate among women, as had his earlier articles on the same subject.[64] The publication in English of Simone de Beauvoir's *The Second Sex* in 1953 did not go unnoticed by feminists. Biographer Alma Lutz thought that the reviews made it sound very feminist and was eager to discuss it with her friends once her copy arrived. ERA activist Florence Kitchelt did not feel she could afford to buy it but spent an hour or two in a bookstore reading it and summed up her impression: "Three cheers for Simone de Beauvoir!"[65] Women could also choose from a number of biographies of feminists, including those written by Alma Lutz. However, for some inexplicable reason, one of the most feminist books, Ruth Herschberger's *Adam's Rib*, received no attention in the correspondence of women's rights supporters.[66]

Other works of interest to feminists published during this period include *Womanpower* (1957), a study of women in the labor force, and *Work in the Lives of Married Women* (1959), both products of conferences sponsored by the National Manpower Council in which a number of individuals and women's organizations participated. Eleanor Flexner published *Century of Struggle*, her pioneering survey of women's history, in 1959, and the AAUW "Status of Women Notes" insisted that it should be on every woman's required reading list.[67] So feminists were not entirely without literature when Betty Friedan's *The Feminine Mystique* hit the scene in 1963. "This really IS a book!" one Woman's Party member exclaimed.[68] Echoing what other women saw in the book, Alma Lutz recommended it to her friends because it gave her "a glimmer of hope that some of the younger generation are waking up."[69]

But even the existence of feminist literature could not obscure the reality of the general mood of antifeminism that prevailed. By creating an inaccurate image of women's lives, by holding women who did not conform to the ideal responsible for the problems of the American family, by focusing attention on individual women at the expense of the issues, and by ignoring the women's movement, the media helped to create an atmosphere in which it was difficult to put forth arguments for women's rights. Feminists understood, at the time, that this media treatment had important consequences.

The Impact of Antifeminism

Feminists active in the late 1940s and 1950s knew that the social climate of antifeminism made their work difficult. Business and Professional Women's Clubs president Margaret Hickey, for example, recognized the danger facing women in the immediate postwar period. She feared that women's "very big part in victory may become just a footnote in the history books" and expressed alarm at the "toboggan in public esteem" women experienced at the end of the war.[70] Another BPW officer noted that "[c]riticism, much of it bitter, most of it unrealistic and unfair, has been hurled at American women." She offered the 1947–1948 program theme, "We Face Tomorrow," as a response to those who believed that women should "restore the old models of the past for their sex."[71] Alma Lutz, a longtime feminist, recognized that magazines preferred articles that poked fun at women. She commented in a letter to the editor of the *Nation* that it had become the fashion, as in 1848, to ridicule feminism and diagnose women's needs as purely psychological. She saw *Modern Woman* as the 1948 equivalent of the denunciation of the Seneca Falls feminists as "aged spinsters crossed in love, wishing to ape man."[72] Along the same lines, a National Woman's Party staff member in 1947 hoped to pass the Equal Rights Amendment "before the dreadful anti-feminist reaction we are facing closes all doors to us."[73] Florence Kitchelt, head of the Connecticut Committee for the Equal Rights Amendment, feared that the members of the Connecticut General Assembly considered the women who appeared before them for the ERA "cranks and queers" or "old maids."[74] Fannie Ackley, a working-class member of the National Woman's Party, found that some women dubbed a woman a "man-hater" simply because she believed in equality of rights.[75]

The media attacks created pressure on individuals and groups to abandon the feminist cause. The negative connotations were by no means entirely new in the post-1945 period; in the 1920s, feminism had come to evoke images of militance, prudishness in sexual matters, and a seriousness out of tune with the Roaring Twenties. But during World War II, prominent spokeswomen had begun to announce that equality would be won without the agitation of "a militant sisterhood," or, as Margaret Hickey, BPW president put it, that "the days of the old, selfish, strident feminism are over."[76] An article published in 1950 stated that "the very fervour of the pioneers has become somewhat ludicrous in modern eyes, nowhere more in evidence than among groups of so-called Emancipated women."[77]

Deliberate antifeminism, then, added its strength to the general atmosphere of conformity and consensus to discourage women from voicing protests about gender inequality in American society. What stood out about this period for one woman active in the civil rights and the women's movement was the ridicule of women trying to stand up and be counted as persons.[78] Ironically, the backlash drove scholar Mirra Komarovsky, "if not into street demonstrations, then into polemical writings and speechmaking," which resulted in her classic work, *Women in the Modern World.*[79]

The paradox of "woman's place" did create the potential for a groundswell of discontent among American women, but that discontent did not find an outlet until the resurgence of the women's movement in the mid-1960s. Sociologist Joan Huber, in writing about her own career as a housewife in the fifties, summarizes particularly well the significance of societal support in defining problems as collective injustice rather than individual misfortune: "A person who maintains a self-definition with no social support is mad; with minimum support, a pioneer; and with broad support, a lemming. Most of us are lemmings. We accept or change our ideas of our own rights and duties only when we perceive social support for doing so."[80]

Post-1945 American society did not provide any significant support for women's claims of injustice, let alone for the efforts of women who would challenge the sources of injustice. But some women remained active on behalf of women's rights. That they were not themselves, for the most part, the women experiencing the work and family pressures of the times helped to limit the appeal of their movement. In order to understand how the women who identified as feminists maintained their commitment in this period, and what impact their activities had, we turn our attention to these individuals and their organizational activities. We begin by looking at the National Woman's Party.

3

The National Woman's Party

In 1978, fifty-five years after the National Woman's Party arranged for the introduction of the Equal Rights Amendment in Congress, a small contingent from the Woman's Party joined the massive march for the amendment in Washington, D.C. The sponsor of the march, the National Organization for Women, was, from the perspective of the Woman's Party, a newcomer in the work for women's rights. The march dramatized the changes in the women's movement, and the role of the Woman's Party in it, since the early years of the ERA struggle. The National Woman's Party could at no time in its history have mobilized 100,000 people to march for the ERA, as NOW did, but that does not belie the significance of the Party's history. Although membership had dwindled and the group's efforts had languished long before the 1970s, the National Woman's Party played a central role in the women's rights movement in the post-1945 period. A small, exclusive single-issue group, hardly popular even within the women's rights movement, it brought together individuals and groups interested in the ERA and other issues of concern to women, issued information on the status of the ERA, attracted publicity, and, perhaps most important, provided a continuous feminist presence throughout the years. The distinctive character of the women's rights movement in the two decades after the end of World War II owed a great deal to the National Woman's Party, so a closer look at this organization is essential.

The 1978 ERA march stood out not only as a symbol of the peripheral role of the National Woman's Party in the movement in the 1970s, but also as a poignant reminder of the origins of the Woman's Party in the suffrage struggle.[1] Born out of the National American Woman Suffrage Association in 1913, the National Woman's Party and its founder, Alice Paul, stood for militant action. While the NAWSA had worked cautiously for suffrage and supported the American war effort in World War I, members of the young Woman's Party (originally called the Congressional Union) took to the streets to demand suffrage and picketed the White House in protest against a government that promised to make the world safe for democracy while denying half of its citizens the right to vote. Although the National Woman's

Party never returned to its militant tactics after the passage of the suffrage amendment in 1920, the heritage of arrest and imprisonment remained crucially important within the Woman's Party, and the reputation of fanaticism clung to the group throughout the years.

The end of militance, however, had not signified satisfaction with the status quo after the suffrage victory in 1920.[2] By 1921, Alice Paul had decided on a constitutional amendment that would guarantee women equal rights, and she never swerved from the course she set for herself and her group, despite the objections of most of the women who had been active in the suffrage movement. Support for the ERA bloomed during World War II as large national women's organizations joined the previously short list of endorsers, and the amendment made headway in Congress.[3] Introduced in every Congress since 1923, the amendment came up for a vote in the Senate for the first time in 1946. To the National Woman's Party, the immediate postwar years looked especially promising. Year after year, the group worked to have the ERA introduced in Congress, questioned candidates on their position on the amendment, built the list of Congressional sponsors, fought to keep the ERA in all party platforms (the Republican platform first included it in 1940, the Democratic platform in 1944), lobbied senators and representatives, sought the endorsements of other organizations, pressured the appropriate subcommittees and full committees, lobbied again— always, until 1972, to no avail, but always without losing hope. Despite its occasional work on other goals, the Woman's Party defined itself as a single-issue organization: "Our work is confined entirely to obtaining the passage of the Equal Rights Amendment."[4]

The story of the National Woman's Party after 1945 is a story of decline and survival, since the group failed to achieve its single goal, passage of the ERA. It lost membership and failed to attract new and young members, suffered serious internal conflicts that hampered its effectiveness, and developed a reputation as a collection of amusingly eccentric and anachronistic old feminists, but it sustained a feminist community throughout the years that made continuing activity in a hostile environment possible.

Characteristics of the Woman's Party

The National Woman's Party has suffered throughout its history from a bad press. Denounced as militant and fanatic for its tactics during the suffrage struggle, it later received bad marks from historians who believed it had jeopardized the suffrage victory.[5] When the Woman's Party threw its support behind the ERA, women reformers who feared that the amendment would harm working women by eliminating protective labor legislation wrote off the group as a bunch of selfish wealthy women unconcerned about the plight of their working sisters, if not secretly financed by the bosses.[6] This is the reputation that has clung to the Woman's Party, which was indeed composed largely of white, middle- or upper-class, well-educated, and older women.[7]

Despite Alice Paul's insistence that the group welcomed all women, that it
was a *"classless movement,"* it was an elite group which brought together
women of differing political views.[8] A Party pamphlet described the organization as a group of "progressive women with differing political opinions,"
which was true with the qualifications that they were "progressive" as a
group only in supporting equality for women. As a member active in progressive groups, particularly the peace movement, noted: "There were some
very conservative women, there's no doubt about that, but there were others
of us who were not."[9]

The National Woman's Party had never been a mass organization,
although it climbed to a membership of approximately 60,000 in the last
years of the suffrage movement. It lost membership quickly after 1920, and
by 1945 claimed a general membership of approximately 4,000, which rose
to 5,500 in 1953 but dropped to 1,400 by 1965. "General" membership
included active members, who paid annual dues of at least ten dollars, as
well as "enrollment" members who paid a flat twenty-five-cent enrollment
fee to the national headquarters. Since the general membership figures are
inflated in this way, it is more revealing to consider the number of active
members. In 1947, the Woman's Party listed 627 active members and in
1952, 200. The organization periodically discussed the problem of membership, with few results, because on the whole the leadership preferred to
keep the organization small and elite. Alice Paul believed that the group
would function best as a cadre of women working to get the existing women's
organizations to endorse the ERA, and other members echoed her conviction that a small group could work most effectively.[10] The leaders recognized,
however, the impressive nature of a large membership and as a result kept
the membership figures secret. As death thinned the ranks, little new blood
flowed into the Woman's Party. At the same time, internal conflicts periodically resulted in the resignations of those on the losing side. So the
organization dwindled in size throughout the post-1945 period, increasing
its exclusiveness and its sense of isolation.

By the end of World War II, the National Woman's Party functioned
almost entirely on the national, rather than on the state or local, level.
Active state branches existed in Washington, D.C., Maryland, New York,
Ohio, and California, but the bulk of the active members lived in the north-
east and most of the activity took place in Washington. The state branches,
which had been active in the 1920s, served the organization primarily as
names on letterheads to use in lobbying senators and representatives. In
1944, Alice Paul introduced a system of regional chairmen to coordinate
the work of the states, but this never became really effective. In some states,
branches consisted of little more than a chairman. Some state and local
branches held meetings and pursued independent projects—although these
might, as was true for the District of Columbia branch, include such things
as flower arranging—but direction for the ERA campaign came from national headquarters. A national chairman, who presided over a national
council, headed the Woman's Party, although Alice Paul, often with no

formal position, exercised as much if not more control than the chairman. As one member put it, Paul "gave the chairman all deference. But if you were a wise chairman, you did what Alice Paul wanted, because she knew what was needed."[11] A national advisory council included prominent women who lent their names to give weight to the group's work. A number of committees functioned sporadically in particular areas such as lobbying, publicity, organization, women's history, and work with churches. What the Woman's Party did primarily, however, was lobby Congress and coordinate letter-writing campaigns to win support for the ERA from politicians and women's organizations.

Despite the Woman's Party's reputation as a wealthy organization, financial problems plagued the group throughout the years after 1945. The legacy left to the group by Alva Belmont, who had been married to both William Vanderbilt and O. H. P. Belmont, had been spent by 1949, and only a few wealthy members remained to send in large sums. Despite occasional large contributions and bequests—including ten thousand dollars in 1967—the Woman's Party depended on its diminished invested funds, dues, contributions, and revenue from the rental of rooms in its headquarters, Belmont House, on Capitol Hill, and two other Party-owned houses. The Belmont House remained a prime asset, but the party, while successful in winning a tax exemption for the house as a historical place in 1960, periodically had to expend energy fighting Congressional attempts to take the property for government purposes.[12] Contributing to the loss of membership, resources, and effectiveness, and at the same time partly a result of the decline, were two major conflicts within the organization in the 1940s and 1950s.

Conflict

The most serious conflict in the Party's stormy history blew up into a legal battle over the name, leadership, and resources of the organization in 1947, while another in 1953 led to the resignation of the chairman. Throughout the history of the organization runs a current of conflict, usually centered on the leadership of Alice Paul. The pattern of conflict both grew out of and in turn reinforced the closed nature of the Party, and thus played an important role in shaping the character of the group.

Conflict was by no means new to the Party in the 1940s; a dispute in the mid-1930s had foreshadowed the 1947 schism.[13] A group of women who wanted to reform the Party in the 1930s had temporarily withdrawn and published a rival journal, although they returned to work with the rest of the membership after two years. The personnel of this rebellion differed from that of the larger schism a decade later, but the issues were very much the same.

The larger conflict began during Alice Paul's chairmanship, which she assumed in 1942 for the first time since 1921. The Woman's Party stepped

up its efforts on behalf of the ERA and, in the wartime atmosphere, passage
of the amendment began to seem an attainable goal. In 1945, Paul announced
that she would not run for chairman again, and the nominating committee
reported that it could find only one candidate willing to serve. Because of
wartime restrictions on travel, a Convention By Mail replaced the regular
convention, and the nominating committee sent out ballots with one slate
of officers only. Dissatisfaction with Anita Pollitzer, the candidate for chairman, and disgust at the lack of choice led to the formation of a small group
calling itself the Coalition Council which proposed an alternative slate of
officers. Not surprisingly, the official slate easily won and Pollitzer took
office.

    With Pollitzer's victory, the Coalition Council began to take the shape
of a true opposition group, and it aimed its criticism at the failure of the
officers to call a regular convention. But the real issue lay deeper—unhap-
piness with Paul's and, by extension, Pollitzer's, leadership. Throughout the
conflict, the opposition—the self-styled "Action Group" or "Constitutional
Group"—saw Pollitzer as a mere puppet of Paul and referred constantly to
"the two A.P.'s." Pollitzer's election, then, really had little to do with the
conflict; even before the Convention By Mail, one of the rebels had expressed concern with the leadership at headquarters and warned that the
situation would "eventually result in a conflagration unless something *is
done.*"[14]

    At this early stage, the opposition based its objections to Pollitzer
primarily on her role in a conflict between the New York City branch, which
she headed, and the New York state branch, led by Jeannette Marks. The
city branch refused to merge with the state branch, as it had, according to
Marks, promised, and Pollitzer received the blame. Complicating this situation was Marks's conviction that the national leadership acted in a high-
handed manner toward the state groups, particularly in the matter of the
allocation of dues collected by the states. The Maryland and Massachusetts
branches joined with New York in complaining about the national leadership's authoritarian approach to relations with the states. In both cases,
members accused Alice Paul of dictating the choice of state chairmen, or,
even worse, of inciting revolts against existing state leaders.

    These issues came to a head at the meeting of the Eastern Regional
Conference, which included the rebellious states, in June 1946. The audience
heckled Anita Pollitzer during her speech and called on the national leadership to hold a convention in the autumn. By this time, unrest in some
quarters had gone beyond anything that could be solved by calling a convention. In June, Doris Stevens, who emerged as one of the three leaders
of the rebellious group, wrote that she had "joined with others to . . . throw
out Paul and her goons and get a sane, neutral chairman who will not destroy
what is left of the party."[15] Stevens, who had been one of Paul's closest
associates during the suffrage struggle, had nursed a grudge against Paul
since the 1930s when the National Woman's Party, rather than Stevens,
received a legacy from Alva Belmont, despite Belmont's promise to Stevens

that she would receive part of the money as a reward for her years of personal service. Apparently Stevens stayed out of the conflict at first, writing that she had "not time nor taste for schisms," despite the fact that she believed that Alice Paul had "purged" her from the national council.[16] But other members unhappy with Paul's leadership drew Stevens into the conflict, and eventually she became one of the leaders of the group, along with Laura Berrien, a lawyer who reportedly fell under Stevens' spell, and Anna Kelton Wiley, who had edited the group's journal, *Equal Rights*, for years and felt that no one appreciated her efforts.

The national council met in New York in September 1946, and there the discontented members proposed a reorganization of headquarters and demanded a convention. Paul and Pollitzer perceived these resolutions, correctly, as an attack on the current administration and walked out of the meeting. In response, the opposition made use of a constitutional provision that, they claimed, allowed ten members of the council to call a meeting without the chairman's approval. They held such a meeting the next month and scheduled a convention for January 1947. Pollitzer, who did not attend the October council meeting, then announced different dates in January for the convention.

At this point the situation degenerated even further. The opposition went ahead and held what came to be known, to the other side, as the "Rump" Convention, despite an attempt to arrange a compromise date for a convention of the entire Party. The Rump Convention chose as chairman Sara Whitehurst, a Baltimore woman whose primary work had been with the General Federation of Women's Clubs, not the National Woman's Party. The possibility of Whitehurst's chairmanship, which had been raised even before the 1945 Convention By Mail, had caused Paul to reconsider leaving her post, since Whitehurst, perceived by Paul as merely a "clubwoman," had never been a member of the Woman's Party. Although some members of the opposition counseled compromise, after the January convention the Woman's Party had two separate sets of officers. The newly elected officers, along with their supporters, went to headquarters and found the door guarded by a male detective who refused them entrance. Outraged, they pushed their way into the house, where a scuffle ensued. Eventually the rebels filed two lawsuits, one over the name, leadership, and resources of the organization, and the other asking for an injunction against the closing of Belmont House. Numerous attempts at reconciliation, one initiated by Jeannette Marks, who had played such a major role in the beginnings of the conflict, failed. Whitehurst resigned as chairman in June 1947, and the opposition group eventually lost both cases.

A number of issues emerged during this conflict, although all ultimately led back to Alice Paul's leadership. One concerned the question of priorities: should national or international work take precedence? Alice Paul had been devoting herself primarily to the World Woman's Party, established in 1938 with headquarters in Geneva and then, after 1945, in New York. The insurgents accused Paul of placing international ahead of national work and of siphoning funds from the National Woman's Party to the World Woman's Party. Another issue involved expansion of membership, which the insurgents favored, and this too focused criticism on Paul. Doris Stevens insisted that "A.P. can't be bothered with members," and the state branches that wanted to increase membership were pitted against the national officers who wanted to maintain a small cadre organization.[17]

The opposition also directly attacked Alice Paul's leadership. The rebels accused her of frightful extravagance and misappropriation of funds. Critics expressed shock at her openhandedness with the group's money in making telephone calls, sending telegrams, and using unnecessarily expensive postage rates for mailings. Others focused on the closed leadership ranks and the autocratic control exercised over the membership. The 1945 Convention By Mail, with its single slate of officers, symbolized for the rebels the lack of democratic procedure in the Party. Wiley, Berrien, and Stevens charged in a letter—characterized by Paul's supporters as "wicked" and the product of "an insane mind"—that, from the time of Paul's return as chairman, "you surrounded yourself with a small clique, and to question your imperious will even mildly was to provoke the clique's attack."[18] When Paul, in the summer of 1947, suggested that the group split, Berrien and Stevens circulated an open letter than expanded on this accusation: "You have made it clear that you consider yourself and the small group around you an *elite* with superabundant intellect and talents, and consider us, in contrast, the commonfolk, although you were generous enough to say that some of us, at least, would attract 'thousands and thousands' of women. Thus, these two parts of a whole would march forward together, you providing the brains, and we, the busy, but witless, hands and feet."[19]

In connection with such attacks on Paul's elitism and dictatorial methods, the opposition suggested that the Woman's Party leadership had failed to keep pace with the changing times. They embodied this criticism in their later designation of the Paul group as the Conservative Group. One woman described the rebels as "the people who want a little more 1945 and a little less 1920 and before in our propaganda."[20]

From the other side, explanations for the conflict lacked coherence. Some of Paul's supporters suggested that the rebels acted out of jealousy of Paul's brilliant leadership. Others saw the rebellion as a simple greedy attempt to seize the valuable resources of the Party. Paul, years later, suggested that it had to do with the fact that Pollitzer was Jewish, although she at the same time denied knowing why it all had happened.[21] As time went on, the Paul group begin to suspect that "some sinister outside group" that opposed the ERA lay behind the lawsuits.[22] Lavinia Dock, a prominent nurse and settlement house worker and one of Alice Paul's mentors, suspected business interests in the service of fascism.[23] Other members, however, immediately looked fearfully to the left. Alice Paul, who later supported McCarthy in his anti-Communist witch-hunts, wondered if the Communist Party wanted to take over the Woman's Party. Other members, too, perceived the conflict as smacking of Communist tactics, and the officers eventually gave evidence concerning the schism to the FBI and the Committee on Un-American Activities of the House of Representatives.[24]

The charges of Communist subversion were, of course, absurd. Alice Paul later remembered the conflict—which she claimed was not important—as originating with Caroline Babcock, one of the rebels who served as secretary and whom Paul described as "very much more to the left than most of the people on the board."[25] Paul's characterization is probably accurate, but what is particularly interesting about this conflict is the mixture of political perspectives on each side. Doris Stevens and Laura Berrien both became outspoken supporters of McCarthy. Anna Kelton Wiley deplored racial integration and later left the AAUW over the admittance of a black woman to the District of Columbia chapter. There is no evidence that basic political differences separated the two sides. One member described the schism as "another of those things that happen in [the] history of movements," and that is perhaps an accurate description.[26] The revolt pitted Paul's critics against her supporters, and it was not the first nor the last time that members chafed under Paul's firm control.

Paul was, reputedly, committed to the point of fanaticism to the ERA, and she attempted to maintain control of the forces working for its passage. She decided when it was time to press Congress and when time to keep a low profile. Even when she was not chairman—and she never served as chairman again after 1945—she kept a tight rein on activities in the Woman's Party. As one woman active in the movement but not part of the Woman's Party observed, "Alice Paul was chairman of the Woman's Party all the time."[27] Her supporters as well as her critics recognized this, but they tended to hold her in such reverential awe that they would never think to criticize her. Those who rejected her leadership rejected along with it the adoration bestowed on her and the exclusive closed nature of the Party.

It is difficult to tell precisely how the membership lined up in this schism. The rebels claimed support from as much as 81 percent of the membership, but this is certainly exaggerated. There seem to have been approximately equal numbers of women actively involved in the conflict on both sides; the insurgents could claim support from the Maryland, New York, Massachusetts, and Michigan branches, but the majority of the membership elsewhere, who got wind of trouble in Washington but did not always know what the issues were, seemed to support the Paul-Pollitzer group. The extremes to which Doris Stevens in particular went in denouncing Alice Paul began to alienate even those who wanted to see reform in the organization.

Stevens wrote to Paul after the opposition decided not to appeal the case and indicated that Paul would have to decide whether she would accept the former rebels as workers for the ERA. Despite this formal attempt at reconciliation, the schism resulted in the withdrawal of many of the rebels. Anna Kelton Wiley resigned, although she returned in 1959. Laura Berrien left and also faced the dissolution of her law firm partnership with member Burnita Shelton Matthews, who had remained loyal to Alice Paul. Doris Stevens resigned and devoted her energies to the anti-Communist crusade of the 1950s. Decades after the schism, rebel Olive Hurlburt continued to carry a grudge and warned a newspaper reporter that "Alice Paul was not the saint everyone thinks she was."[28]

Many of the rebels rallied around the Connecticut Committee for the ERA as an alternative to the National Woman's Party. Anti-Woman's Party groups formed in Maryland, New York, and Washington, and a number of the rebels corresponded or met together to discuss the organization of a new national network of state groups, although little came of these plans. Feminist author and activist Alma Lutz, who had sympathized with the rebels but opposed the filing of the suit, stated her disgust with the Woman's Party as a "closed corporation with no desire for any fresh viewpoints," but insisted that she would always cooperate with it or with anyone working for the ERA, even if she had no desire to associate actively with the Party.[29] She did, however, rejoin the national council in 1959. Most of the opposition expressed a desire to continue to work for the ERA even if they could not go back to the Woman's Party. One of the leading rebels worried that Wiley, Berrien, and Stevens were proving the truth of the accusation that the rebels were trying to wreck the organization and the ERA.[30]

There is no question but that the schism had serious consequences for the Woman's Party. Although membership figures are not available for every year, members reported a large number of resignations, and the figures for active members show a downward trend after 1947. Furthermore, the schism made the leadership suspicious and inclined to shut the door against new members. Even more important, work for the ERA nearly ground to a halt while the two groups sought support from the membership and tried to raise funds to cover the costly legal fees.[31] Attention from the newspapers reinforced the reputation of the Woman's Party as a group of eccentric and quarrelsome women.[32] And, in spite of all these serious consequences, none of the issues had been resolved, so they broke out again in the 1950s.

The Woman's Party, in the wake of the lawsuit, elected a chairman who, it was hoped, would avoid controversy. From 1949 to 1951, Agnes Wells, retired Indiana University dean, chaired the Woman's Party in a low-key manner that led the rebels to dismiss her as inarticulate, dull, and old. In 1951, Ethel Ernest Murrell, described in contrast by one member as a "live wire," became chairman and was determined to rejuvenate the Party.[33] Murrell was a wealthy lawyer who headed the Florida branch; she was married and had two children, and Party publicity and correspondence among members frequently commented on her attractiveness. Murrell almost immediately ran into trouble. As in the 1947 schism, the vision of reform and reorganization came up against the "old guard's," and especially Paul's, notion of the proper role of the organization. The same issues emerged, although this time with a slightly different and very significant twist.

Murrell started right in with a commitment to reorganization, expansion of membership, and solution of the recurrent financial problems. She planned a large-scale publicity campaign to precede a membership drive.

One member who became her right-hand woman enthusiastically described this as "a kind of Burma Shave campaign," something that was certainly foreign to the Party's normal style.[34] Even more out of character was her decision to hire a man to take charge of publicity and membership. Murrell and Mary Roebling, a New Jersey banker whom Murrell appointed to head a new Ways and Means Committee, hired Hiram Serkowich to raise funds. He then brought in a public relations firm with whom Murrell contracted for the publication of a slick magazine to replace the existing journal *Equal Rights*, the setting up of a National Woman's Party radio corporation, and the establishment of a tax-exempt foundation.

These changes threatened the character and structure of the Woman's Party on a number of fronts. Murrell applied for tax-exempt status, which could only be granted if the organization ceased to devote itself primarily to attempting to influence legislation. Although the sole purpose of the Woman's Party was passage of the ERA, Murrell ordered a halt to lobbying. In order both to make the Party eligible for tax-exempt status and to attract large numbers of members, Murrell expanded the program to include such things as strengthening the American home by stressing the dignity of marriage, endorsing the recognition of the power of God in national and personal life, and opposing collectivism in American government.[35] This new program, smacking of right-wing politics as it did, prompted Alma Lutz, always sympathetic to but also critical of the Party, to wonder, "Has Mrs. Murrell forgotten the Equal Rights Amendment?"[36] Even Florence Kitchelt, head of the Connecticut Committee for the ERA and no friend of the Woman's Party at this point, deplored what she saw as the demotion of the group to "an ordinary woman's organization" with a "middle-of-the-road" program.[37] Furthermore, Murrell's decision to hire a man to take charge of fund-raising went against Party practice. Men could not join the group, since Alva Belmont's bequest specified that her property would revert to her heirs if the organization ever allowed a man to hold an office or paid position. Some members even feared that the Belmont heirs had somehow engineered the Serkowich contract in order to get their hands on the Belmont property.[38]

It is not surprising, considering how few years had passed since the resolution of the lawsuits, that the events and issues of 1947 kept cropping up in the early fifties. At first an incident between Murrell and one of the former rebels led Murrell to associate herself with the Paul faction and to suspect a revival of the old rebellion.[39] But soon Murrell began to see herself in the role of the rebel fighting the entrenched interests and antiquated ideas of Paul and her cronies. Although the opposition from 1947 stayed informed about the events of Murrell's term of office, none of them played any role in the new conflict. But all of the old issues reemerged.

Alice Paul, on behalf of the World Woman's Party, offered to assume joint responsibility with the National Woman's Party for Belmont House, leading Murrell not only to decline but later to accuse Paul of attempting to suck the National Woman's Party into her international organization. Murrell's eventual resignation, in fact, followed on the heels of her defeat on the issue of national versus international priorities. More important, Murrell's plans raised once again the issue of transforming the Party into a mass membership organization. As the conflict developed, one of the public relations people wrote bitterly to Alice Paul that the Woman's Party apparently "prefers to remain a small and hard-pressed organization," an accusation that the old-guard leadership could not deny.[40] As one young member who deserted Murrell explained the old-guard attitude: "no mass appeal will ever bring into the Party that type of woman who can best carry forward our particular aims. We are an 'elect body' with a single point of agreement."[41] Just as before, all of the issues came together in an attack on Alice Paul's leadership. Although Paul tried to conceal her interference by working indirectly through other members, Murrell took her on firmly and directly. When Murrell refused to stand for reelection in 1953, she suggested that either Paul or Anita Pollitzer run for chairman, since "Miss Paul founded the organization, [and] she rules it."[42]

As in the 1947 schism, criticism of Alice Paul and the old guard involved the charge of anachronistic methods. Murrell's emphasis on publicity, and especially her employment of a public relations firm, placed her squarely on the side of those who wanted "less 1920" in the Party's operation. Murrell, however, went beyond this to attack the old guard as a group of old unmarried women without clout in Congress. She consistently asserted the need for young women in the organization, and apparently expressed privately the conviction that the group consisted of "a bunch of senile old women" and referred to its committees as "a screaming mob" and "the lunatic fringe."[43] In her biennial report in 1953, she recommended the recruitment of young married women and the elimination of the "archaic" notions of "exaggerated feminism." Her statements made clear her fundamental rejection, not only of Alice Paul's leadership, but of the membership of the Woman's Party: "Men and women complement each other. Any doctrine which loses sight of this truth wars on nature. Worse; harried or misfit women, who approach Congress, preaching such a doctrine, either by their words, or by their attitudes, will hamper instead of helping any legislature [sic] they endorse. Therefore I recommend: Let NWP send to Congress women who have shown in marriage or in business that they know how to get along with men. After all, Congress is largely composed of men. You want a large body of men convinced. Attempt then to persuade—not to harass."[44]

With these words, Murrell dismissed the National Woman's Party lobbyists as eccentric old man-haters unable to work effectively for the ERA. It was a matter of principle for Alice Paul and others that women work together to demand their rights. As Paul had expressed her conviction in relation to the vote, it was more dignified to ask it of other women than to beg it of men.[45] Murrell, however, seemed to see the members as pitiful old women with little in their lives besides the Woman's Party. She wrote to Paul that if she were circumvented in her plans, she would have to conclude that "there is truth in what has been said to me so often by those friends of ours on Capital [*sic*] Hill, and other friendly, if unkind, critics who laughed: 'those dear ladies do not want this Bill passed. It would take their reason for being alive away from them.' "[46]

But Murrell could not wrench the Party away from its traditional goal and approach. When she resigned her chairmanship, she took with her the Wyoming branch (she and her husband owned a ranch in Wyoming and she based her efforts after her chairmanship there) and part of the Virginia branch as well. She established the American Woman's Council with the help of a few of her supporters within the Party, and she tried to implement her ideas through the new organization. This Council supported the Hunt Resolution, an alternative to the ERA introduced in the Senate by a Wyoming senator. The Hunt Resolution stated that whenever the Constitution used the terms "person," "persons," "people," or any personal pronoun, it would be interpreted to include both sexes. The Woman's Party, steadfast in its dedication to the ERA, opposed the Hunt Resolution, as did most, but not all, women active on behalf of the amendment. In addition to the Hunt Resolution, the American Woman's Council took up many of the positions Murrell had tried to foist on the National Woman's Party, lending the new council what Alma Lutz characterized as a "reactionary" and "superpatriotic" tinge.[47] It did not last beyond the mid-1950s.

The Murrell conflict had far less serious consequences for the Woman's Party than the earlier split, but it revealed the persistence of the issues that had divided the Party in 1947. Perhaps most threatening to the group's existence was the attitude that Murrell expressed when she lost patience with the organization—that the whole approach of the Woman's Party suffered from "deplorable wrongness."[48] Alice Paul and the old guard opposed Murrell primarily because she sought to put a great deal of responsibility in the hands of outsiders, even men. Alice Paul's later evaluation of Murrell's term of office revealed not only Paul's anti-Semitism but also her hostility toward Hiram Serkowich as a man: "So one Jewish man who was a professional lobbyist, persuaded her to put him in charge of all the lobbying and every activity practically of the Woman's Party."[49] Alma Lutz stated more directly what others who opposed Murrell suspected: "I do not feel that Mrs. Murrell is at heart a true feminist."[50]

Why did such conflicts periodically split the organization? Some observers— both reflecting and contributing to the media image of the women's rights movement—blamed women's inability to cooperate. But such a charge disregards the conflicts and schisms that rend all social movements. Many National Woman's Party members believed that the momentous events of the twentieth century, especially the Russian revolution and the rise of fascism, had so altered the social environment that all organizations responded in troublesome ways. Jeannette Marks, for example, believed that "[w]omen are sceptical nowadays about 'leaders' and leadership, for they have seen all too well for too long a time what male leaders can do! . . . the day of the leader is almost past."[51] In this context, Alice Paul's leadership proved troubling to many members.

For many members, Alice Paul simply *was* the Party. She seems to have been a charismatic leader—fanatically devoted, powerfully appealing to her followers, and able to exercise control through the force of her personality.[52] The evaluation of a psychiatrist who examined her during her time in prison during the suffrage struggle became part of her legend; he compared her to Joan of Arc and reported that she would die for her cause but never give up.[53] Followers often referred to her in religious terms, describing her as a "martyr," admiring her devotion to her "mission" or "vision," and even suggesting that God had sent her to the Woman's Party. But the women who worshipped her could also become disillusioned and begin to see her as the devil. Florence Armstrong, who credited Alice Paul with helping her "enter into a new life," left Party work because of disappointment in Paul, noting that many other women had suffered the same "terrible experience."[54] Several years after Armstrong withdrew, she felt "like weeping with perplexity as to what went wrong" to make Paul hostile and implacable.[55] Women frequently left Party work because of an inability to get along with Paul or described being "dropped" or "snubbed" by her.[56] Doris Stevens, who had "affectionately" dedicated her 1920 book about the suffrage struggle to Alice Paul, noting her "brilliant and devoted leadership," described her in 1946 as the "high priestess of the National Woman's Party" and "a venomous specimen of the snake pit."[57]

The fact of the matter is that both adoration and denunciation of Alice Paul tended to focus on the same qualities. Doris Stevens compared Paul to an alcoholic character in "The Lost Week-End," noting that the film "[s]tripped naked . . . the moving force of one who is completely absorbed in one thing."[58] But then other members lavished praise and admiration on Paul as "a thoroughly dedicated person," "an ideal to women all over the world," "the motive force behind the whole movement," or "the greatest woman living."[59] Inez Haynes Irwin described in her 1921 history of the Woman's Party how Paul offended women by forgetting to thank them for their work, explaining that Paul probably "had an innate conviction that it was egregious personally to thank people for devotion to a cause."[60] Unquestionably, women became disgruntled at the lack of expressed gratitude. As Anna Kelton Wiley put it, "One could die working for the NWP and no one would care; one does not work for gratitude to be sure but human understanding is the least one does expect."[61]

Paul's fanatic devotion must have seemed a reproach to many of her followers. Thinking of her as godlike, they did not expect weaknesses, and when faced with a lack of "human understanding," they saw her as the fallen angel. Because so many members sought her approval, some always suffered when she seemed to "play favorites."[62] Unfortunately for the Woman's Party, rejection of Paul's leadership came to mean, for the most part, rejection of the organization. The rebels in the 1940s and Ethel Ernest Murrell in the 1950s tried to change the Party instead of leaving, but they failed in both cases. As one astute observer wrote in 1946, "it is very difficult for a woman who has really lived a cause for a great many years, and as its leader, finally to distinguish between herself and the cause itself."[63] Neither Paul nor the other members could easily separate the leader from the organization.

Paul's single-minded devotion to the ERA kept the party doggedly working throughout the years, at the same time that her domination caused periodic eruptions. One member loyal to Alice Paul admitted that Paul's personality had everything to do with the lack of recruitment in the 1960s. "She was a rather forbidding person, you know. Young people were really afraid of Alice, they didn't know how to talk to her."[64] As membership dwindled in the 1960s, she used whatever resources she had to keep the ERA alive. In the 1970s, Paul maintained that she had only run the ERA campaign while she was national chairman from 1942 to 1945, but then she admitted that since she cared very much about the amendment and often had more time than the chairman, she had regularly "helped out."[65] Paul never returned to headquarters after 1972, when she quarreled with the then president, Elizabeth Chittick, who favored participation in a coalition of women's groups working for ratification of the ERA. Paul's loyal followers described Chittick's election as a "coup," and Paul resented and criticized the new leadership.[66] Nevertheless, even after Paul's death in 1977, the Woman's Party continued to list her on its letterhead as the founder of the Party and to advertise "Alice Paul Jail Jewelry"—a replica of the famous jailhouse door pin proudly worn by the imprisoned suffragists—as a symbol of Paul's dedication and commitment to her goal. The women's movement in the 1970s took up Alice Paul as a symbol, and the woman whose leadership sparked so many conflicts eventually received recognition throughout the movement.

There is special irony, then, in the events of the 1981 Woman's Party convention, at which the Congressional Union, a militant ERA group named in honor of the predecessor to the Woman's Party, sought to take over the Party by electing its own members to office. In a manner reminiscent of the 1947 schism, Woman's Party president Elizabeth Chittick tried to deny CU members admittance to the convention on the grounds that there was no record of their Woman's Party membership, and she ruled the CU slate of candidates and resolutions out of order on technical points. The Woman's Party leadership was in some sense a victim of its own membership drive, since most of the CU members had joined just before the convention. The CU members wanted to restore what the masthead of its newsletter proclaimed as the "non-violently militant spirit of Alice Paul and Lucy Burns," but the longtime Woman's Party members insisted that militance was inappropriate in the 1980s.

The irony lies in the fact that the CU members used the name of Alice Paul to criticize Chittick for many of the things that the 1947 rebels held against Paul herself. For example, when the Nominating Committee offered only one slate of candidates for office, having disqualified the alternate slate, a CU member cried out, "We have not been able to vote in our party! What would Alice Paul think of that?" And Chittick had come to sound very much like Alice Paul when she proclaimed that "For 10 years I've worked night and day and weekends. I've given my life, my blood for this Party." What is clear from this struggle—which ended with defeat for the CU and rejection of militance by the Woman's Party—is that the history of the Woman's Party continues to make the organization vulnerable to what its members can only perceive as a takeover.[67]

Community

The picture of the National Woman's Party that emerges from an examination of its major internal conflicts is one of a small, relatively closed, tightly knit organization held together by personal as well as political ties. Alice Paul's presence remained important even when she held no formal administrative position in the Party. Her conception of the organization as a small elite group that would mobilize other women on behalf of the amendment held sway throughout the years despite the periodic challenges of reformers within the group. The conflicts reveal not only the forces that tore the organization apart, but also the bonds that held it together throughout a period especially hostile to feminism. For the members of the Woman's Party constituted a community held together and characterized by five elements: (1) a shared bond of participation in the suffrage struggle; (2) feminist identification in the face of hostility from the larger culture; (3) activity at the Alva Belmont House which served as organizational headquarters, a feminist hotel, and a surrogate home for some members; (4) a high level of commitment to the feminist cause; and (5) close personal ties.[68]

It should not be surprising that the bond of shared participation in the suffrage struggle could hold together a community of women for so many years. Many of the most active members, and especially the officers, had joined during the suffrage struggle and carried their memories and experiences into later work. By the post-1945 period they were no longer young women, and they often reminisced about the good old days of the suffrage movement. "I felt as I read your letter the warm glow that always comes when old Suffrage ties are renewed," one member wrote in typical fashion.[69] Others commented on their vivid recollections of a first suffrage meeting or parade. Having been a suffragist, and especially having been jailed and force-fed, gave one status within the Woman's Party. One member wrote that people in her city flocked to see and hear a speaker because she had been in prison: "the work for suffrage is now far enough in the past to cast glamor and appeal on those who worked for it."[70] Women running for office who had served a jail term for suffrage activities proudly listed this as a prized accomplishment. Members too young to have been in the suffrage movement adopted a humble attitude toward the pioneers; one wrote, for example, "I was too young to enter the brilliant and successful campaign . . . but you must admit I tried to make up for it after I became a member of the National Woman's Party."[71] Lena Madesin Phillips, founder of the

International Federation of Business and Professional Women's Clubs and a National Woman's Party member, confessed to envy of the jailed suffragists and to a slight sense of shame that she had never gone to prison because of the strength of her convictions.[72]

What publicity the Woman's Party attracted inevitably mentioned the suffrage past. There is irony in the fact that the media tended to describe the Woman's Party as militant, since the organization had employed militance for only a brief period in its history. The leadership apparently never considered engaging in militant actions on behalf of the ERA and even opposed them once the National Organization for Women, in the early 1970s, took over leadership of the ERA struggle.[73] Members often, however, applied lessons learned in the suffrage past in working for the ERA. The decision to draft and pass a constitutional amendment came naturally to women who had fought for the suffrage amendment. The many similarities between the suffrage and the ERA fights—especially the nature of the opposition and the arguments raised against the amendments—were evident to Woman's Party members, who took courage from the long years of the suffrage struggle, rarely losing hope over the years. Perhaps the experience of the ex-suffragists with the victorious culmination of the seventy-two-year fight gave them a unique and optimistic perspective. In any case, they often hoped that they would not have to wait as long for passage of the ERA.

The Woman's Party used suffrage history quite consciously to further its current program. The Committee on Woman Pioneers kept in contact with women throughout the country working to win recognition for suffrage leaders Susan B. Anthony, Elizabeth Cady Stanton, and Lucretia Mott. The Woman's Party regularly celebrated birthdays of suffragists and anniversaries of historic events, such as the Seneca Falls convention and the ratification of the Nineteenth Amendment. Members like Anita Pollitzer, who told a friend how much she enjoyed browsing through her six-volume *History of Woman Suffrage*, believed that suffrage history had shaped their perceptions, could teach them useful lessons, and would inspire women all over the country to join the ERA fight.[74]

A second and particularly powerful characteristic that held the Woman's Party community together was the explicit feminist identification of most of the members. National Woman's Party members, as part of the only national organization to identify publicly as a feminist organization, were especially likely to claim the label of "feminist" for themselves and their coworkers. The group's journal, *Equal Rights*, carried a column entitled "A Feminist Thinks It Over" and reported women's achievements as "Feminist Firsts" and "Feminist Milestones." The librarian described the Florence Bayard Hilles Library, part of Belmont House, as a "feminist library."[75] In private correspondence, members complimented other members and friends by describing them as "good feminists," "ardent feminists," "active feminists," "staunch feminists," "strong, articulate feminists," "real feminists," or "militant feminists." Alice Paul, herself with a reputation as a "super-feminist," described women as "born feminists."[76] The correspondence of the Woman's

Party shows that the term "feminist" was central to the group's identity, despite the prevailing societal conviction that feminists were "kooks," "freaks," or "eccentrics."[77]

Feminism meant different things to different women. Perhaps most important in organizational terms was the Woman's Party commitment to work by women for women. Members sought endorsements of the ERA primarily from other women's groups. One especially interesting indication of this commitment to working with women emerges from the notes of members involved in lobbying. Members frequently reported talking to secretaries in order to get information and win support for their cause. A Woman's Party staff member, for example, received a call from an ERA supporter who worked for an anti-ERA senator and used her position to whip up support for the ERA in the senator's home state.[78] As one officer reported, "this is the way the women worked, they could always find a back door to get in. . . . You see, we had women in every department in Washington."[79]

The preference of most members for working with women, along with their explicit feminist identification, made the group unique in this period. Women who joined the Party found the comfort of community support for their beliefs. One woman who came to the group in the 1940s as a relatively young woman recalled in an interview the excitement of finding other women who were feminists: "I had this marvelous feeling that I was no longer alone. . . . [A]s soon as I became a member of the National Woman's Party, I felt this wonderful backing. I was not alone. . . . And I was not quite as peculiar as I thought I was. I was not the only one. . . . I said, 'Oh my God! . . . this is terrific, this is terrific.' "[80]

The Woman's Party community centered around the Alva Belmont House, national headquarters in Washington, D.C., which served not only as an office but also as a permanent residence, feminist hotel for visitors coming to Washington, and clubhouse. A number of women lived at Belmont House and in two other Party-owned houses, and others stayed there for periods ranging from days to months while engaged in lobbying. Although some rooms in the houses were rented to nonmembers, even men, Belmont House played an important role in maintaining the women's community. The woman-centered character of the house's environment is suggested by the fact that one member wondered if being married would prevent one's acceptance there.[81] Another member who traveled to Washington to serve as house manager emphasized the importance of the house in referring to its "big contribution to the national and world woman-movement."[82] Some of the members made their home there; one woman wrote that she was "looking forward with joy to my return home, and *Home* to me now, means the dear Alva Belmont House."[83]

The women who lived and worked at Belmont House became, for some members, the "Woman's Party family." Members occasionally developed ties that they described as mother-daughter or sister-sister relationships. In fulfilling their functions as family members, women provided support of various kinds for each other. For example, Alice Paul, concerned about suffrage cartoonist Nina Allender's loneliness, gently encouraged her to come to Washington to live: "You would be surrounded by loving friends and by people who know and appreciate the very great gift you have made to our movement."[84] The daughter of another recently deceased member wrote to thank the members at Belmont House whose kindness and consideration had made it possible for her mother to continue working as long as she did.[85] Members sometimes tried to provide financial support for other members in need; one woman, for example, collected a fund to provide another with dentures.[86] Often members offered emotional support in times of stress and took a personal interest in each other's welfare. Members worried about Alice Paul's health in particular, urging her to rest, telling her what to eat, encouraging each other to take care of her.[87] By the 1960s, one member reported in an interview, the house had become a home for old women with problems.[88]

If Belmont House served as a surrogate home for some members, it also functioned as a sort of feminist hotel. Visitors who stayed there—often women scholars engaged in research in Washington—sometimes ended up hard at work on the ERA at Alice Paul's urging. A staff member at the house, for example, wrote to another member that a woman staying there while researching seventeenth-century British actresses at the Folger Library "has always been a feminist, but only since living here . . . did she take an active part in what we are doing."[89] Bringing friends to Belmont House, in fact, seemed to be one of the ways that women recruited new members. Martha Souder, the house manager, took to headquarters a friend who became so interested and appreciative of Alice Paul that she decided to try to help. Another member wrote about a friend who wanted to stay at Belmont House, adding, in what seems to have been a hopeful vein, "She has *our* point of view and is a top news woman."[90] Several members we interviewed noted with amusement that Alice Paul would immediately try to recruit any woman who visited Belmont House.[91]

For all National Woman's Party members able to travel to Washington, Belmont House served as a clubhouse and center of organizational activities. Since the group's main activities were lobbying Congress and garnering support from other organizations, much of the actual political activity took place at Belmont House. National conventions were originally held in different locations throughout the country, but by the post-1945 period they were scheduled only sporadically and only in Washington at Belmont House. Organizational decisions were made not at national conventions but by the officers or the national council, which met periodically at headquarters. Belmont House was also important as the site of feminist celebrations that were designed to win publicity for the organization and its cause—teas to honor women politicians or sponsors of the ERA, parties on the anniversary of Susan B. Anthony's birthday or the passage of the suffrage amendment, or celebrations of some victory in the ERA fight. Whatever the function, Belmont House served as a female world on some occasion for all members of the Woman's Party and as a feminist space for women who identified with the women's rights movement.

The symbolic as well as practical importance of Belmont House became clear in the 1947 schism. The height of the conflict came with an "invasion" of Belmont House by the opposition group, and one of the lawsuits sought an injunction against the closing of the house to members. The opposition group sought the house not only because of its considerable monetary value but also because of its history and its association with the women's movement. In an attempt to guard against such a fate, the Woman's Party after 1947 continued to be suspicious about new members and finally, in 1975, rewrote the bylaws to separate the house and property from the organization by setting up a corporation to take over ownership of the assets. By the 1980s, the Woman's Party was as interested in the just- completed restoration of the house, which had become a national historic landmark, as in its political battles. Elizabeth Chittick, the president since 1972, put a great deal of effort into saving the house from the ravages of time and the threats of government confiscation; in the early eighties, it once again opened as a feminist museum, filled with portraits and busts of women, a living monument to Alice Paul and her feminist sisters.[92]

The final two elements of the Woman's Party community, which are discussed in Chapter 5, need only be mentioned briefly here. Commitment was important throughout the women's rights movement, but especially in the National Woman's Party. As Ethel Ernest Murrell put it when appealing for contributions at Christmas time, "At this season of each year, we rejoice in gifts to those we love. National Woman's Party is truly the beloved of every member."[93] Close personal ties and the friendship networks that resulted from them also played a vital role throughout the women's rights movement. Within the National Woman's Party, friendships grew out of and in turn supported the work for the ERA. Correspondence among members—and there was little separation between official and personal concerns, a manifestation of the intensely personal nature of the group's political work—rang with expressions of intimacy and friendship. The intensity of emotion expressed during the internal conflicts is understandable in light of the centrality of friendships and intimate relationships in the group's everyday life. There were those who hoped during the lawsuit "to be able to differ on these questions and be the best of friends." Laura Berrien, not long before she signed her name to a series of vicious attacks on Anita Pollitzer, wrote to her, "If I send my love to you I hope it will be reciprocated."[94] Even Alice Paul, never renowned for her tact and diplomacy, wished Laura Berrien and Doris Stevens "all possible happiness and peace" on New Year's Eve of 1946.[95] But, while friendship could help to limit conflict, it could also aggravate it. The bitterness of the lawsuit was itself in part a result of the intermingling of personal and political ties for many of the members. When the court ruled against the rebels, they lost a friendship circle and a gathering place as well as a case.

The community described here helped make it possible for women in the post-1945 period to work for the feminist cause in a society either indifferent or hostile to feminist concepts and values. It provided a supportive environment that gave women a place in the historical tradition of the women's movement, valued feminism, gave meaning to members' lives by calling for commitment to a cause, and facilitated and nurtured warm friendships and intimate relationships. But the very elements that tied the community together—suffrage bonds, feminist identification, the centrality of Belmont House, the level of commitment, the intensity of relationships—could lead to serious conflicts and also worked to further the elitism and exclusivity of the group. Not surprisingly, as a more mass-based women's movement began to grow in the 1960s, the National Woman's Party faded into obscurity.

Conclusion

Despite the exclusivity and penchant for conflict of the Woman's Party, it played a role in the women's rights movement out of all proportion to its size or even its political effectiveness. It managed this in part because it established a reputation as a feminist organization with a militant past at a time when other women's organizations were disassociating themselves from feminism, and in part because it simply persevered in a "very one track" way.[96] The structure and functioning of the Woman's Party were shaped by the organization's past history, and they were well-suited to survival in a social climate of antifeminism.

Participation in the suffrage movement affected all aspects of the Woman's Party—its structure, membership, ideology, goals, and strategy. Because the Party took shape during the suffrage struggle and many of its leaders and core members had been suffragists, the organization remained committed to bringing about change through passage of a single piece of legislation. Although the Woman's Party in the 1940s and 1950s never considered using the militant tactics for which it had become famous, it followed the path laid down in the early years in every other way.

Furthermore, the exclusivity and hierarchical structure of the organization served its interests in the antifeminist atmosphere of this period.[97] The organization had originally lost much of its diversity of membership in the early 1920s, when Alice Paul forced through the commitment to the ERA. In the 1947 schism, a faction within the organization proposed a nonmember "clubwoman" for chairman, and in 1951 a relative newcomer attempted, as chairman, to change the character of the group in fundamental ways. In response, the leadership became even more cautious about seeking new members. From the perspective of the leadership, the commitment to the ERA had to be protected from outsiders or new members who might attempt to change the goals and strategies of the group. An exclusive organization could accomplish this.

The Party's hierarchical structure also helped to keep the organization firmly on the track of support for the ERA. A hierarchical structure is useful for attaining an immediate, tangible goal, but is less useful for changing individual viewpoints, experimenting with new strategies, and dealing with diversity and conflict.[98] The history of the Woman's Party in the 1940s and 1950s shows that its hierarchical structure did allow the organization to continue to work for the ERA, but at the expense of converting women, trying new approaches, or building a diverse organization.

If one were to evaluate the Woman's Party on the basis of results, the record would not look very good. Nor would the Woman's Party receive good marks from feminists of the 1980s, since its elitism, to say nothing of the conservatism and racism of some of its members, violates the principles of participatory democracy and the commitment to progressive politics that have come to be associated with the women's movement. Yet the tenacity of the Woman's Party quest for equal rights and the community it created demand recognition.

The National Woman's Party stood out for its feminist identification and perseverance, but it was not alone in fighting for women's rights in the late 1940s and 1950s. In fact, it pursued its goals in a context of women's rights activism on the part of a variety of organizations and individuals. We turn now to the larger women's rights movement.

4

The Women's Rights Movement

The National Woman's Party took the lead in the late 1940s and 1950s as the center of a loosely structured women's rights movement. What does it mean to consider these women as participants in a social movement? According to John McCarthy and Mayer Zald, the opinions and beliefs in a population that favor change in the social structure or reward system of a society constitute a social movement.[1] Social movements are generally comprised of a number of organizations that compete and cooperate in pursuit of specific goals.[2]

A wide variety of groups and individuals in the post-1945 period contributed to the set of opinions and beliefs that favored change in the status of women. But the organized women's rights movement had come to mean a relatively small group of organizations and individuals that saw themselves as the heirs of the suffrage movement, focused narrowly on what had traditionally been understood as "women's rights," and accepted an explicit or implicit feminist ideology. Three criteria set apart those groups and networks of women that identified as part of the women's rights movement from those that did not: commitment to fighting for women's rights, work in concert in pursuit of common aims, and the existence of connections, if only on occasion, with the self-identified feminist center of the women's rights movement, the National Woman's Party. (See Table 4.1.) In a period in which feminism had come to be seen as anachronistic and abrasive and feminists as "more hysterical than historical," groups and individuals who remained committed to women's rights were seen as peculiar, if not deviant.[3] In "What Has Happened to the Feminist Movement?," published in an anthology on leadership in 1950, the authors explicitly contrasted those women who were a part of the old-fashioned, separatist, militant feminist movement with the modern "shoulder-to-shoulder stand with American men" characteristic of other 1950s women leaders.[4]

As the only self-proclaimed feminist organization devoted exclusively to winning women equal rights, the Woman's Party had an influence out of all proportion to its numbers. This was not because it was well-liked. To the contrary, most other women's rights organizations detested the Party,

45

TABLE 4.1. Women's Movement Organizations, 1945 to the 1960s

	Core	Periphery
Preexisting	National Woman's Party Business and Professional 　Women's Clubs (BPW) National Association of 　Women Lawyers American Medical Woman's 　Association Pioneers of the women's 　movement network	General Federation of 　Women's Clubs National Council of Women American Association of 　University Women (AAUW) National Association of 　Colored Women Women in politics network Zonta Soroptimist Altrusa Federation of Women 　Shareholders
Emergent	Connecticut Committee for 　the ERA Massachusetts Committee for 　the ERA Lucy Stone League Women's Joint Legislative 　Committee Industrial Women's League 　for Equality St. Joan Society	Women in World Affairs Multi-Party Committee of 　Women Assembly of Women's Organi- 　zations for National Security

especially if they opposed the Equal Rights Amendment, or at best distrusted it because of its proprietary attitude toward the amendment. It seemed sometimes as though the Woman's Party would rather see the ERA go down to defeat than relinquish leadership of the struggle. But as a consequence of its persistence and perseverance, the Woman's Party could not be ignored by any group or individual concerned with women's rights. From the perspective of its members, the Woman's Party was "the only organization that has consistently worked over the years in behalf of women."[5] Even an ambivalent observer admitted that it "is of course the spearhead of the movement."[6]

A core of organizations, clustered around the National Woman's Party, supported its central goal, the Equal Rights Amendment. As a result of the Party's persistence, the ERA had, in fact, come to symbolize the primary definition of "women's rights" that characterized the women's rights movement. Many of these groups had existed prior to the 1940s and 1950s and had purposes broader simply than passage of the ERA. Still, their activities on behalf of the amendment were central to the functioning of the women's rights movement in this period, and they engaged in cooperative and competitive interaction in the struggle for equal rights. Preexisting organizations that formed the core of the movement included groups of professional women such as the National Federation of Business and Professional Women's Clubs (BPW), the National Association of Women Lawyers, and the American Medical Woman's Association. Although not all BPW members would have considered themselves part of a women's rights movement, it is significant that a study published by the BPW in 1954 described the members as "emancipated women."[7]

In addition, a loose network of women dedicated to winning recognition for the pioneers of the suffrage movement can be considered part of the core. This network included a Minnesota woman, Rose Arnold Powell, who belonged to both the National Woman's Party and the General Federation of Women's Clubs; the head of the Susan B. Anthony Memorial in Rochester, New York; the director of the Susan B. Anthony Memorial Library in California; and the membership of the Committee on Pioneers of the Woman Movement of the National Woman's Party, chaired by Ethel Adamson.

Not all of the groups interested in women's rights could be traced to the suffrage movement. Several new or emergent groups in the 1940s and 1950s enlarged the core of the movement. The Connecticut Committee for the Equal Rights Amendment, founded in 1943, was a state group active on the ERA and other feminist issues. It spawned other state groups, the most successful of which was the Massachusetts Committee, established in 1955. The Lucy Stone League, a local women's rights group organized in New York in 1950, was a revived and broadened organization based on a 1920s forerunner that had fought for the right of a married woman to retain her own name. In its new incarnation, it sponsored luncheons and talks and put out a newsletter on the status of women.

The National Woman's Party also set up several new organizations intended to demonstrate wide and diverse support for the ERA. These included the Women's Joint Legislative Committee established by Alice Paul in 1943 as a coalition lobbying group; the Industrial Women's League for Equality, organized in 1944 to counter arguments that women workers did not support the ERA; and the St. Joan Society a group of pro-ERA Catholics founded in 1943 to influence Catholic opposition to the amendment.

On the periphery of the women's rights movement were a variety of preexisting and emergent groups that supported some of the goals of the movement and cooperated periodically with core organizations to pursue common aims, but not with the same level of interest and consistency as the core groups. Included among these were the General Federation of Women's Clubs, which had endorsed the ERA during World War II; the National Council of Women, a coalition of women's organizations that supported some efforts to improve women's status but was relatively inactive in the 1940s and 1950s; the American Association of University Women (AAUW), which dropped opposition to the ERA in 1953 and worked to win women positions in policy-making in conjunction with other organizations; the National Association of Colored Women, which endorsed the ERA and belonged to the Women's Joint Legislative Committee; the Federation of Women Shareholders, an organization dedicated to placing women on the boards of directors of every large corporation; service organizations of executive women in business and the professions such as Zonta International, Soroptimist International, and Altrusa International; and a network of women in politics, including women active in the women's divisions of both the Democratic and Republican parties, as well as some women politicians. Two new groups, Women in World Affairs and the Multi-Party Committee of Women, brought together women's rights activists and members of other women's organizations to support the appointment of women to policy-making positions. Another coalition, the Assembly of Women's Organizations for National Security, emerged in response to the Korean War.

State and local groups, individuals, networks, and national organizations joined together to promote women's rights. Particularly important was the participation of national leaders who could command a following within their organizations. Often state and local branches of large national organizations, such as the National Federation of Business and Professional Women's Clubs and the American Association of University Women, remained uninterested in and uninvolved with the feminist issues identified by the national leadership.

For example, one small-town Ohio member of the local AAUW branch remembered that her group always had a Status of Women chair because the national leadership required it, but the chair rarely did anything and the post became merely a way to introduce potential new leaders to the board. The national organization, she noted, was always ahead of the local branches since program development took place at national meetings and it was the activists who attended conventions.[8] But activity at the local level also depended on the interests of members. In Columbus, Ohio, for example, the local AAUW branch launched a campaign to get women appointed to the board of trustees of Ohio State University and published a booklet on outstanding women of Franklin County in response to the national program on women in policy-making.[9]

All of the diverse organizations and networks of women, some of which owed their existence to the suffrage struggle and others newly formed in the period, made up a loosely connected women's rights movement. As one National Woman's Party member described it, "the movement is a huge mosaic [sic]" in which "every little fragment of work contributed by some one" was important.[10]

What held together these different organizations and networks was a commitment to women's rights and interaction on behalf of particular goals. In contrast, a number of other groups worked to improve women's lives but were generally hostile to the groups dedicated to women's rights. The League of Women Voters, for example, explicitly disavowed feminism and even denied any particular concern with women's issues. Anna Lord Strauss, president of the League from 1944 to 1950, complained to her office staff: "If I hear much more about women's rights I am going to turn into a violent anti-feminist."[11] Throughout the 1950s, the League asserted that it was a citizen's group rather than a women's organization and even considered changing its name to the League of Active Voters.[12] Throughout the late 1940s, League officers struggled to decide whether or not to drop out of the International Alliance of Women, a federation of women's groups they considered too feminist. One leading member contrasted "those primarily interested in women (the old line 'feminists') and those primarily interested in seeing women play their part in world affairs."[13] The League wanted to see the International Alliance "put less emphasis on fighting for the rights of women and more on doing the job that needs to be done and winning recognition through their accomplishments."[14] By 1949, Anna Lord Strauss had decided that it was time to withdraw. She explained to the head of the Alliance that "the feminist approach has little appeal" to League members, who "think of themselves as citizens first and as women incidentally" and had developed a national program that had "no subject which is primarily of interest to women."[15] In 1951, the League severed its connection with the International Alliance of Women.

In addition, the League left the Women's Joint Congressional Committee, a coalition lobbying group that opposed the Equal Rights Amendment, because it did not want to be affiliated solely with women's organizations, and even dropped opposition to the ERA from its program in 1954 because it did not seem important enough to oppose and its membership knew little about the amendment. The League was a solidly liberal organization that, like the country's most prominent liberal woman, Eleanor Roosevelt, rejected feminist identification while serving as an example of what women could do in leadership roles. Although the League had roots in the nineteenth-century women's movement, trained generations of women leaders, and included women who were feminists, whether or not they identified as such, it chose in this period to disassociate itself from the women's rights movement. Rejecting identification with feminism and even with any characterization as a women's organization, League officers sharply distinguished themselves from the "feminists or 'equal righters.' "[16]

In the same way, the Women's Bureau of the Department of Labor worked hard to improve the conditions of working women but distinguished its activity from that of working for "women's rights." In 1947, for example, director Frieda Miller testified before a Senate subcommittee on behalf of jury service for women by claiming that "I am speaking today not for women who are merely grasping a 'right' as part of a feminist program."[17] Making a similar distinction, the assistant director in 1954, in planning a conference on womanpower, sought a title that would "avoid the old battle of the sexes idea and pitch a conference around the idea of men and women working together and maintaining the Nation's economy."[18]

The Women's Bureau had specific class interests, and these, like the race or religious interests of organizations such as the National Councils of

Negro, Jewish, and Catholic Women, tended to prevent identification with the women's rights movement. "Women's rights" had traditionally meant property rights, the right to vote, and other legal and political issues of relevance to the white middle-class women who comprised the bulk of the membership of the movement. Reforms of particular interest to black women or working-class women, such as protection of the right to vote, anti-lynching laws, the abolition of segregation, or legislation affecting the conditions of labor, had been pushed aside as racial or class matters. Organizations such as the Women's Bureau worked on behalf of women, but they often perceived their interests as antagonistic to those of the self-identified women's rights movement, largely as a result of the long tradition in the American women's movement of distinguishing gender from race or class issues.

Membership

The membership of the women's rights movement was, in the post–1945 period, quite homogeneous, being overwhelmingly composed of white women.[19] The only black woman publicly identified with the movement throughout this period was Mary Church Terrell, founder of the National Association of Colored Women, who belonged to the National Woman's Party and fought to integrate the Washington, D.C. chapter of the AAUW. Pauli Murray, a black lawyer involved in the civil rights movement and active in the founding of the National Organization for Women (NOW) in 1966, participated in organized feminist activity in the early 1960s but did not really see herself as a part of the women's rights movement in earlier years. In her opinion, however, her lack of participation was a result of the general pattern of segregation that characterized women's organizations; a white woman in her position would likely have belonged to some of the organizations active on behalf of women's rights.

Most feminists in this period were by birth, marriage, or occupation middle, upper middle, or upper class. Some, like Perle Mesta, the famous Washington hostess and diplomat, came from wealthy families and married within their class. Others, like Mary Roebling, a New Jersey banker, came from middle-class origins and married into wealth. Roebling, after her husband's death, took over the management of a bank and by 1957 had become one of the ten richest women in the United States. Extremely wealthy women were the exception, but few members had to struggle to survive. Some working-class women did participate, although often with difficulty. Fannie Ackley, for example, a retired linotype operator, threw herself into work for the National Woman's Party but could not play a national leadership role because she lacked the funds to travel to Washington and to national conventions. Sufficient income was important since women paid their own expenses for most of their feminist work.

The majority of women involved in women's rights work fell in between the extremes of upper-class wealth and working-class limited income, and most were well-educated and employed. Few of the women about whom information is available had less than a college education, and many had received advanced degrees. Many of the women's rights organizations, such as the AAUW and the BPW, had educational and occupational requirements which affected the status of women in the movement as a whole. Of the 108 core members identified for purposes of this analysis, 81 percent (88) could be identified as employed. Of the twenty whose occupations are unknown, some were retired, some wealthy women who did not work, some house-wives, and some full-time activists in the movement. Lawyers made up the largest occupational category: 23 percent (20) had been trained as or were active as lawyers or judges, 10 percent (9) taught school below the college level, 6 percent (5) were college professors, and 7 percent (6) were college administrators. Nine percent (8) were in business or public relations, 10 percent (9) were writers or journalists, and 11 percent (10) were politicians or civil servants. Five percent (4) worked in industry or the trades, two of them as linotype operators. Other occupational categories included librarian, reformer, labor union official, engineer, nurse, physician, psychiatrist, sculptor, factory counselor, economist, geologist, historian, and student.

Perhaps the most striking characteristic of women's rights activists in this period is their age. Few of the women were young, and a large proportion were well beyond middle age by the post-1945 period. Most were born in the last quarter of the nineteenth century. Of the fifty-seven women in the core membership for whom information could be found, 2 percent (1) were born in the 1850s, 12 percent (7) in the 1860s, 25 percent (14) in the 1870s, 30 percent (17) in the 1880s, 19 percent (11) in the 1890s, 10 percent (6) in the 1900s, and 2 percent (1) in the 1910s. In short, 88 percent were over fifty in 1950. The few women who came to the movement in the 1940s and 1950s as relatively young women seemed to feel out of place. Indicative of this, one woman who joined the National Woman's Party at forty was actually considered a youthful asset to the organization.[20] This age pattern—a problem recognized by women throughout the movement—resulted from the fact that most women seemed to have developed their commitment to feminism in the early years of the twentieth century, particularly in the last years of the suffrage struggle. By the post-1945 period, the elderly age of women's rights activists drew barely concealed ridicule from the media and others and undoubtedly contributed to the portrayal of feminism as an anachronistic cause.

Almost as striking as the age of the members is the breakdown on marital status. In a period during which a larger proportion of women than ever before were married, 41 percent (36) of the eighty-eight women in the core membership whose marital status is known were not married. A third of these women lived with other women. Of the 59 percent (52) of women who identified as married, about 23 percent (12) were widowed. Marriage could make participation in the movement difficult, and children increased the barriers to participation. Since so many of the women involved in feminist activity were past middle age, few had children living at home. Some of the most active married women never had children.

A survey of BPW members published in 1954 shows that the membership characteristics outlined here match the general profile of the BPW membership in the 1950s. A third of all BPW members had graduated from college, almost 40 percent worked as professionals or semi-professionals, six out of ten were at least forty-five years old, and 44 percent had never married.[21]

One final characteristic of the membership is important here. Women's rights supporters did not come predominantly from one political party, nor did they all share a common political outlook. Women prominent in both the Republican and Democratic parties participated in the women's rights movement and worked together easily without setting aside their commitment to party politics. Some women were liberals active in a variety of progressive causes while others held right-wing views. In short, participation in women's rights activities had little relationship to an individual's general political views. As one Woman's Party member commented, the fact that some members were conservative and others were considered rather radical "just shows that these women were conscientious in their point of view about equality of rights for women."[22]

Feminist Ideology

What held together this group of women of elite social status with disparate political views was a commitment to feminism. In the public view, feminism had already become by the postwar period "as quaint as linen dusters and high button shoes."[23] To be sure, the social climate of antifeminism limited the number of women willing to identify publicly as feminists, but it did not eliminate them altogether. The National Woman's Party continued to identify itself publicly as a feminist organization. Members regularly used the term "feminist" to describe themselves, their friends, and their organization. Women outside the Woman's Party were less likely to use the term "feminist," but its use was by no means unknown among other women. One woman who advertised as a professional lecturer described herself in her brochure as an "ardent feminist."[24] Rose Arnold Powell, the Minnesota woman who devoted her life to winning recognition for Susan B. Anthony, regularly described herself as a feminist, even when writing to individuals outside the movement. Lena Madesin Phillips, founder of the International Federation of Business and Professional Women's Clubs, described herself in her unpublished autobiography as a feminist. BPW leader Lucy Somerville Howorth wrote in her college alumnae bulletin in 1948: "Briefly, I am a lawyer; I was a politician; I am a feminist."[25]

Other women used the term to describe friends and acquaintances. Doris Stevens, who left the Woman's Party in 1947 and busied herself with anti-Communist activity, described an acquaintance as a "life-long feminist."[26] Her friend Jane Grant, who revived the Lucy Stone League in New York in 1950, used the term to describe acquaintances to her friends. The head of the Connecticut Committee for the ERA, Florence Kitchelt, referred to the "feminist cause" and described other women as feminists, but wrote good-naturedly to the editor of the New Haven newspaper when an article described her as an "ardent feminist" to inform him that she was "a little skeptical about its applicability."[27]

The term "feminist" had come to be associated in the 1920s and 1930s with support for the ERA, so women who opposed the amendment but worked on other women's issues would have been less likely to use it to describe themselves. Some women avoided the stigma of identifying as a feminist by coining new words. Florence Kitchelt, who doubted that she was an "ardent feminist" and avoided the term "feminism" "because it is so widely misunderstood," thought the author of an article she liked sounded as if she might be a "good Equal Righter."[28] National Woman's Party member Perle Mesta, appointed U.S. Minister to Luxembourg in 1949, claimed to have invented the term "Equalism" as a substitute for "feminism."[29] Several women used the term "Equalitarian" instead of "feminist."

Women who did identify as feminists recognized that the public at large did not share their pride in this label. BPW leader Lucy Somerville Howorth noted that "too many women shudder at being called feminist," but proclaimed that "I glory in being a feminist."[30] Other women also remembered the hostility directed against them as feminists.[31] Some recalled total avoidance of the label; one woman active in the early 1960s discussed how carefully sympathetic women used the word "feminist" among themselves—she described it as an "in-house term"—which suggests that some women who described themselves and their friends as feminists in their correspondence would have been loath to do so in public.[32] It is perhaps revealing that another woman active on the local level recalls that she began calling herself a feminist "whenever that label became an accepted label."[33] As a result of the pressure, some women, like Susan B. Anthony II, grandniece of the suffragist and a leftist activist in the 1940s, found the experience of being a feminist "very lonely."[34] But others, like optimistic feminist biographer Alma Lutz, believed things would improve and pointed out that "feminists have always been a minority."[35]

What did the women in the movement mean when they spoke of feminism? There is no simple answer to this question, although the term usually did suggest support for the ERA. But not all women who supported the amendment claimed the label, and in any case feminism suggested much more than simply support for one piece of legislation. The fifth edition of *Webster's Collegiate Dictionary*, published in 1942, defined feminism as the "theory, cult, or practice of those who advocate such legal and social changes as will establish political, economic, and social equality of the sexes."[36] The most interesting word here is "cult," which suggests a conception of feminism peculiarly fitting for this period. For our purposes, we define feminism as a world view that ranks gender as a primary explanatory factor and/or category of analysis for understanding the unequal and unjust distribution of power and resources in society; integral to this world view is a commitment to changing that unequal distribution. Feminists may hold differing views of the origins of inequality that lead to entirely different solutions, but feminism provides elements of a common conceptual framework and a basic perception of women as a disadvantaged group.

An exploration of the feminism of various women's rights supporters shows the range of ideas encompassed by the term "feminism" in this period. Nora Stanton Barney, the granddaughter of suffrage leader Elizabeth Cady Stanton and the daughter of militant suffragist Harriot Stanton Blatch, defined feminism in a letter written with an eye to publication to Communist Party member Bella Dodd: "To me, a feminist is one who thinks that women are primarily human beings with the same minds, ambitions, ability and skill, consciences, and power for evil and for good, as men. . . . The feminist believes that any differentiation of treatment of the sexes can only find excuse in the protection of the health and interests of the child."[37] Barney's definition speaks to an important point of controversy in the women's rights movement. Some women believed that the way to win equality was to deny gender differences; since they wanted equal opportunity, they reasoned, they should assert the basic humanity of women. Florence Kitchelt, for example, echoing the feminist critique offered by theorist Charlotte Perkins Gilman in her 1898 work *Women and Economics*, lamented that American women were "over-sexed," by which she meant that society paid too much attention to sex differences. She emphasized, in contrast, the common interests of women and men, and preferred words such as "people, citizens, companionship, community."[38] Historian Mary Beard agreed that Americans placed too much emphasis on sex and longed for women to understand that "they are first human beings, with the dignity therein implied."[39] Katie Louchheim, head of the Women's Division of the Democratic Party, spoke at the 1960 Democratic Convention to emphasize that women were first of all people.[40] These women and others like them recognized that society distributed rewards unjustly on the basis of gender, but argued for change by insisting that women be treated like men rather than that the system as a whole be transformed.

In contrast, other women denied that women were persons first and women second, although they insisted that women should nevertheless have equal rights with men. They put women's gender identity first because they believed that women were in fact superior to men. These women carried on the tradition of the nineteenth-century feminists who believed in the moral superiority of women, especially with regard to their peace-loving and life-giving nature. Florence Allen, the first woman to hold a federal judgeship, suggested at the 1946 convention of the Federation of Business and Professional Women that women had the responsibility to save the world from the moral bankruptcy that had led to World War II.[41] In the same vein, Rose Arnold Powell wondered on July 4, 1949, whether, if the Founding Fathers in 1776 had given women equality, the world might not have

escaped two world wars.[42] Lena Madesin Phillips corresponded with a number of European women in the International Federation of Business and
Professional Women who believed that women, the transmitters of life, could
govern the world better than men, and she agreed that "Women are the
great adherents of a peaceful world."[43] In the 1970s, Alice Paul told an
interviewer that: "Women are certainly made as the peace-loving half of
the world and the homemaking half of the world, the temperate half of the
world. The more power they have, the better world we are going to have."[44]
Some women asserted that socialization made women different and better,
as did politician India Edwards, who stated in her memoirs that she believed
women were more honest than men, although that might not be true after
women had been in the political arena for awhile.[45] Other women said
nothing about socialization but simply asserted women's superiority. What
all of these women shared was the conviction that women were fundamentally different from and superior to men and that the movement of women
into positions of power would transform society in basic and desirable ways.

Feminists of all stripes, however, insisted that women were discriminated against in American society. Rose Arnold Powell complained of "the
inferiority of women and their duty to remain in silence and subjection
enwrapping every baby girl in her cradle, like an invisible straitjacket."[46]
Florence Kitchelt believed that women "have always been a subject class,"
and that the "whole world functions on women's subservience."[47] Vivien
Kellem, a flamboyant right-wing manufacturer and feminist, insisted, according to a profile in the *New Yorker*, that "the female sex is forever being
done out of its rightful desserts by underhanded male machinations."[48] In
"Confessions of a Feminist," Lucy Stone League member Jane Grant traced
her realization of discrimination to her marriage and subsequent loss of
name. Once launched on the struggle to regain her name, she realized that
there must be other things to fight for as well.[49]

Women throughout the women's rights movement spoke out on different aspects of gender-based discrimination. Some came to feminism
through the experience of employment discrimination. One National Woman's Party member reported to Alice Paul that many women bore the brunt
of job discrimination and that she had talked to many women in government
jobs in her own building who complained bitterly but were too afraid to
protest openly.[50] Other women noticed discrimination against women in the
family, although the movement as a whole kept silent on issues such as
abortion and birth control, child care, and the unpaid labor of women in
the home. Florence Kitchelt, whose husband shared in the domestic duties
of their home, believed that day nurseries should be funded as part of public
education. Kitchelt believed that "young husbands are accepting homemaking as a 50–50 job," but ultimately she envisioned less a revolution in
the division of labor than the availability of choice for women. She wanted
every woman to have the opportunity to combine marriage and career, but
had no intention of telling women what choices to make: "I hope, and am
sure, great numbers would choose the domestic role."[51]

Kitchelt stands out among women's rights supporters for her broad analysis of discrimination. Not only did she support public child care, she also attacked the consumerism of the period for its impact on women. She congratulated the founder of a Dallas group, the "Little Below the Knee Club," for her stand, covered in the *New York Times*, against changing hemlines. She saw it as a protest by "self-respecting women who resent being treated as puppets, as an idle class whose chief occupation is self-adornment."[52] She carried this theme further in letters to a radio broadcaster, to writer Ashley Montagu, to UN General Assembly President Vijaya Lakshmi Pandit, and to her friends, complaining about the exploitation of women by the fashion and cosmetic industries. Kitchelt expressed horror, for example, over a *Life* magazine photograph of a fourteen-year-old girl dressed in a strapless chiffon ball gown for a formal dance, blasted a Connecticut labor newspaper for printing an exploitative picture of a woman in a bathing suit, and chided Eric Sevareid for his frivolous coverage of Perle Mesta's swearing-in as Minister to Luxembourg.[53] "[I]f Mr. John Doe had been the new envoy," she wrote to Sevareid, "you would not have selected certain guests for mention because they had been reducing. . . . There was a breath of masculine disdain in your allusion to the National Woman's Party without a word of explanation of its objective, equality under the law."

Like Kitchelt's critique of fashion and consumerism, the attention of other women to the issues of language and names foreshadowed the protests of women in the resurgent movement of the late 1960s. Rose Arnold Powell wrote to the chief of the U.S. Weather Bureau to object to the use of women's names for hurricanes and tornadoes.[54] Fannie Ackley objected to the use of masculine pronouns in the generic sense, pointing out that if "one sex has to mean all humanity, surely it would be more logical to use 'woman' instead of 'man,' since it is womankind that bears and rears all humanity."[55] Other women commented on the unfairness of the terms "Miss" and "Mrs." when men were not identified by marital status. Several women, in the tradition of Lucy Stone, refused to change their names when they married, leading one Woman's Party member to explain to a correspondent that she was "a Miss, a genuine one—I have no husband in the background as some of us have."[56] In 1950, fifteen women met to resurrect the Lucy Stone League, even though some members feared that the issue of married women's names would frighten the "timid souls" among younger women.[57] But the issues of language and names, which to many symbolized discrimination against women, remained important to some feminists in the post–1945 period.

Women's rights advocates spent little time dwelling on the origins of gender-based discrimination or on the responsibility of men as a group for the social system, but they did on occasion express anti-male sentiment, which may have had something to do with Fannie Ackley's lament that the word "feminist" had come to mean "man-hater." Rose Arnold Powell was particularly vehement, complaining in her diary of "Masculinity running rampant *all over the earth!*," rebelling at the "utter man-mindedness" she saw around her, determining to break down the "autocracy of sex so firmly imbedded in the subconscious minds of men," and writing of her soul's loathing of her treatment by men.[58] Fannie Ackley herself, who resented the charge of "man-hater," denounced a male antifeminist author for writing an article "saturated with masculine egotism, sex bigotry, jealousy, and prejudice—supported by false and vicious propaganda—which I think you will agree are among the world's worst evils."[59] A California Woman's Party member wrote to praise the work of her fellow members in Washington, remarking that "anyone who has worked with politicians has no doubt in his mind why his Satanic Majesty is a male."[60] Another woman, a lawyer, stated plainly that if she were not smarter than most men, she would not even ask for equal rights; she thought that most women felt the same way.[61] A woman who left the Woman's Party in the 1940s reported that a member of her state branch of the Party was furious when a man certified the accuracy of the treasurer's book since the male signature "defiled" the book.[62] Perhaps it is not surprising that the issue of man-hating became the central one in the contested will of a woman who left her estate to the National Woman's Party in 1945.[63]

What all of this suggests is simply that feminism was, for women's rights supporters, a world view. Although they usually worked on specific issues that do not seem to us today a comprehensive attack on the patriarchal order, and although they did not develop a well-articulated and systematic theoretical position, they did identify the manifestations of women's inequality in institutions and interactions throughout American society. Women's rights supporters, and especially National Woman's Party members, were usually acutely conscious of gender in all situations. Rose Arnold Powell, for example, burned with indignation when she learned that the publication of Alma Lutz's biography of Susan B. Anthony had been delayed because of the appearance of another biography: "Bosh! That isn't the real reason," she fumed in her diary. "No delay if it were a man."[64] While Powell was unusual in the vehemence of her denunciations of discrimination, her feminist world view was typical of the most dedicated women's rights supporters.

Women's rights advocates believed in the existence of discrimination against women, whether they used the word "feminist" or not, but perhaps even more important, they believed in the solidarity of women. For many, collective action by women—that is, a separatist organizational strategy— seemed the only way to bring about change. The very use of the term "woman" rather than "women"—in the name of the National Woman's Party, in the use of the term "woman movement," and in other usages— expressed a commitment to the unity of women.[65] Alice Paul characterized women's rights advocates as sharing a "feeling of loyalty to our own sex and an *enthusiasm* to have every degradation that was put upon our sex removed."[66] The National Woman's Party, like other groups, acted on this feeling of loyalty by turning to other women's organizations for support in attaining its goals. For most women, activism meant working with women

for women, either with other individuals or with exclusively female organizations. But some, believing with one labor woman that "[t]his *was* a
woman's rights movement; now it is an *equal rights* movement," sought to
bring men into partnership in achieving their goals.[67]

There were those, even strong feminists, who thought that men could
be more effective than women. One old suffragist, for example, had her
brother-in-law write a letter on behalf of the ERA because "a letter from
a *MAN* would be more effective."[68] Florence Kitchelt praised an article in
the *Saturday Review*, noting that its male authorship was "much in its favor."[69] The head of the Susan B. Anthony Memorial in Rochester tried to
get men to write letters supporting Anthony for the Hall of Fame because
she thought that they would count for more than women's letters.[70] A group
of women arranging a debate in the American Civil Liberties Union on the
ERA made sure to have a man speak for the pro-amendment side.[71] Usually
women who sought men's participation did not believe that men could do
a better job, but based their approach on the conviction that society paid
more attention to men.

Florence Kitchelt, who herself worked in happy partnership with her
husband Richard throughout their long and companionable marriage, was
one of the main advocates of bringing men into the women's rights movement. She wrote to one friend that the idea of a Woman's Party did not
appeal to her and that she would prefer to see a National Committee for
the ERA that welcomed men into membership.[72] In her own Connecticut
Committee for the ERA, Kitchelt eagerly sought male members, trying to
get men for 50 percent of the national board positions. Other women who,
like Kitchelt, left the National Woman's Party during the schism, also rejected the all-woman composition of the Party. Even Jeannette Marks, with
her strong feminist identity, wanted to recruit men for her New York group.[73]

The National Woman's Party did not admit men. In fact, Alva Belmont,
the wealthy benefactor of the Party in the early years, included a clause in
her bequest that revoked her legacy if men ever joined or participated in
the organization. As a result, much of the agitation about men in the wom-
en's rights movement centered around the Woman's Party, despite the fact
that men could not join most women's organizations. Although Alice Paul
never seemed to waver on the idea of a separatist organization, the issue
arose periodically. One member of the Finance Committee responded ve-
hemently in 1949 to an inquiry concerning the hiring of paid male fund-
raisers. Not only would it be contrary to the spirit of the Belmont money,
but it would represent an "injury to the prestige of our wonderful membership. I can scarcely think," she wrote, "of any move that could reverse
our position more ostentatiously, and completely."[74] The 1949 Party convention defeated a resolution to admit men. But the issue came up again
during the conflict-ridden chairmanship of Ethel Ernest Murrell in the early
1950s, when she hired a public relations man and sought to organize mixed
membership groups that endorsed the ERA, although she carefully avoided
referring to them as branches of the Party. The chairman of the southern

California branch of the Party wished to enroll men; one of the California members noted that "scores of women *just do not like* the idea of a special political party for women only." As a result of consultation with Murrell, the group established a Council of Legal Equality which admitted men as associate members.[75] No other branches seemed to follow the lead of the southern California one, however, and Murrell's plans eventually led to her resignation from the chairmanship. In 1959, a woman chemist who described herself as a "friendly stranger" in favor of the ERA suggested to the Woman's Party convention chairman that the organization change its name to the National Women's and Men's Party and that men be brought in to hold half the offices. The Convention chairman dismissed the proposal as "fantastic."[76] Despite the convictions of some women that men should be brought into the movement, the women's rights movement retained a basically separatist organizational strategy.

Goals and Strategies

Three major issues received the most attention in the women's rights movement: the ERA, women in policy-making roles, and advocacy of women's history.

The Equal Rights Amendment

The Equal Rights Amendment was the foremost, as well as the most controversial, feminist issue in the 1940s and the 1950s. Drafted by Alice Paul immediately after the suffrage victory in 1920, it had won little support until after World War II. The Federation of Business and Professional Women endorsed it in 1937, the General Federation of Women's Clubs in 1943, and a host of smaller organizations joined the Woman's Party in favoring the amendment from the 1930s on. The League of Women Voters opposed the ERA from the beginning, as did the majority of women's organizations that had grown out of the social reform tradition. At stake was the fate of protective legislation, for which most women's organizations had battled long and hard in the late nineteenth and early twentieth centuries. The anti-ERA forces believed that the Woman's Party and other pro-ERA organizations consisted of wealthy women and professional women, probably financed by the National Association of Manufacturers, who cared not a whit for the well-being of working women who desperately needed the protection of social legislation to allow them to compete in the labor market. Pro-ERA groups, on the other hand, believed that protective legislation in fact discriminated against women, and that only through complete legal equality would women ever advance in American society.

Although some opponents of the amendment sought to keep women in their "place," most of the women's organizations that fought the ERA in this period sought to improve the lives of women. They opposed the amendment because, as one letter from a coalition of opposition groups stated, they believed that "its feminist ideology, holding that identical treatment would result in equal treatment, is both fallacious and obsolete."[77] Those opposed to the ERA emphasized the differences in the biological and social functions of men and women, which in their eyes made the concept of equality something that could not be achieved by the ERA. Mary Anderson, the first head of the Women's Bureau in the Department of Labor, stated this clearly in a letter written to Lena Madesin Phillips in 1937. She believed that the word "equality" represented a myth and that, because "there is no equality among men," equality for women "will mean little in our everyday life."[78] Historian Mary Beard agreed that the ERA could not establish "any conception of 'equality' as a fixed reality."[79] Mary Van Kleeck, a reformer who had served as the head of the agency that eventually became the Women's Bureau, explained to Alice Paul that "I do not believe that you and I have ever differed by a hair line in our clear objective for women." She attributed her anti-ERA stand to the fact that she saw "economic and industrial causes of women's inequalities," rather than legal ones, as basic, but expressed her respect for Paul's devotion to her cause.[80] Judge and prominent ERA opponent Dorothy Kenyon would have agreed with Van Kleeck's analysis. She believed that the two sides parted company only in "the means of securing equality" and insisted she was "passionately in favor of equal rights."[81]

ERA supporters in general could not understand how anyone in favor of equality could oppose the amendment. Florence Kitchelt described Dorothy Kenyon with disgust as someone "who doesn't know the meaning of equality."[82] Edith Goode, a National Woman's Party member, admitted that she understood why women had supported protective legislation when women were exploited in the sweatshops of New York and barred from joining unions, but could see no reason in the 1950s for protecting men by union contracts and women by legislation.[83] A number of ERA supporters, including the rare working-class members of the National Woman's Party, had no doubts about why labor favored protective legislation for women. The "nut in the shell," as one woman saw it, was that men feared competition from women.[84] One woman lawyer who represented Arizona waitresses fighting an eight-hour law that barred them from employment wrote angrily that "when any man or class of men shout of 'protecting' women, either biologically or economically, it is high time for the female of the species to look for a cyclone cellar and proceed (even the hard way) to not only *protect* but *defend herself*."[85] She insisted that protective laws were never enforced unless women offered competition to men. She and women like her could look back to the precedent of the Depression, when the public outcry focused on the inappropriateness of married women teachers but entirely overlooked married women in domestic service.[86] Emma Guffey Miller, president of the Woman's Party in the 1960s and a figure in Democratic politics, raised this very point and accused the "ladies of leisure" in the League of Women Voters of telling their less fortunate sisters how and when to work.[87] For

ERA advocates, the issue was crystal clear: "Equality of rights means precisely what it says. Applied industrially, it means that you cannot take a group of women in a given industry employing both men and women and apply to them laws which do not apply to their men competitors.... Whatever protection is decided upon must be applied to the job and the industry and not the sex of the workers."[88]

Just how confusing the whole issue of protective legislation and the meaning of equality could be is made clear by a controversy within the women's rights movement in 1945. Florence Kitchelt, who had switched from opposing to supporting the ERA in 1943, began to solicit and distribute legal opinions to the effect that the ERA would not eliminate protective legislation. This infuriated her allies, who moved to dissuade her from what they saw as her confused and irresponsible position. Her coworkers in the Connecticut Committee managed to pass a resolution ending the distribution of opinions on protective legislation by the Committee and Kitchelt seemed to back down, but in July 1945 she published an article in the New York *Herald Tribune* that stated that the ERA would not abolish protective legislation.[89] Ella Sherwin, head of the Industrial Women's League for Equality, a group of working-class women created by the National Woman's Party, informed Alice Paul that she had learned through mental telepathy that Kitchelt was a traitor and tool of the League of Women Voters.[90] Despite such dramatic charges and a great deal of concern among active women, the "Kitchelt affair" ended with no more serious consequences than a lingering distrust of Kitchelt in certain circles. But what is important here is that even an ardent, if newly converted, ERA activist like Kitchelt could fail to understand the arguments about protective legislation. Kitchelt never realized that, from the perspective of her allies, the difference between what she termed "so-called protective legislation" (meaning discriminatory legislation) and "protective legislation" (meaning legislation that would benefit women) was not at all clear and would in any case have to be interpreted by the very legal system that had for so long kept women in their place by "protecting" them.

In retrospect, there seems to have been some validity to both sides in the controversy. The anti-ERA forces were probably correct in asserting that the amendment's supporters often cared little for the well-being of women in industrial jobs. On the other hand, the amendment's backers seemed to have been accurate when they insisted that protective legislation often hurt women and especially when they argued that the old reason for supporting such legislation for women—that the Supreme Court had declared protective legislation constitutional only for women and thus such legislation could not be passed in sex-neutral terms—had become obsolete. ERA advocates, for example, pointed out that the New Deal's Fair Labor Standards Act created minimum wage and maximum hours laws for men as well as women.[91] The fact that more women belonged to unions, and thus could protect themselves through union contracts, also swayed some anti-ERA women to change their minds. Eleanor Roosevelt, a longtime opponent of the amendment, changed her mind in 1951. When archfoe Esther Peterson, director of the Women's Bureau and the power behind the 1961 President's Commission on the Status of Women, changed her position on the ERA much later, she mentioned the Fair Labor Standards Act, the Equal Pay Bill, and Title VII of the Civil Rights Act as measures that really made protective legislation that singled out women discriminatory. "So it seems to me that the question of protective legislation for women workers only is no longer an issue and that we might better spend our time and efforts on extending good labor standard laws to both men and women and eliminate those that benefit one sex only," she wrote in 1973.[92]

Although the ERA controversy involved a basic ideological difference on the meaning of equality, it also, by the 1940s, had a great deal to do with personal and organizational antagonism. Florence Kitchelt, herself a major figure in the League of Women Voters, traced the League's opposition to disapproval of the tactics and strategy of the National Woman's Party during the suffrage struggle and to dislike of League leaders for Alice Paul. Women did change their minds—the most exciting conversion for the Woman's Party in the post–1945 period was that of Alice Hamilton, a prominent opponent in the 1920s and 1930s—but imagine the difficulty of endorsing a position defended by one's longtime enemies. At times, positions on the amendment seemed to have little to do with the merits of the legislation itself. In fact, the amendment's sponsors and supporters in Congress included, in the 1950s, such figures as Richard Nixon, Barry Goldwater, and Strom Thurmond, all men who later switched to opposition. Woman's Party President Emma Guffey Miller decried the about-face of "Sir Richard, the Nimble," who sponsored the ERA as a member of the House and Senate but as a presidential candidate refused to acknowledge his past support and hid from members of the Woman's Party.[93] Since it is difficult to imagine Nixon, Goldwater, or Thurmond as ardent supporters of women's rights, it seems likely that their support had to do with the fact that most liberals opposed it and that the ERA did not become associated politically with issues such as abortion or gay rights until the 1970s. That the conservative position on the amendment may have come in part from simple opposition to liberals is suggested by the comments of an anti-ERA individual confused by the politics of the amendment: "It seems that all the wrong people are against the Amendment this year so we wonder if we ought not to be for it."[94] The issue is, of course, more complicated than this, but the position of one's friends (or enemies) did seem to have some impact on the issue.

In 1944, opponents of the ERA established the National Committee to Defeat the Un-Equal Rights Amendment, which became the National Committee on the Status of Women in 1947. This group supported the Women's Status Bill, an alternative to the ERA which called for an end to discrimination on the basis of sex except where reasonably justified by differences in physical structure or biological or social function.[95] The Committee also called for a presidential commission to review the status of women and make legislative recommendations, something that became a reality almost fifteen years later during John F. Kennedy's administration. Despite this kind of organized opposition, and the fact that many liberals believed that "nearly all the people in the United States whose opinion is of any value" opposed the ERA, the amendment received consideration by the Senate in 1946, 1950, and 1953.[96] In addition, Truman and Eisenhower endorsed the ERA and it received support in every Democratic platform from 1944 to 1956, and in every Republican platform from 1940 to 1960. This kind of support was, in fact, more a nod to women than a serious political position. Although in retrospect the amendment realistically had little chance of passage in the fifties, it is not surprising that supporters fought so hard for it and believed so fervently that it would soon become law. Their belief in the imminent passage of the ERA was not irrational in the context of the amendment's history.

Women's rights supporters worked for the ERA by lobbying Congress, seeking endorsements from candidates for political office and other organizations, establishing organizations specifically to support the amendment, and educating the public through newspaper and magazine articles, letters to the editor, and radio and television appearances. But this gives no idea of the magnitude of the efforts involved in the continual struggle on behalf of the amendment. Even if a politician or an organization endorsed the amendment, advocates had to maintain support by winning a new endorsement every year. There was also the fear, as the sympathetic male editor of the *Saturday Review* warned, that candidates with their eyes on the women's vote would promise to support the ERA at election time but promptly forget it afterwards.[97] ERA supporters not only had to win new and maintain old allies, they also had to guard the amendment against attempts to revise it in ways they believed negated it. The most serious threat came in the form of the Hayden rider, brainchild of powerful ERA foe Senator Carl Hayden. The Hayden rider, tacked on to the ERA, exempted any legislation that benefitted women from the provisions of the amendment and thus in effect nullified it. The passage of the amendment with the Hayden rider in 1950 and again in 1953 put ERA supporters in the uncomfortable position of having to oppose the amendment for which they had so long been working. They feared that the amendment in its new form would pass both the House and the Senate and force them to oppose ratification, so they worked to delay the vote until Congress adjourned. The Hayden rider succeeded in stopping the pro-ERA forces, and the confusion over strategy led some observers to conclude that the ERA activists did not want the amendment to pass and deprive them of their life work.

Some ERA supporters considered alternatives to the ERA, afraid that opposition had grown too entrenched. Rose Arnold Powell, who tried to persuade Alice Paul to reword the amendment, also drafted a Declaratory Act that would, if passed by Congress, guarantee that the generic "man" in the Constitution included women. The Hunt Amendment, introduced in the Senate in 1953, took the same approach and won support from some women's organizations, although the National Woman's Party vigorously opposed it. Woman's Party members, committed for so many years to the ERA, seemed increasingly hostile to any tampering with "their" amendment. In 1959, for example, a lawyer and Woman's Party member drafted a new amendment that granted women equality of status rather than rights, leading other members to wonder if she had been "planted" to foil passage of the amendment and to condemn her as "a pest" who was "somewhat lacking mentally."[98]

The ERA remained the foremost feminist issue in the post–1945 period, even though it could not pass Congress. The 1940s and 1950s saw more and more women's organizations begin to change from opposition to support, thus laying the groundwork for the women's movement's solid support of the ERA in the 1960s and 1970s. Although many women, especially in the National Woman's Party, fought solely for the ERA, ignoring all other feminist issues, not all women's rights advocates believed that the ERA alone would put an end to women's disadvantaged status. One woman told a correspondent how anxious she was to finish the amendment so she could go on to "the next thing."[99] But many women, engrossed in the long struggle, could see no further than the next step toward victory for the ERA. Anna Kelton Wiley, for example, believed that the ERA would help to change attitudes toward women, that it would "tend toward making women into finer beings, depending more and more on their mental and spiritual capacities and less and less on personal appearance, glamour and charm."[100] With such hopes riding on the amendment, it is not surprising that so many women put it first on the feminist agenda.

Women in Policy-Making

Both supporters and opponents of the ERA worked on a second feminist issue, the struggle to win for women policy-making roles in American society. Staunch ERA advocates tended to see this as a second priority item. One National Woman's Party member, for example, stated forcefully that "*After* we get *equality* I want full opportunity to have women get a chance at the top jobs!"[101] But to many women, the fight for policy-making jobs seemed a practical approach to fighting for women's rights. They believed that women would bring special qualities to top posts that would change the course of American politics and that, more important, recognition of individual women would benefit all women. This perspective on change is not surprising, given the many high-powered women professionals who participated in the women's movement. But what saved the program on policy-making from pure elitism was the sincere conviction that women in top jobs represented and benefitted all women. As Molly Dewson, a major figure in Franklin Roosevelt's administration, put it, "I am a firm believer in progress for women coming through appointments here and there and a first class job by the women who are the lucky ones chosen to demonstrate."[102]

Individuals who ran for office or won appointments often stated their convictions that they merely served as representatives of all women. For example, when Margaret Chase Smith became the first woman to win election to the Senate solely on her own merit, the BPW honored her at a luncheon at which she accepted "the unofficial responsibility of being senator-at-large for America's women to the extent they desire."[103] Anna Kross, Commissioner of Corrections in New York City, expressed her sadness at losing an appointment not personally, but "because of the women's angle."[104] Democratic politician Genevieve Blatt, after winning an election in Pennsylvania politics, expressed her pride, "not for myself, but for the women who made it possible."[105] Alice Paul congratulated Burnita Shelton Matthews on her appointment as a U.S. District Court judge with assurances that the appointment would "bring honor and glory to the Woman's Cause," and would give "a big upward lift to the Woman Movement."[106] Although Matthews felt obliged, as a judge, to stop working for the ERA, she worked for women by mentoring a long series of women law clerks.

Much of the impetus behind the effort to win policy-making roles for women came from the experiences of World War II. Franklin Roosevelt had been the first president to bring women into government, thanks in large part to the prodding of Eleanor Roosevelt and Molly Dewson, and women vigilantly guarded their hard-won advances during the war. A protest by organized women against the all-male composition of the War Manpower Commission had led to the appointment of a Women's Advisory Committee to the Commission. This nod to women, although unsatisfactory to all concerned, at least showed that concerted action could bring results.[107]

Women's rights supporters perceived a dramatic change with Roosevelt's death. Lucy Somerville Howorth, one of the women particularly concerned with the atmosphere in Washington, described the beginning of Truman's administration: "One of those curious waves went through the Government, petty persecutions of women began, they were denied access to phones, they were pulled off interdepartmental committees, they were denied promotions."[108] Howorth mentioned some scattered agitation by women, but she decided that the problem was lack of access to Truman and set out to open a channel of communication. With a group of other women, she persuaded a personal and political friend of Truman to speak to him, organized conferences, published articles, and met regularly with women who shared her concern. In contrast, India Edwards, the top woman in Truman's administration, gave her boss, whom she continued to support loyally throughout her life, high marks. She claimed to have asked Truman for the appointment of women as a reward for her work in the campaign and claimed credit for what few appointments the president did make.[109]

Despite Edwards' efforts and Truman's endorsement of the ERA, however, some women continued to make an "anguished protest against the low political estate to which women have fallen in the Truman administration." Journalist Doris Fleeson, writing in the Washington *Star* in 1945, described the influence of women in Washington as "virtually nil."[110] Twelve years later, she indicted Eisenhower for the lack of women in policy-making positions in his administration.[111] No president, in fact, satisfied women's rights advocates. Emma Guffey Miller, National Woman's Party member and longtime Democratic committeewoman, expressed her "grievous disappointment" that so few women during the Kennedy administration received worthwhile positions, and politician Genevieve Blatt commented that "Things do seem to get worse and worse, so far as women in Washington are concerned, don't they!"[112] Johnson did little better at pleasing activist women. Burnita Shelton Matthews, herself a federal judge, complained that "President Johnson is not making a better showing."[113]

But women did more than complain to the presidents about the poor representation of women in government. Believing—or perhaps merely hoping—that government officials did not appoint women because they were unaware of qualified women, organizations and individuals set out to compile rosters of women for use by the government in making appointments. During the war, women from a variety of women's organizations met with Eleanor Roosevelt to discuss the appointment of women to agencies dealing with postwar problems and succeeded in planning a White House conference on women in policy-making. This conference, which included representatives of working-class and black women's organizations as well as white professional women, compiled a Roster of Qualified Women, which it sent to the State Department.[114] That same year, the BPW, which participated in the White House conference, set up its first Roster of Women for Policy-Making Posts.[115] The BPW, especially under the leadership of Margaret Hickey, put a great deal of effort into campaigns to place women in high positions.[116] Other organizations, too, used the roster approach. The National Woman's Party sought to list names of women who were strong feminists as well as qualified in more traditional ways, and the American Association of University Women also focused on compiling rosters.[117] In 1961, the National Council of Women undertook a National Survey of Women Leaders by hiring a women's firm to send out questionnaires to women leaders.[118] In addition to these projects by individual organizations, women came together in coalition groups to support the appointment of women to high-level posts through the compilation of rosters.

Another tactic women used was the endorsement of women for particular offices. In 1946, the BPW took a big step by deciding to endorse specific women for office rather than simply to support the general principle of women in office. Margaret Hickey saw this as a historic departure for women's organizations, since groups such as the League of Women Voters had traditionally avoided any appearance of partisanship by supporting only issues, not individuals.[119] Women activists worked hard on behalf of other women and rejoiced when their candidates won elections or received appointments. When both India Edwards and Sarah Hughes received nominations for vice president at the 1952 Democratic convention, for example, Edwards responded to inquiries as to why she allowed Hughes's name to be placed before the delegates with the comment that the more women recognized in public affairs, the better.[120] Although women active in partisan politics worked for women of their own party, organizations like the BPW took care to support both Democrats and Republicans. In 1952, for example, the BPW endorsed Margaret Chase Smith for vice president on the Republican ticket and Sarah Hughes for vice president on the Democratic ticket. Without any real hope that either woman would receive the nomination, the Federation worked nevertheless to win recognition for women in the political world. Even India Edwards, a staunch Democrat, spoke to non-partisan and bipartisan groups of women to urge them to work for whatever party they supported. She remarked wryly in her autobiography that she might have done her work too well, since experts claimed that women were responsible for Eisenhower's election in 1952.[121]

A third tactic favored by activists interested in policy-making posts for women was the mobilization of women in politics in order to build a real constituency that would have to be acknowledged by the men in power. Much of this effort emanated from the women's divisions of the Republican and Democratic parties. Bertha Adkins of the Women's Division of the Republican National Committee attempted to mobilize Republican women and managed to institute a series of women's breakfasts—dubbed "hen breakfasts" to match the "stag breakfasts" which already existed—with Eisenhower in 1955.[122] Democratic women also kept the pressure on their male colleagues. Emma Guffey Miller worried about the fate of the Women's Division of the Democratic National Committee (DNC) and reminded the Democratic men that women worked hard in national and state campaigns.[123] Molly Dewson, the first head of the Women's Division and the person responsible for bringing women into government during Roosevelt's administration, kept informed about political work even after her retirement, and she praised the efforts of those who sought a larger role for women in politics.[124]

Expressing her support in a unique way, India Edwards sent out her 1959 Christmas card bearing the printed message: "May there be more women in politics and more 'little women' in Girls Clubs in 1960!"[125] She believed that the chairman of the DNC, one of her enemies, had outwitted her by "integrating" women's activities into the larger committee in 1952. Theoretically she approved, but in practice she worried that this merger would mean that women would lose control over their own activities.[126] These women and others like them believed that the political involvement of women would eventually lead to their greater participation at all levels. As one political woman wrote to Emma Guffey Miller: "Maybe if you and I live long enough we will accomplish what we are fighting for: political equality."[127]

Whichever of the three tactics individuals or groups favored—compilation of rosters of qualified women, endorsement of women for office, or mobilization of women in politics—they believed that women in policy-making posts would make an enormous difference in society and would transform the status of women. Most women active on this issue saw the appointment of women to high-level posts as a potential cause, rather than a result, of an improvement of women's status. For example, Margaret Hickey wrote to Lucy Somerville Howorth and commented: "Well, Lucy, the old battle cry was 'votes for women,' and now it ought to be 'politics for women' if we are to get any place."[128]

Some women blamed women themselves for their lack of representation in policy-making posts, but more often they commented on men's resistance to change. At a luncheon celebrating the 100th anniversary of the 1848 Seneca Falls women's rights convention, Mary Donlon of the New York Workman's Compensation Board insisted that the problem lay in the "unchanged attitude of men who do not want either women or women's thinking at policy levels."[129] Sarah Hughes, a judge and national leader in both the BPW and AAUW, favored a project to get more women on policy-making boards, but feared that the women had "run up against a stone wall" and could make "very little impression . . . on the powers that be."[130] Two members of the National Woman's Party wrote to urge the confirmation of Marion Harron as judge of the tax court and insisted that the criticism of Harron voiced by male lawyers "is not well founded but is the result of a discriminatory policy or prejudice of such critics against women judges."[131] Just as the debate over the ERA exposed different feminist world views, so too did the convictions of women working to bring more women into policy-making positions.

All of the women's organizations in the movement supported women for high-level positions in government, but some—in particular Woman's Party members—distrusted those who put this issue first. For example, one member of both the BPW and the National Woman's Party denounced the Federation's emphasis on getting women into policy-making jobs as a "betrayal" and "repudiation" of its program.[132] It is clear that Woman's Party founder Alice Paul, although she refused to talk about it on tape, believed that some women's groups were more interested in getting their officers into highly paid government positions than in the ERA, which she believed would bring about fundamental change for all women.[133] The suspicion that some female activists worked only to advance their own careers was voiced by other women as well. It is not hard to see why such charges would be made against those who worked to attain high-level positions for women, especially because the very women who promoted this tactic often received appointments. Nevertheless, the struggle to win a place for women in policy-making positions remained a central one in the women's rights movement throughout this period.

Women's History

A third feminist issue—advocacy of women's history—was also popular in the post-1945 period. Women active in this area believed that recognition of women's past would serve to improve the status of women in the future.

Interest in and support for women's history took many forms. One was the celebration of feminist holidays, particularly the anniversary of the suffrage victory on August 26 and the anniversary of the first women's rights convention at Seneca Falls in 1848. Every year the National Woman's Party held a ceremony on August 26 in the crypt of the Capitol and invited honored guests and the media. In 1948, women's rights advocates celebrated the 100th anniversary of the first women's rights convention with a commemorative program in Seneca Falls. Controversy over the celebration reflected the conflict among women's groups over the ERA—a clear indication that women's history was, for women's rights advocates, closely tied to contemporary issues. Members of the National Woman's Party worried that the celebration would be dominated by anti-ERA women and found their fears confirmed when Dorothy Kenyon gave the keynote address. Nora Stanton Barney, granddaughter of Elizabeth Cady Stanton and a strong ERA supporter, felt slighted because she was not asked to speak despite her having had an illustrious grandmother. Feminist biographer Alma Lutz, who attended the celebration, found the attendance disappointing and suspected that the local group responsible for planning had only timidly advertised the event out of fear of other women's groups barging in and taking over.[134] The National Woman's Party, left out of the planning and offended by the participation of anti-ERA women, sent no official delegation.

Another form of advocacy of women's history was support for commemorative postage stamps of women. The National Woman's Party had been instrumental in securing the Susan B. Anthony stamp in the 1930s and continued to buy no other three-cent stamps in the post–1945 period. One Woman's Party member, who prided herself on her "women only" stamp collection, worked to win support for a new feminist commemorative stamp honoring the women of Seneca Falls.[135] Ethel Adamson, head of the National Woman's Party's Committee on Pioneers of the Woman Movement, worked with other feminists to persuade the post office to design a stamp picturing Lucretia Mott and Elizabeth Cady Stanton, the two responsible for calling the Seneca Falls convention. Emma Guffey Miller used her political clout with Truman and her personal influence with her brother, a U.S. senator, to win approval for the stamp, to be issued in 1948. How important this work seemed is clear from Miller's comment that "I have never taken part in anything that has made me any happier than this for I feel sure it will help the cause tremendously."[136]

It is not hard to imagine the anger and dismay of Woman's Party members who believed that the stamp would help the cause of the ERA when it appeared with a portrait of Carrie Chapman Catt, the anti-ERA suffrage leader, alongside those of Mott and Stanton. Outraged women wrote to the post office to complain of the historical inaccuracy of the stamp, since Catt had not even been born at the time of Seneca Falls. Ethel Adamson believed that the post office "richly deserves all the lambasting they are getting" over the stamp, since they failed to consult experts who KNOW [the] history of the Woman Movement." It was clear to her that the post office officials "are neither historians nor feminists!"[137] Woman's Party members saw the "ruined" stamp not as the product of ignorance, however, but as the result of an anti-ERA conspiracy. One woman reported that a post office official showed her a file of letters asking that Catt be included, and

Alice Paul believed that "someone with political influence," probably Eleanor Roosevelt, had requested the inclusion of Catt.[138] Another member involved with the stamp accused the Women's Bureau of the Department of Labor of getting to the post office and changing the stamp.[139] Since the Woman's Party believed in the stamp as a means of educating women throughout the country on the history of the women's movement, they took the changed stamp very seriously, illustrating the significance of women's history to many feminists.

Women in the movement also sought to honor pioneer feminists by promoting their election to the Hall of Fame. Ethel Adamson wanted to see Lucretia Mott, Elizabeth Cady Stanton, and Susan B. Anthony take their places alongside the great men already enshrined in the Hall of Fame, but eventually the efforts of women's rights supporters focused on Anthony. Rose Arnold Powell, an ardent devotee of Anthony whose fondest dream was to see Anthony's face on Mount Rushmore, tried to whip up support for her heroine. Florence Kitchelt, too, devoted the summer of 1950 to writing to the electors and to friends and acquaintances in support of Anthony. Finally, in 1950, Anthony won her place in the Hall of Fame, causing Kitchelt to exclaim: "If we were a little younger, we might dance a ring around the rosy!"[140]

Rose Arnold Powell believed absolutely that winning recognition for the pioneers of the suffrage movement would contribute to the fight for women's equality, and she devoted her life from the 1920s until her death in 1961 to the memory of Susan B. Anthony. She tried to win support for her Mount Rushmore project, fought to have schools named after her heroine, urged calendar companies to mark Anthony's birthday, worked to make Anthony's birthday a national holiday, and, in fact, succeeded in making it a state holiday in Minnesota. All of this she did out of the conviction that "for true democracy—democratic balance, there should be instruction in our public schools on the Woman Movement and its pioneers, especially Susan B. Anthony."[141]

Powell believed in education, and so did other women who sought to educate the public through courses or publications on women's history. Mary Beard, probably the foremost early women's historian, contacted college presidents to urge the teaching of women's history and the acquisition of library materials on women. In 1949, she called together a group of women to propose a women's research institute that would serve as a resource for a revision of the curriculum.[142] Although the idea came to nothing, a few universities did offer courses on women's history in this period. Florence Kitchelt reported that in 1948 Syracuse University sponsored a course on the Status and Responsibilities of the American Woman, and Eugenie Leonard taught a course on the History of Education of Women at Catholic University in 1949.[143] Historian Miriam Holden believed that a course in women's history "should awaken women generally to a renewed consciousness of their broadest role in society" and congratulated Catholic University for offering such a course.[144] Alma Lutz taught a seminar on the Role of

Women in American History at Radcliffe in 1953, since she believed that it was important to teach young women about the achievements of their foremothers.[145]

A number of women's rights advocates wrote early works on women in history, convinced that this knowledge would help to bring about change in society. Alma Lutz wrote biographies of a number of feminists. Writer and publisher Marjorie Barstow Greenbie sought to put behind women "the authority of a really adequate recognition of the achievements of women in history," angered as she was by the way "women are erased from the record, and men put in their places."[145] Miriam Holden, in response to the omission of women from the public record, built her own collection on women's history into the largest library on women in the country. Mary Beard had worked to set up a World Center for Women's Archives in the 1930s, a project that she believed provoked "many kinds of college efforts to bring women into education."[147] Even though this particular project collapsed, the idea eventually succeeded in an altered form as the Radcliffe Women's Archives, now the Schlesinger Library. Beard wrote *Woman As Force in History* in 1946 to show that women had contributed actively to all of history, but her denunciation of the early feminist movement for its emphasis on the oppression of women alienated her from the women's rights movement.[148] She was amazed by the number of women who said they had read her book and then expressed "their inviolate convictions in the old-style feminism," which seemed to indicate that they had not understood her point.[149] For Beard, her approach to women's history was "my firm belief; my 'cause,' and I work at it all the time, in various ways."[150]

Most of the women who devoted their energies to women's history supported other feminist causes as well and perceived their work as contributing directly to the improvement of women's status in society. Education of women, they believed, was the key to mobilizing them to fight for change.

Interaction Among Organizations

The organizations involved in work on these different issues were loosely tied to one another through overlapping membership, cooperation, competition, and coalition-building. Overlapping membership facilitated contact and communication among different organizations. Although it is not possible to quantify the full extent of overlapping membership, it is clear from the records of women's organizations and from the papers of individual women that multiple memberships were common. Many women belonged to both the BPW and AAUW or combined membership in a professional or executive service organization with membership in some other women's group. Many National Woman's Party members belonged to the BPW and the AAUW and some even to the anti-ERA League of Women Voters. Although some women participated actively in two or more groups, most seemed to have had a primary commitment to one. As judge Sarah Hughes, a national figure in both the BPW and the AAUW explained, "it is very difficult to be active in more than one organization."[151] Agreeing that leadership in more than one group was difficult, the president of the AAUW wrote that she understood Lucy Howorth's reluctance to take on the leadership of a new organization: "You have your personal life, your professional life, and your AAUW life, and that's lives enough for any woman."[152] Many women did find active participation in one organization more than "enough for any woman," and thus maintained multiple memberships in name only. For example, some BPW leaders, like president Marguerite Rawalt, belonged to the National Woman's Party but never participated actively.

Overlapping memberships proved particularly useful for advancing the interests of the group with which a woman identified most strongly. National Woman's Party members were especially likely to use their multiple memberships in this way. For example, some Woman's Party members "infiltrated" the BPW and other groups in order to win or maintain endorsements for the ERA. At one point, a number of Woman's Party members who also belonged to the BPW attended a BPW convention and worked carefully to conceal their Woman's Party membership, mingle with the delegates, and gather at intervals in a central room to map out strategy.[153] Woman's Party members regularly reported working within the AAUW, the League of Women Voters, and other organizations. Within local branches, they would speak for the ERA and attempt to create a groundswell of support that would change the group's position on the national level. Florence Kitchelt, who had helped to establish the League of Women Voters in Connecticut, worked on fellow League members but complained that "[e]ven Hitler could not devise a more perfect system for thwarting discussion," since local Leagues could not discuss issues not on the national program, and the national board refused to act without demand from the local Leagues.[154]

Other organizations, too, used overlapping membership to support their own work. When the four founders of the Connecticut Committee met in 1943 to discuss what could be done in their state to help pass the ERA, Florence Kitchelt spoke in favor of a "rainbow division" of individuals and organizations that would include pro-ERA members of the League of Women Voters and the American Association of University Women.[155] Women working on issues other than the ERA also tried to win support by appealing to organizations to which they belonged. Rose Arnold Powell sought support for her Mount Rushmore project from the National Woman's Party, the General Federation of Women's Clubs, and the BPW, as well as the Susan B. Anthony Memorial Library of California and the Susan B. Anthony Memorial in Rochester.

Overlapping membership provided a communications network— women regularly reported on the meetings of one organization to the officers of another—and allowed individuals and organizations to try to co-opt the resources of other groups for their own ends. Interaction among organizations also took the form of cooperation. Groups working on the ERA, in particular, sometimes tried to coordinate their efforts. Occasionally, for example, representatives from the National Women's Party and the BPW met to discuss strategy for the ERA campaign. Sally Butler, president of the BPW in 1948, invited the presidents of forty-two women's organizations to a luncheon in New York to discuss the political, economic, and social future of women.[156] In 1952, Butler proposed the formation of a group, consisting of the Woman's Party, the General Federation of Women's Clubs, and the BPW, to work for the passage of the ERA. Although this committee never got off the ground, later BPW officers continued to plan joint campaigns. In 1960, Libby Sachar suggested that the National Woman's Party, the National Association of Women Lawyers, and the BPW raise a chest of money to hire public relations people to launch a media campaign for equal rights.[157]

Florence Kitchelt of the Connecticut Committee, like other ERA supporters, felt ambivalent about the National Woman's Party but nevertheless sought to cooperate on the ERA campaign. She sought advice from Woman's Party leaders, relied on *Equal Rights*, the Woman's Party journal, for information on the ERA, and met with Party officers to map out strategy. In 1946, the Connecticut Committee passed a resolution calling for cooperation on ERA work among the National Woman's Party, the BPW, and the General Federation of Women's Clubs. Kitchelt believed that the time had come for "the formation of a campaign committee of women of organizing ability with a chairman of wide national reputation, with a large personal following, and with qualities of efficiency and political strategy, who could give her whole time to the campaign . . . Can we not," she asked, "gather together our forces to make a mass movement that is irresistible?"[158] Periodically Kitchelt considered merging the Connecticut Committee into the National Woman's Party, despite her ambivalence about the national leadership, but her friend Alma Lutz, herself a Woman's Party member, warned against such a move. Lutz believed that the Woman's Party was "the real center for the ERA," but advised the Connecticut Committee to remain "an organization working harmoniously with the NWP but separate and with no interlocking members."[159]

It is clear from the cautious approach to cooperation illustrated in these examples that interaction among movement organizations also took the form of competition among groups working on common goals. The National Woman's Party and the BPW often competed for leadership of the ERA effort. Women active in other organizations sometimes encouraged this competition, hoping that one or the other group would win and push ahead with the Amendment.[160] Kitchelt placed great trust in Sally Butler, who became BPW president in 1946 and urged it to "strike out under our own colors to accomplish this goal, not to swing on some other group's coattails."[161] When the BPW opened its Washington office in 1947 and Butler declared it the "campaign headquarters for the passage of the Equal Rights Amendment," it could only have seemed a challenge to the National Woman's Party, whose own Capitol Hill headquarters one member described as "a center for the legislative work of various women's organizations."[162]

When Florence Kitchelt took over the leadership of the Connecticut Committee shortly after its founding, she took care to disassociate it from the National Woman's Party by informing her correspondents that it was not a branch of any national organization, but rather a "rallying-point for all persons and group[s], men and women," interested in endorsing the amendment.[163] She left the National Woman's Party in 1947 at the time of the schism, although she maintained contact with the group. Her friends warned her against Nina Horton Avery, a woman suspected of "infiltrating" the BPW for the Woman's Party, and they advised her when Alice Paul, having moved to Connecticut, wanted to join the Committee.[164] Many of the Woman's Party members who had left it at the same time as Kitchelt joined the Connecticut Committee. When Alice Leopold, director of the Women's Bureau, changed her mind on the ERA and prepared to endorse it, despite the Bureau's long and ardent opposition, Kitchelt worked desperately to ensure that Leopold would announce her conversion to the BPW rather than to the Woman's Party.[165]

The National Woman's Party, throughout its history, believed that no other organization could be trusted to devote itself wholeheartedly to the ERA. A struggle with the BPW in 1953 and 1954 over strategy in the face of the Hayden rider simply confirmed this suspicion. Alice Paul and the Woman's Party worked to have the amendment dropped, fighting the BPW for pushing ahead in support of the amendment, oblivious, as they saw it, to the dangers lurking in the rider. Paul complained that she could not get an appointment to see Helen Irwin, president of the BPW, and the two organizations worked at cross-purposes. Irwin criticized the Woman's Party for thwarting her efforts at every turn and reported that "a great many people are saying that they have come to the conclusion that certain leaders in that organization do not actually want the Equal Rights Amendment passed because they would no longer have a 'cause' for which to work."[166]

While it is not true that National Woman's Party members would have sabotaged the ERA in order to keep their "cause," it is true that they met any suggestion that they give up control of the campaign with hostility. In 1954, for example, a state officer of the BPW wrote to the Woman's Party on her own initiative to suggest that they pass on the torch to the BPW, since the Party was "not thriving," had not made "visible progress," and "many of your number are growing old."[167] She suggested that *Equal Rights* be published as part of the Federation journal, and that the Party will its headquarters to the BPW. One Woman's Party member responded with amazement and hostility, suggesting that enemies of the Woman's Party stood behind the letter.[168] Alice Paul replied, somewhat patronizingly, that she hoped that the BPW would give "active service" to the cause of the ERA and would "stand firm against all compromises." If the Federation were to do this, she wrote, it would "undoubtedly soon have a position of real leadership in the amendment campaign."[169] On the whole, Woman's Party members would probably have agreed with one of their number who wrote Alice Paul to urge her to retain her leadership despite the jealousy of other women and women's organizations: "the women of this nation should be on their knees begging you to go on with your work."[170]

Even competition within the movement shows how closely tied the groups were, since those engaged in social conflict have strong bonds of connectedness.[171] Though various organizations tended to work individually toward common goals, women's rights supporters and groups also engaged in coalition-building to work on specific projects. The most long-lived, although not necessarily the most effective, was the Women's Joint Legislative Committee for Equal Rights, organized in 1943 by Alice Paul as part of a plan to revitalize the ERA effort. At first, only the Woman's Party and the American Medical Woman's Association belonged, but by 1945 the coalition claimed thirty-one member organizations. Each organization sent two representatives to meetings and paid yearly voluntary dues for expenses. The Joint Legislative Committee worked exclusively on the ERA. The idea behind it was both to coordinate lobbying and letter-writing campaigns among national organizations and also to use the organization's name to convince Congress that thousands of women throughout the country favored the ERA.

The biggest problem facing the Women's Joint Legislative Committee (WJLC) was the accusation that the National Woman's Party dominated it. Alice Paul created the organization. Moreover, the Woman's Party paid most of its expenses, handled most of the secretarial work, and its leaders and many, if not most, of its members belonged to the National Woman's Party. The suspicion that the group was an arm of the NWP was, in part, justified. Both the chairman, Nina Horton Avery, and the convenor, Katharine Norris, belonged to the Woman's Party, although Avery was also a BPW member and in fact served on the Joint Legislative Committee in her BPW capacity. At some meetings, all the representatives present from organizations, ranging from the General Federation of Women's Clubs to the National Association of Colored Women, also belonged to the Woman's Party. Yet Alice Paul denied that the Woman's Party controlled the organization, and convenor Katharine Norris, who left the Woman's Party in 1947, tried to steer an independent course. "When I organized the WJLC," she wrote, "I wanted to make the members feel that it was *their* committee, not merely the tool of the NWP, although I realized how much we are dependent in many ways and how indebted."[172] Despite Paul's denials and Norris's determination, however, women's rights advocates for the most part continued to see the Joint Legislative Committee as a puppet of the Woman's Party. Florence Kitchelt reported that people doubted its independence, and Alma Lutz put little faith in it because "always in the past NWP would not play unless they controlled it."[173]

Whatever the truth about the coalition in its first years, it is clear that the conflict that rocked the Woman's Party in 1947 put the Joint Legislative Committee out of commission. Both sides in the conflict sent partisans to the Joint Legislative Committee meetings claiming to represent the Woman's Party, and the two leaders, Avery and Norris, stood on opposite sides in the struggle. After the lawsuit, the losing side threatened to destroy the Joint Legislative Committee. Katharine Norris, who, according to one member, did all the work of the coalition, died in 1949 and the Committee, although it continued to meet occasionally, left little further record of any activity. In 1954, Alice Paul responded to a suggestion from Florence Kitchelt that a coalition take up work for the ERA with the information that the Joint Legislative Committee had been reactivated under the leadership of Avery, who had been appointed by the BPW.[174] The BPW, however, withdrew from the organization, leaving the group more than ever under the domination of the Woman's Party. Although the organization still existed on paper in the 1980s, it long seemed to have served simply as a name used in lobbying. It never functioned as a real coalition, but it facilitated communication and occasional cooperation among women's organizations working for the ERA.

Women like Florence Kitchelt, who worried about the domination of the Women's Joint Legislative Committee by the National Woman's Party, sought to establish alternative coalitions to support the ERA. As early as 1945, while still a member of the National Woman's Party, Kitchelt approached Jeannette Marks, chairman of the New York state branch of the Woman's Party and one of the women who eventually left the group, with her idea of developing an "equalitarian movement" separate from the Woman's Party.[175] By the end of 1945, Marks had begun to consider leaving the Party and establishing an "Inter-State Coalition Council for the Equal Rights of Men and Women."[176] Kitchelt continued to try to build a coalition of the existing national organizations, but the National Woman's Party leadership consistently met her suggestions with the reply that a coalition, the Women's Joint Legislative Committee, already existed.

Those who left the Woman's Party after the 1947 schism established new state groups and attempted to build a national coalition. Caroline Babcock, the former executive secretary of the Woman's Party, suggested that the Eastern Regional Conference of the Woman's Party withdraw and form the U.S. Committee for the ERA, with state committees and membership in the Women's Joint Legislative Committee.[177] Alma Lutz urged Kitchelt to expand the Connecticut Committee to take in all of New England and favored the formation of a National Committee for the ERA to act as an advisory and coordinating group for existing organizations.[178] Former Woman's Party members in Baltimore formed the Mayflower Club, which held meetings and raised money but did not seem to last long. Jeannette Marks formed a Federated Committee for Equal Rights in New York in the hopes that a national committee would get off the ground, but it amounted to little. Alma Lutz encouraged Frieda Ullian, a past president of the Massachusetts AAUW, to set up a Massachusetts Committee in the mid–1950s. While Lutz chafed at the slowness with which the planning progressed, Ullian began to gather together women, many from the AAUW, into a correspondence committee similar to the Connecticut Committee. In 1956, Lutz wrote on Massachusetts Committee letterhead to a friend in order "to show you what a good front we put up." Although the Committee did not do enough to suit her, she took comfort in the thought that "at least Massachusetts Senators etc. will know some women in Massachusetts want the amendment."[179] The Committee continued to exist, at least through its letterhead, into the 1960s.

Women already involved in ERA work stood behind these local efforts to build state groups into a national coalition. Alma Lutz, in addition to her efforts in Massachusetts, wrote to a Rhode Island woman referred by a friend to request that she think about setting up a Rhode Island Committee for the ERA and recommended that Kitchelt contact a Chicago woman who had left the Woman's Party years before and formed an independent group.[180] A group of ex-Woman's Party members met in Washington in 1949 to discuss the formation of a committee there.[181] Kitchelt's contacts extended as far as Montana, where she encouraged a friend to establish a state committee.[182] Although these scattered and short-lived groups never united in a national coalition, what is important is the conviction among some women that only such a coalition could bring victory for the ERA. One woman's comments suggest why the formation of state groups was so important. Describing the situation in Massachusetts, she noted: "When we make a convert, usually middle-aged or more, there is no activity in which the convert can participate regularly and at once."[183] Local organizations could direct activities aimed at achieving national goals.

Women attempted to form coalitions around issues other than the ERA as well. Those interested in increasing women's participation in policy-making came together in a group known during World War II as the Committee on the Participation of Women in Postwar Planning, later called the Committee on Women in World Affairs. Established by educator Mary Woolley and run in the postwar period by historian Emily Hickman, it brought together representatives from organizations as diverse as the BPW, the National Council of Women, the League of Women Voters, the National Women's Trade Union League, and the National Council of Negro Women. The committee tried to keep the need for appointing qualified women constantly in the minds of government officials and also to compile a roster of women for use by the State Department in making appointments to international conferences.[184]

At the same time that the Committee on Women in World Affairs pushed to take advantage of the early postwar atmosphere, a group of women in New York formed the Multi-Party Committee of Women to work for the election and appointment of qualified women for public office. Anna Kross, the first woman Commissioner of Correction in New York, suggested that the BPW help to establish the Multi-Party Committee nationwide, and she contacted Emily Hickman to work out some kind of cooperative arrangement with the Committee on Women in World Affairs.[185] Hickman's death in 1947 seemed to put a halt to the Committee, but activity resumed in 1951. In 1953, Dorothy Kenyon reported to the board members that the "perennial question was again discussed as to whether the luncheon should be our 'swan song' or whether we should keep Women in World Affairs alive a little longer."[186] Lucy Somerville Howorth resigned from the organization in 1954, although she believed that "there is a need for some central clearing house and cooperative medium in securing appointment of women."[187] Dorothy Kenyon agreed with her that the group had reached the end of its usefulness and that a new approach was needed.[188] Organizations such as the AAUW continued to compile rosters of qualified women, as discussed earlier, but the dissolution of the Committee on Women in World Affairs marked the end of a broad coalition of working-class, black, and white middle-class women's organizations working to bring women into powerful positions in government.

One last coalition that deserves mention focused on the role of women in national security. At the start of the Korean War, the BPW, responding to wartime experience with the mobilization of women, called for all women's organizations to meet in Washington to discuss the utilization of womanpower.[189] In October 1950, leaders from thirty-eight national women's organizations met to consider how they might ensure the full partnership of women in all phases of civil defense and economic and military mobilization. In her keynote address, Margaret Hickey called for a standing conference of all women's organizations present, which included the BPW, League of Women Voters, AAUW, National Woman's Party, National Association of Women Lawyers, the National Councils of Catholic, Jewish, and Negro Women, National Association of Colored Women, YWCA, and the Junior League, as well as patriotic and labor groups. The conference set up a steering committee of the nine original sponsoring organizations, mostly middle-class women's or church groups, with the exception of the Associated Women of the American Farm Bureau Federation, to plan for a general clearinghouse of women's organizations. In response to prodding, the women present added the National Council of Negro Women to the steering committee, and then included patriotic, labor, and service organizations as well to make the steering committee more representative. This body established the Assembly of Women's Organizations for National Security, a permanent organization with a clearinghouse structure.

That it was not easy to hold together such disparate organizations is clear—the League of Women Voters withdrew shortly after the organizing meeting, perceiving conflict between those who wanted women to be an integral part of national mobilization and those who wanted to push to get women into key positions; the National Woman's Party withdrew early in 1951; and the National Association of Colored Women voted against the establishment of the clearinghouse because it was not in a financial position to contribute to such an organization.[190] By 1953, the Assembly considered two possible directions; it could continue to hold one or two meetings a year to discuss women's participation in the defense program, or it could become a "mothball program," ready to be reactivated in the event of hostilities.[191] Lucy Somerville Howorth later evaluated the organization positively because it represented one of the few efforts of women's organizations to work together and because the leaders became acquainted through the Assembly's activities.[192] Since the Assembly had little or no impact on government policy toward women during the Korean War, Howorth's analysis of the latent effects of coalition-building is probably accurate.

It is clear that women's organizations in the post-1945 period never managed to establish any lasting and well-functioning coalition. Nevertheless, the various organizations active on women's rights were linked by overlapping membership, cooperation, competition, and coalition-building. Part of what distinguishes groups and individuals considered to be a part of the women's rights movement from other groups and networks of women whose efforts might also have had an impact on women's lives in this period is that their actions and ideas were undertaken with the sense of being part of a joint effort of women to attain equal rights. To be sure, there were groups of American women in this period concerned with numerous other causes that the women's movement came to adopt in the 1960s—birth control, child care, peace, labor, racial equality, and socialism—but women in these groups did not identify specifically with the women's rights movement. Women's rights continued to mean primarily legal equality and the opportunity to share in the leadership of the nation.

Attempts to Mobilize Constituents

Women involved in different organizations worked in various ways to attempt to broaden the base of support for women's rights. Although the women's rights movement was unsuccessful in these attempts, it created some publicity for its activities and touched the lives of some individual women.

Feminists sought publicity and support from a number of different media. They set great store in newspaper coverage; Florence Kitchelt, for example, regularly wrote letters to the editors of the *Herald Tribune* and the *New York Times*, and the papers of most feminists include occasional letters to the editor on women's issues. Women working for the ERA won a great victory in 1943 when both the New York *Herald Tribune* and the *Christian Science Monitor* began to support the amendment. National Woman's Party members from New York met with *Herald Tribune* publisher Helen Reid and convinced her of the need for such legislation, and Alma Lutz, prominent Christian Scientist and frequent contributor to the *Monitor*, in conjunction with other Christian Scientists persuaded the editorial board of the newspaper to support the ERA. Both papers continued to do so, not always with the same vigor, until the amendment passed Congress in 1972.[193] ERA supporters rejoiced when the ERA or other feminist issues made headlines. In 1946, Katharine Norris, the convenor of the Women's Joint Legislative Committee, took confidence from the fact that three days of debate on the ERA in Congress "has put the Equal Rights Amendment on the map, with front page notice in all the big metropolitan papers."[194] Similarly, Alma Lutz believed in 1954 that feminist letter-writing was "stirring up a bit of interest and that's all to the good."[195]

Feminists also looked to popular magazines, especially middle-class women's magazines, for access to the public. They sought to rebut antifeminist articles and believed that they could have a significant impact by reaching the audiences of these magazines. When Perle Mesta wrote an article for *McCall's*, her friend Emma Guffey Miller informed her that "it is evidently influencing a lot of women for we have had several requests for information about the Equal Rights Amendment."[196]

Some women believed that they needed to make use of new techniques to reach women, and their thoughts often turned to radio and television as media that reached farther than the printed word. Even before World War II, some members of the National Woman's Party had put together a "Footlights Committee" that planned to use Hollywood stars in a radio campaign for the ERA, but Alice Paul had squelched that idea.[197] At the end of the war, however, the Woman's Party did establish a Radio Committee that prepared a fifteen-minute dramatic sketch on the history of the women's movement and attempted to distribute it through its branches.

Women's rights advocates managed periodically to get radio and television coverage, particularly on women's shows such as "Tufty Topics," a radio show hosted by commentator Esther Van Wagoner Tufty, and the Mary Margaret McBride radio show.[198] Woman's Party leader Anita Pollitzer discussed the ERA on a television program, "The Power of Women," in 1952, and announced presidential hopeful Nelson Rockefeller's support for the ERA on Dave Garroway's television show in 1959.[199] One of Groucho Marx's guests on his famous television show, "You Bet Your Life," announced that her hobby was working for the ERA; Marx argued that women already had equal rights and promised to ask future contestants what they thought. In response, National Woman's Party members determined to get on the program to discuss the amendment.[200] In 1962, CBS produced a show, "The Women Get the Vote," which touched on the post-suffrage movement and incorporated appearances by Anita Pollitzer and Nora Stanton Barney of the Woman's Party. Woman's Party members appreciated the opportunity for publicity, and anti-ERA women from the League of Women Voters and the Women's Bureau worried because the show depended so heavily on the "Equal Rights people."[201]

In all of these ways women's rights supporters attempted to use the media to inform the public on particular issues and win women to the cause. Individual organizations also issued their own publications, especially the National Woman's Party's *Equal Rights* and the BPW's *Independent Woman*, which they hoped would reach nonmembers as well as members. One editor of *Equal Rights* believed that the journal "does more good than almost anything we do. . . . [W]e sit back and crow over our 'fan mail.' "[202] Despite the general belief in publicity, Alice Paul argued on occasion that the ERA forces should try to get the amendment through Congress with as little notice as possible, in this way throwing the opposition off guard.[203] But Paul was unusual in this, and her policy often raised resentment from Woman's Party members as well as women outside the organization. Most ERA supporters probably would have agreed with a Woman's Party member who, in 1960, read *The Hidden Persuaders* and longed for money for advertising, which, she believed, would lead to certain passage of the ERA.[204]

Women's rights advocates made a few more direct attempts to recruit women for their cause. As older women, they looked especially to the responses of younger ones. Some, like one National Woman's Party member, found the younger generation "hopeless," "not interested in anything but cosmetics, T.V. and modern amusements."[205] Others complained that younger women took for granted the rights that had been won for them by the women's movement and failed to understand that they still lacked full equality. The social climate of antifeminism, after all, portrayed feminism as old-fashioned. Alma Lutz commented that "young women of today don't want to have anything to do with any project which they think even hints at a 'battle of the sexes.' "[206]

But most feminists did not despair, even in the face of indifference. Even the woman quoted above who found the younger generation "hopeless" went on to hope, believing that "[i]f a thing is *right*, it is bound to come to pass eventually!" Throughout the years, activists placed their hope in younger women. One, a public relations consultant, believed that "there is a generation of younger women who can really take up the torch as it drops from some of our willing but weakened hands."[207] Another woman wrote to the National Woman's Party and excused her lack of activity with the conviction that "new blood—in the form of the younger women, many of whom are coming to the rescue of we older crusaders—makes me feel sure eventually the men who hold the power will succumb."[208] Rose Arnold Powell anticipated help from younger women, too, but had no intention of giving up her own part in the struggle: "Younger hands must uphold the banner of women's recognition but I'll work as long as I can."[209] Whether they maintained hope or not, the older women's rights supporters turned their eyes to younger women, realizing that equal rights activism could not go on without "new blood."

Organizations and individuals sought on occasion to bring young women into equal rights work in specific ways, although they never enjoyed much success. The National Woman's Party attempted periodically to stimulate the formation of groups specifically for young women. A junior organization, made up of "young business women . . . active in defense work"—described by one member, in a manner revealing her own class bias, as "not to the manor born but definitely well informed"—sprang up in California in 1945.[210] In 1949, the Woman's Party bookkeeper organized a box social and square dancing exhibition to raise money and attract young women to the group.[211] The national chairman wrote to state chairmen in 1950 to try to spark the formation of young people's groups, and Ernestine Bellamy, an active young member, took the title of Young Adults Chairman to work in a systematic way to attract other young women. She organized a square dance and went on a nationwide speaking tour in 1950, and in 1951 formed a group in Massachusetts that she wanted to bring into the Woman's Party as a Junior Council.[212] Eventually Bellamy's energies went to her college work and then her marriage, but another Woman's Party member in 1961 reported that she had taken the first steps toward organizing a junior group.[213] In 1969, the Woman's Party had a Student Action Committee made up of young college women "dedicated to the renewed feminist movement" and designed "to bring young hands and energies to the front to relieve shoulders too long burdened with responsibilities other[s] refused to help them carry, and to utilize the training and talents given them by past feminist[s] so their sacrifices shall not have been in vain."[214] By the late 1960s, however, young women formed their own new feminist organizations, and the Woman's Party had little success recruiting members from among those newly committed to the feminist cause.

Other organizations, too, sought to reach out to younger women in the post–1945 period. Jane Grant suggested to Florence Kitchelt that the Connecticut Committee, in conjunction with the AAUW and BPW, produce material for college newspapers that would appeal to young women.[215] Grant's revived Lucy Stone League tried from the start to attract younger participants. The members planned to hold a luncheon during Easter week, when students came home from college, hoping that young women "will really take charge" since "it is now their problem and we should not hold the bag indefinitely."[216] In the same spirit, Lena Madesin Phillips rejoiced in 1948 that the International Federation of Business and Professional Women's Clubs had attracted younger women.[217]

Supporters of women's rights not only tried, on occasion, to recruit young women, but also to make individual contacts with the next generation. Some spoke to groups of students or sent material to individuals who requested it for their school work. Others sought out contact in unusual ways. Florence Kitchelt, for example, wrote to four young women whose letter to the editor had appeared in *Life* magazine to ask them to tell her whether they agreed with her about women's issues.[218] When Florence Armstrong, a National Woman's Party member, received a letter from her Sunday school class asking her to continue as their teacher, she proposed an unusual exchange. Explaining that she wanted to spend the rest of her life doing things no one else would do, and that relatively few women really exerted themselves for the vital matter of raising the legal status of women, she offered to continue teaching, which required eight hours of preparation weekly, if her class gave her in exchange eight hours of work for the National Woman's Party.[219] Whether they accepted her offer or not, Armstrong's interest in recruiting young women is evident.

Despite such efforts, and despite media coverage, the women's rights movement never succeeded in attracting young women to its ranks in the 1940s and 1950s. Women's rights advocates for the most part recognized that the majority of women knew nothing about their activity and had no interest in their cause. Sometimes they used the metaphor of women sleeping through the decades to describe their apathy. "Women who are still asleep, not conscious of the dignity and responsibilities of citizenship, are not stirred about the proposed amendment," Florence Kitchelt wrote in 1945.[220] "We must wake up our dormant women," Anna Kross urged the next year.[221]

Others complained directly of women's apathy. One found it "almost impossible to realize how little the average woman knows (or cares)" about the ERA.[222] Another found it "Incredible—but true" that three of her friends were surprised to learn that women did not have all their rights.[223] A National Woman's Party member from Cleveland reported that her small group there worked up a luncheon from time to time, "to which the faithful and a few of their friends will come, but we have found it very difficult to broaden our base and enlist the interest of enough women to make a letter-writing campaign of any value."[224] As late as 1966, one woman found it discouraging that some of her best friends were totally uninterested in the ERA: "You can expound to your heart's content," she complained, "and they just smile sweetly and say: 'It's wonderful to see your enthusiasm.' "[225] But even recognizing how apathetic others were about women's rights did not cause women to give up. Alma Lutz, who complained occasionally about the indifference of women, admitted that "I too often lose my patience over women's apathy, but though the world moves slowly, it does move."[226]

Women's rights supporters offered different explanations for other women's lack of interest in feminism. Some gave regional explanations; one woman from Wyoming and another from North Dakota believed that women in the west enjoyed greater equality and so showed little understanding of the need for a struggle for equal rights.[227] A professor at Newcomb College in New Orleans wrote that "Louisiana women are dedicated to Mardi Gras, and coffee parties and debutantes and all such jazz—an all year affair."[228] Other women reported that gains already made in employment opportunities and voting rights satisfied most women. But even when women declared, as Alma Lutz did, that women "are their own worst enemies," they tended, implicitly at least, to blame socialization and societal structures for the situation.[229] Rose Arnold Powell consistently criticized women who were "mired in masculinity," suggesting that, in her eyes, adherence to male values was the culprit.[230] Perhaps most activists would only have criticized women so harshly to other women. One National Woman's Party member deplored the fact that the men "are up to their old trick: playing one woman against another." She saw women as "a political football" and, extending the analogy, regretted that some of them "*do need* a good *kicking* around." But her closing lament is revealing: "Oh dear—sometimes I run and hide to keep up my faith in them, but I'll never tell *one man* that—and you are the only woman."[231]

Despite all of this gloom about the responsiveness of the majority of women, advocates of women's rights reported finding support for their goals in surprising places. One National Woman's Party member spoke at a Temple of Education on women's rights and reported that many of the women in the audience came to speak with her afterwards to thank her for her message.[232] Another told of a "nice married woman neighbor" about to go to work in a clothing factory who was "an ardent Equal Righter."[233] When one longtime member in 1945 accompanied her husband to a church luncheon and spoke briefly of the need for equality, she was surprised and pleased that so many women there knew of the ERA.[234] Another woman spoke at a Soroptimist convention where the lieutenant governor of California was "impressed by the great demonstration the women made when I brought up in the course of my address the Equal Rights Amendment."[235]

Such experiences must have helped to make it possible for some women to continue to try to mobilize potential supporters, both through the media and through direct recruitment of young women. Despite such efforts, however, women's rights supporters never spent much energy and never succeeded in convincing young women to join their ranks in the 1940s and 1950s. The majority of American women remained indifferent to, or uninformed about, women's rights until the resurgence of the movement in the mid–1960s.

Conclusion

Throughout the post-1945 period, women actively working to win equality conceived of their efforts as part of a women's rights movement. These women, mostly white, older, well-educated, professional women, belonged to a variety of autonomous organizations tied loosely together by a common interest in fighting for women's rights and by their interactions at individual and group levels. Although not all would have used the word "feminist" to describe themselves, and although they did not all hold the same ideological convictions, they shared a commitment to reforming American society and a world view that made gender primary. Women's rights supporters worked on one or more of three major issues: the Equal Rights Amendment, advocacy of women in policy-making positions, and recognition of the importance of women's history. Different organizations and networks came together around these issues by sharing members, cooperating, competing, and building coalitions with one another. What few efforts they made to mobilize potential supporters relied on publicity and personal contacts.

The women's rights movement remained a small one of elite older women committed to a cause the media portrayed as old-fashioned. Yet older feminists did not give up on the struggle for women's equality. Instead, they held on, maintaining their commitment to women's rights within a supportive community of like-minded women.

5

Commitment and Community

Since we have been conditioned to think quantitatively, feminists often begin the Journey with the misconception that we require large numbers in order to have a realistic hope of victory. This mistake is rooted in a serious underestimation of the force/fire of female bonding.

MARY DALY, *Gyn/Ecology*[1]

Women's rights activists in the post-1945 years might have disagreed with much of Mary Daly's "metaethics of radical feminism," but most of the core membership would have agreed with the sentiments expressed in the above quotation. As militant suffragist and peace activist Mabel Vernon put it in an interview in the 1970s: "we had very active, devoted people. It didn't make any difference whether there were few or many."[2] Other women might have longed for a mass mobilization of women, but the fact remains that throughout the postwar years, the women's rights movement did seem to attract "very active, devoted people." How could it have been otherwise in a society so hostile to feminism? Although women might casually join some of the organizations that concerned themselves with women's issues, identification with feminism and the women's rights movement required commitment to an unpopular cause. Commitment and the bond of personal relationships were vital not only to the National Woman's Party community but to the survival of the women's rights movement as a whole.

Commitment is essential for the survival of any organization or community, especially when the group is not successful in attaining its stated goals. In the case of the women's rights movement, the years from 1945 to the mid-1960s saw little progress toward the goal of establishing equality for women in American society, yet women continued their activism. What motivated them was commitment, which has at its core a reciprocal relationship between the individual and the group in which the individual benefits from the process of giving to the group. As Rosabeth Kanter, in her sociological study of nineteenth-century American utopian communities, put it: "Commitment thus refers to the willingness of people to do what will help maintain the group because it provides what they need."[3] Individuals join and remain with groups in response to a variety of incentives: material incentives such as money, goods, or tangible rewards; specific solidary incentives such as status, power, or other intangible benefits; collective solidary incentives such as prestige, friendship, fun, or other rewards from being part of a group; and purposive incentives such as value fulfillment or satisfaction in working for worthwhile causes.[4] The women's rights movement offered few material or specific solidary incentives, although some women made a career of feminist work, achieved a kind of status as spokeswomen, or made a home for themselves within women's organizations. But for the most part, commitment resulted from the rewards of being part of a group and the satisfaction of dedication to an important cause.

Commitment of Individuals

The different women's organizations and their members ranged along a spectrum of commitment, with the core National Woman's Party members who devoted their lives to "the Cause" at one end and the women who participated minimally in some aspect of feminist work at the other. The model for lifelong commitment was Alice Paul, who had by 1945 established an almost legendary reputation as a dedicated, iron-willed super-feminist, a fanatic and a martyr. Although trained as a lawyer, Paul devoted herself completely to the NWP and World Woman's Party, living and working only for the women's rights movement. In 1965, a Woman's Party member reported that eighty-year-old Alice Paul "still works from six A.M. to twelve P.M."[5] In an interview conducted in the 1970s, Paul seemed unable to understand why the interviewer thought the ERA campaign must have been discouraging. When the interviewer commented that the Woman's Party could not have survived without someone like Paul, Paul replied, "But that's all any campaign needs, isn't it?"[6] This kind of simple faith and unquestioned commitment seemed to characterize Paul's involvement in the women's rights movement from suffrage days until her death in 1977.

Other National Woman's Party members approached the level of Paul's commitment. When Fannie Ackley, a member described by a friend and coworker as a "zealot," died, her twin sister wrote to Alice Paul that "EQUAL RIGHTS was Fannie's life work."[7] After her retirement in 1945, Ackley devoted herself to the ERA campaign, remaining "a brave spirit faithful to the cause with her last breath."[8] In the same vein, another member, who suffered a serious and eventually fatal automobile accident, could, according to the woman with whom she lived, "bear the broken bones but the breaking up of her life's work was far harder to take."[9] Anna Kelton Wiley, one of the original Woman's Party members who led the rebels in the 1947 schism, wrote that she had given the "very best years of my life, the years when I could have been happy and carefree, to the cause." She felt that she had sacrificed her mother, her husband, and her sons to the ERA, and she wanted to be remembered for her work. Her sacrifices weighed heavily on her mind as she grew older. In her seventies, she longed to see the ERA sent to the states for ratification before she died. She made clear what her sacrifices had cost her in explaining her opposition to a piecemeal approach to legislative reform: "It is too much to ask of anyone to give up her life just to get one bill passed."[10]

These examples suggest why the Woman's Party gained the reputation of fanatical devotion to the cause of the ERA. Florence Kitchelt of the Connecticut Committee, for example, reported to one of her correspondents that she found the Woman's Party "ingrown to the point of fanaticism."[11] Another woman active in the Woman's Party distinguished her approach from that of the other members by noting that she refused "to be fanatical about one idea or one organization."[12] And another member, explaining her lack of activity, wrote: "In my heart I know I'm just not the type to carry on as you would like. . . . I am interested but not with that dedicated drive I so admired at the meeting Monday night."[13] It would seem that the expectation, if not always the reality, of Woman's Party membership was intense commitment. State chairman Lillian Hulse confronted this expectation after returning from a visit to national headquarters. A colleague expressed the conviction that the visit to headquarters would change her life. As Hulse reported it, "Her eyes just beamed, and she said, 'Now you have done something,' and added, 'The rest of your life must be dedicated to the National Woman's Party.' "[14] In response to this level of commitment, outsiders and newcomers to the Woman's Party found it "a revelation . . . to meet women who are working with such dedication" or believed that "the secret of the ability of a small group to do so much in the face of such odds is that it can attract just such devotion and loyalty."[15] Even opponents of the ERA, such as Senator Hubert Humphrey, commented on the devotion and dedication of the Woman's Party.[16]

Some women outside the Woman's Party also devoted their lives to the women's rights movement. Rose Arnold Powell, for example, dedicated her life to winning recognition for the great women of the suffrage movement, particularly Susan B. Anthony. Although a member of the National Woman's Party, she worked for her own goals in her own ways. Powell threw herself into her work with such intense commitment and persistence that she seemed eccentric. She expressed her willingness to die for her cause: "I have done what I could," a newspaper interview in 1949 quoted her as saying, "I would have given my life, if it would have helped."[17] She claimed her broken health as a sacrifice to the cause, since two accidents in the course of her work plagued her throughout the years. She kept on with her "life work," her "great mission," her "child," until her death at eighty-five. Her own self-concept is revealing. She alternately envisaged herself as Martin Luther, weary of the world and prepared to depart; Jesus in the Garden of Gethsemane, forsaken by all; the Little Red Hen, left to take all the responsibility; and "a tired little mouse gnawing at the ropes of the net that imprisons women."[18] Working almost alone, Powell could only contribute what she did because the strength of her commitment carried her through. Women who were part of the correspondence network dedicated to winning recognition for Susan B. Anthony acknowledged the intensity of Powell's commitment. The head of the Susan B. Anthony Memorial in Rochester, New York, for example, wrote to Powell: "You are in a class by yourself, so devoted to the CAUSE."[19]

Lena Madesin Phillips, the founder of the International Federation of Business and Professional Women's Clubs and a power in the national federation, also devoted her life to feminist work. Trained as a lawyer, she began work with the YWCA at the time of World War I and participated in the founding of the BPW in 1919. Her election at the founding convention to the post of executive secretary launched her in what quickly became her real career. As she herself put it many years later, "my avocation ran away with my vocation," and, like Alice Paul, she concentrated on organizing women rather than on the law.[20] As a result of her efforts, the BPW became, in 1937, the first national women's organization to join the Woman's Party in support of the ERA. Phillips traveled widely in Europe prior to 1930 to set up the International Federation, and she served as president from 1930 to 1947. In fact, she died in the course of her feminist work, struck down by a perforated ulcer in France on her way to Beirut, where she planned to attend a conference on organizing professional women in the Middle East. Phillips gave, as she saw it, "the best efforts of my life . . . to organized womanhood and the causes which women espouse."[21]

Women like Alice Paul, Fannie Ackley, Anna Kelton Wiley, Rose Arnold Powell, and Lena Madesin Phillips devoted their lives to work for women's rights. The term "life work" as a description of feminist activism crops up again and again in the correspondence of involved women. But even women who did not make feminism their "life work" could be committed to women's rights. Although they might not have devoted their lives to the cause, they did give generously of their time and money.

Time and Money

Commitment to feminist work meant that women attended meetings and national conventions, wrote letters and articles, traveled to speak to existing groups or to organize new ones, and lobbied Congress. The National Woman's Party depended on the volunteer labor of members at its national headquarters. A summons from Alice Paul or the national chairman often resulted in an extended visit in Washington devoted to office work or lobbying. Members might come to handle correspondence over a weekend or for the summer, or the national chairman might move to Washington to devote full time to her post. Members who responded to urgent pleas for help recognized how much time feminist activity demanded. "Since 1917 I have devoted all my spare time to feminism both here and abroad," one member wrote in 1952.[22] Another had given thirty-nine years of her life to the work outlined by Alice Paul.[23] Yet another assured the national chairman in 1949 that "wherever I am I will be working to the best of my ability for the cause. I have been at it quite consistently—suffrage and then the Amendment, since 1917, so count on me for whatever assistance I can give."[24] Members worried about the health of Anita Pollitzer, the controversial chairman in the 1940s, and warned her about "dashing around 'giving your all to the cause' "; in the mid-1960s, Pollitzer still felt as if "I'd signed my life away—but I do it with affection!" Although her doctor warned her not to attend the Democratic Convention on behalf of the Woman's Party as she had planned, she dismissed his advice since "he couldn't possibly understand my inner commitments to the Woman's Party"[25] Another member asked for a "cure to giving too much of one's time to one thing" but continued to work for the ERA.[26] One member active in the 1960s remembered that she often worked into the early hours of the morning at Belmont House: "we worked all kinds of hours that nobody who wanted to lead a normal life could hope for."[27]

Women outside the Woman's Party also devoted their time to women's rights. Dorothy Kenyon, a prominent feminist opponent of the ERA, reluctantly accepted a new committee assignment in the International Alliance of Women, "appalled" as she was "at the prospect of serving on any more committees."[28] An admirer described Lucy Somerville Howorth of the Business and Professional Women as "a subject of wonder and admiration" for her "almost super-human powers" in accomplishing so many things so effectively.[29] In the BPW, the rules specified that a national officer had to be "actively engaged" in business or the professions, but the job of president required so much time that the organization relaxed the rules to allow the president to take a leave of absence. According to the official history, one president "represented the Federation at dozens of functions entailing travel, energy and absence from her regular employment."[30] One prominent AAUW member, in listing the qualifications she wanted to see in a chairman of the Status of Woman Committee, noted that the person's "personal affairs" would have to be such that she would have sufficient time and be sufficiently accessible to do the important work of the Committee.[31] BPW leader Margaret Hickey, experiencing the constraints of time in her own life, apologized to a friend for her inability to see her during a quick trip to New York, since she always boarded her plane to return "with a stack of unanswered telephone calls, notes and messages neglected, and sadly, never a chance to see my friends."[32] Perhaps the demands on time are best summed up by one Woman's Party member, Dorothy Shipley Granger, founder of the St. Joan Society (a group of Catholic ERA supporters), who complained that "I could not possibly be giving any more time to a cause if I were a cloistered nun!"[33]

In addition to, or as a substitute for, time, women gave money to the movement. Participation in national organizations usually depended on having sufficient financial resources, since women paid their own way when they traveled to meetings or to Washington to lobby for the ERA. "I wish I had the money to devote more of my time to the cause," one Woman's Party member lamented.[34] As one Woman's Party member commented in an interview: "We spent our own money. You know, it cost me a lot of

money really."[35] Margaret Hickey sometimes spent her own money to speak
to BPW groups all over the country. Another woman, an AAUW member,
remembered that she could never attend national conventions because of
the expense.[36]

Women sometimes gave money because they had no time to participate.
One member sent the Woman's Party twenty dollars, noting: "I can do so
little and I feel so grateful to those of you who are giving so much of your
lives in my interest and the rest of the women of the world."[37] Gifts of
money might mean great sacrifices. Fannie Ackley, after her retirement,
wrote and published at her own expense booklets of pro-ERA doggerel that
she sent to senators and representatives, once spending 350 dollars—an
enormous sum for a retired worker, if minuscule for a powerful lobby—on
a single session of Congress. Making a similar sacrifice, the national secretary
of the Woman's Party in 1948 sold her car in order to raise the money to
go to Washington and work at Belmont House. The next year, the national
chairman, a retired woman of seventy-three, gave such large sums to the
organization that the bookkeeper expressed her anxiety to a coworker.[38]
Other women, less constrained financially, chose to remember the feminist
cause in their wills. Jane Grant, head of the Lucy Stone League in the 1950s,
drew up a will in 1963 in which she left her estate to a college or university
willing to set up a chair in Feminism in the Sociology department.[39] Doris
Stevens drafted a will that directed her trustees to use her estate "to bring
the experience and inspiration of the past to women of leadership in the
years ahead."[40] In the 1980s, the estates of Grant and Stevens gave large
bequests to women's studies at the University of Oregon and Princeton,
respectively.

The giving of time and money might mean that women had to shape
personal relationships around the cause. Women whose personal lives
meshed with their feminist activity found it easier to commit themselves
than those who felt a conflict between their feminist work and their personal
lives. The AAUW member quoted earlier who believed that an officer's
personal affairs would have to allow her time to do the job recognized this
reality. One National Woman's Party member complained because her sisters and friends were filling up all her time and she was being "razzed
unmercifully about always talking about the National Woman's Party and
boring everyone with my endless arguments." Her response to the pressure
was straightforward: "If they are my friends they have to listen—that's all!"[41]

But even women whose personal lives centered around other women
involved in women's rights could find the work disruptive. Lena Madesin
Phillips, who lived and shared her work with Marjory Lacey-Baker, wrote
to a friend that "[t]he last few years have been filled with hard work for
Miss Lacey-Baker and me and we have had to put aside our personal life
and a number of things which we wished and needed to do."[42] Some of the
rebels in the National Woman's Party came to believe that any personal life
at all conflicted with feminist activity. One angry member who left even
before the schism wrote to Doris Stevens "I have been made to realize for

a great many years that most of the best workers in the N.W.P. very definitely put the object of their organization ahead of any human relationship that might exist in the work." "They seem to put their goal before human life," she noted in another letter.[43] One of the leading rebels apologized to a friend for not writing, noting that "I can no longer make the double grade of NWP and personal relationships."[44] Not many women felt this way about the costs of feminist work, but the expenditure of time and money could mean sacrifices of various kinds for highly committed women.

They sometimes complained about the constant sacrifices. A worker at Belmont House, for example, commented on the unremitting struggle to get the ERA through Congress: "You know how it is here, grind, grind, grind."[45] Likewise, a national chairman, thankful for the breathing spell between elections and the opening of Congress, referred to the work of collecting sponsors and lobbying as "the old familiar grind."[46] Charl Ormond Williams, an educator active in the leadership of the Business and Professional Women, gave it up in 1946: "During all my life, I have had one cause after another weighing heavily on my heart and conscience. . . . Since I was eighteen I have spent my entire life serving others. The situation has come to such a pass that I cannot even rest at home when I get there for the thought of things that must be done on the morrow and on the morrows ahead."[47] Doris Stevens, several years after her resignation from the Woman's Party, expressed the same perception of women's rights work when she wrote that she found it "difficult to return to [the] 'feminist harness.' "[48]

Some of the women who left the Woman's Party after 1947 seemed to have lost their commitment. "Who cares about Equal Rights when you have mountains or the sea!," one member wrote.[49] She, like her fellow rebel, Dorothy Shipley Granger, would not make the ERA her "life work." Granger longed for the day when a chairman she respected would take over so she could "do nothing but work, take care of my husband, and *play* all the time. No more causes for me."[50]

Although Granger devoted a great deal of time to women's rights, she seemed to lack the kind of intense commitment that characterized women like Alice Paul. But commitment is not simply a question of individual psychology, for structural conditions determined to a large extent whether women could act on their ideals.

Obstacles to Commitment

Because women's rights supporters were mostly older women, advanced age kept some of them from active feminist work. Lavinia Dock, a pioneer in the nursing profession and one of Alice Paul's mentors, sent a check to the Woman's Party in 1948 with a note that stated flatly: "Shall be 91 in Feb. Tired of life."[51] Another member promised Alice Paul that she would try to get to Washington as soon as possible: "And if I do not get there at all— just charge it to old age maybe—for the old girl 'ain't up to what she used to was' right now."[52] Mary Murray, a working-class member of the Woman's Party, wrote that "[i]t just seems impossible for me to realize that I am an old lady now past the days of usefulness."[53] A woman involved with the Connecticut Committee, too, felt useless as she grew "older and less competent."[54]

Disabilities associated with age—blindness, deafness, arthritis, broken hips—plagued many of the older women. Florence Bayard Hilles, for example, an extremely active Woman's Party member in the 1920s and early 1930s, dropped out of active participation as she grew deaf and developed arthritis in both knees. The eighty-one-year-old head of the Susan B. Anthony Memorial Library of California suffered from a broken hip and failing eyesight. Both Florence Kitchelt of the Connecticut Committee and Clara Snell Wolfe, an active Woman's Party member from Ohio, dreaded going to conventions because they could not hear.

But some women remained active in their eighties and nineties. Emma Guffey Miller, for example, despite her sometimes disabling arthritis, became national chairman of the Woman's Party at the age of eighty-five. Another Woman's Party member wrote that she had just passed her ninety-third birthday and remained as interested in world affairs as when she was twenty.[55] For some women, what one Woman's Party member assured another was true: "Eighty-seven years is not 'old' when one is interested in life and their fellowman."[56] Women throughout the movement continued to lead active public lives long past the traditional age of retirement.

Illness, sometimes but not always associated with age, also limited the participation of some women. One Woman's Party member and former suffragist wrote that "[i]t *tires* me only trying to keep up with you in spirit, for I am ever faithful—really—but I could not take it *physically*."[57] Echoing this sentiment, another member who had left Belmont House and suffered "a sort of nervous collapse" explained: "All the rest of you seem to ignore physical weaknesses, and I tried to."[58] Workers at Belmont House, in particular, periodically collapsed in illness or decided to take a vacation before breaking down. Women in other organizations, too, suffered from overwork. One BPW member complained about the internal politics of the organization to a friend, adding: "Frankly, I think some of the sinus trouble is too much women!"[59] Illness could result from, and make it difficult to continue in, feminist work.

The demands of employment, too, could stand in the way of women's activism. As one Woman's Party member, a university professor, explained: "It is naturally far more difficult for us who are employed . . . to give the time we would like to the Woman's Party."[60] Without some flexibility in her work life, a woman's participation could be seriously circumscribed. Professional employment could provide flexibility—as it did for teachers and college professors who could devote their summers to feminist work—but it could also make demands that limited available time. Rose Arnold Powell left the Internal Revenue Service in 1930 in order to work full time for recognition of the great women of the past. Ernestine Powell, a national chairman of the Woman's Party, has to resign her office in the 1950s because of the heavy demands of her law practice. Another attorney in the Woman's Party struggled to arrange her practice so that she could spend a month at headquarters in Washington. One of the young women in the Party lamented her inability to participate more actively while in college. A college administrator complained that the ERA "has had to take a back seat while less important things claimed all my time. But that is the kind of situation that working for your living involves."[61]

Professional women might have demanding careers, but at least those who were self-employed or worked in education, unlike women in non-professional jobs, might be able to arrange their time so that they could participate in women's rights activities. Retirement might also free women. For instance, several Woman's Party members hoped to be able to devote themselves to ERA work after retirement. But, since feminist work required not only time but money as well, retirement that gave women time might deprive them of the money necessary to play an active role.

Fannie Ackley, for example, could not play a leadership role in the Woman's Party because she could never afford travel from her home in Spokane, Washington, to Washington, D.C. Another working-class member faced the same kind of difficulties. Mary Markajani, in response to a request to appear before a Congressional subcommittee on the ERA as a worker opposed to protective legislation, apologized for her inability to travel from New York. She explained that she had been unemployed, making it impossible for her to pay for the trip, and that she currently had a job so could not take time off to travel.[62] The Woman's Party usually offered to pay the expenses of working-class members, whose testimony could help refute the charges leveled against the organization by women's groups in favor of protective legislation, but often the women, like Markajani, expressed reluctance to take the Party's money.

Family responsibilities, even more than the demands of employment, could tear women away from full-time commitment. The needs of husbands, children, grandchildren, parents, or women companions often had to come before women's rights work. Clara Snell Wolfe, a Woman's Party member from Ohio, could not respond to an emergency call from national headquarters because her husband was in the hospital. In 1951, an AAUW member explained to a friend that she had been out of circulation for almost a year due to her husband's hospitalization.[63] A Connecticut Committee member, who remained interested in work for the ERA, was tied down for two years to the care of her mother.

Women with children often felt overwhelmed by their responsibilities. One AAUW member active on the local level felt unable to attend national conventions, where policy was set, because she had three young children.[64] Other women had to limit their involvement while their children were young. Laments sound throughout women's letters: "I am sorry to be of so little help but I can't do more than family duties"; "The truth is, that while I am interested, I have a family of children"; "Our lives are just too full of FAMILY right now."[65]

Women might assert that family came first, but that did not keep them from feeling guilty about their inactivity. One Woman's Party member wrote that she felt guilty for not helping out with the ERA work, but that she also felt guilty at meetings when she thought of all the things she should be doing at home.[66] Throughout the 1950s, Anita Pollitzer wrote a series of defensive and guilt-ridden letters apologizing extravagantly for her failure to attend meetings in Washington. Another longtime member apologized to Alice Paul for failing to get to Washington and remarked: "How you must dislike the interference of personal matters. . . . It is the price one pays for children, I guess."[67]

There is no evidence that most feminists with family responsibilities questioned the sexual division of labor in their own lives, although Florence Kitchelt, head of the Connecticut Committee, did. Kitchelt wrote in 1948 that she and her husband considered the home a "cooperative institution, not to be maintained at the cost of the development of one of the partners."[68] The Kitchelts' commitment to task-sharing in housekeeping—they had no children—seems to have been highly unusual. One League of Women Voters member active in the 1956 Conference on Womanpower remarked on the gender difference in attitudes toward family responsibility. The reason she had not been in contact with the chair of the conference sooner, she explained, "relates to a sex difference. My husband has been in the hospital and I have spent most of my time with him instead of attending to my work. Had I been a man perhaps I would have put more of my work first and the hospital second."[69]

Even though women with family responsibilities took pains not to neglect them, active feminist work must have been hard on family life. Anna Kelton Wiley claimed, a bit overdramatically perhaps, that her husband's death "had made me realize too late that my strenuous activities in the late 1920s and the early 1930's for the National Woman's Party were all a mistake. Had I concerned myself solely with his needs, when his father died, at that time, Harvey would be alive and happy today."[70] Although feminist work probably did not kill husbands, it no doubt did place strain on marriages and family life. Margaret Hickey, when she left office as president of the BPW, wrote a friend that her husband had bought her a new Packard in order to "encourage me to become a lady of leisure even to the point of wanting me to learn to drive. And then there is considerable pressure brought for me to take up golf again."[71] Charlotte Leyden, president of the National Council of Women, noted that she had never been able to attend meetings of the International Council of Women, in part because her husband "can't see married women taking trips unless their husbands are along!"[72] A National Woman's Party member who had organized a campaign to win ERA endorsements from religious leaders wrote to a friend: "I am working under difficulties for my husband is opposed to my doing all this & I must work while he is not around or we have bad arguments."[73]

He was not the only husband to object to work with the Woman's Party. Another member commented that she "listened to friend husband for once and stayed home."[74] Dorothy Shipley Granger wrote jokingly that she would come to a meeting unless "my lord and master decides to shackle me to prevent any further participation in the affairs of the NWP!" A month later she reported that her husband was about to send her to a desert island to get her away from "the most harmful influences of my life." Granger confided to a friend that she had "got all the members of my dear family up in arms because this is our 25th wedding anniversary and they had planned a wallopping big party for us, and what do I do? Well, you can see that it does appear that a little thing like the ones I love can be shoved aside. Harry doesn't feel that way, but all the others are ready to use their influence to poison the NWP!"[75]

The barriers raised by family responsibilities could be lowered by supportive family members. Some women lived separately from their husbands for periods of time while engaged in feminist work. Burnita Shelton Matthews, for example, a Woman's Party member appointed to a federal judgeship in 1949, lived in Washington and functioned independently while her husband lived sometimes with her and sometimes elsewhere. When Margaret Hickey became the head of the Women's Advisory Committee to the War Manpower Commission during World War II, she commuted to Washington weekly from her home in St. Louis. Anita Pollitzer's husband traveled from New York City to spend weekends with her at Belmont House while she served as national chairman.

Some women saw their husbands as partners in their work. Helen Hunt West, at one time a national chairman of the Woman's Party, described the role that her husband had played: "Byron and I have always felt so much a part of the Woman's Party family. . . . As you know Byron and I spent four happy years there, and I am sure he worked about as hard for our cause as I did, and certainly he was heart and soul in it."[76] Marguerite Rawalt, active in the BPW and the National Association of Women Lawyers, wrote of her husband that his "constant interest in and approval of her activities contributed significantly to her success and satisfaction in organization work."[77] Margaret Hickey found her husband's support for her work essential throughout their thirty-nine-year marriage.[78] Mary Miller, a woman involved in a variety of women's and progressive organizations in Columbus, Ohio, told how her husband supported her in her activities even when he did not agree with her. When she took a stand on a controversial issue, he encouraged her to go ahead, saying: "If you go to jail, you'll need me, and if you go to heaven, I'll go on your coattails."[79] One woman still active in the women's movement in the 1980s insisted in an interview that active participation was only possible for a married woman if she had a supportive husband.[80]

Women's rights advocates were not unaware of these realities. National Woman's Party members sometimes approached recently widowed women with requests to take on offices or particular kinds of work, assuming that they would have more time to devote to feminist work. Women's Joint Legislative Committee convenor Katharine Norris complained that a meeting she had called was poorly attended because so many married women went out of town with their husbands for holiday weekends.[81] In 1965, in a break with precedence, the Woman's Party scheduled a convention on a Sunday and Monday for the convenience of members who were housewives with families and could not leave over the weekend. Pauli Murray, a black woman lawyer involved with the President's Commission on the Status of Women in the early 1960s, noted that feminists at that time tended to be self-supporting women who did not have to answer to anybody at home.[82] Rose Arnold Powell, Susan B. Anthony's advocate, chose Charl Ormond Williams of the National Education Association as the executrix of her estate because she, like Susan Anthony, was free from the domestic ties that took so many leaders out of the struggle. Powell herself believed that "[t]he devotion which would naturally have been bestowed on husband and children and home turned to my own sex and the Woman Movement and I can not feel that it has been in vain."[83]

Rewards of Participation: Satisfaction

We turn now to the other side of the reciprocal relationship at the core of commitment in order to examine the benefits that participants received. Since the women's rights movement did not achieve many concrete results in the post-1945 period, the work itself had to supply its own rewards.

In the most concrete sense, women might benefit from the experience they gained working in women's organizations. Margaret Hickey, for example, remarked in a speech that "[a]ll of us have watched the timid, retiring woman come into her own after a few stimulating years of experience in a woman's club."[84] Other women found these organizations intellectually stimulating. One small-town AAUW member commented that "I'm not sure that it [the AAUW] did anything for the community but it kept my mind from going to mush."[85] Another AAUW member believed that her organizational work kept her from atrophying when her son died in an automobile accident.[86] A woman stepping down as president of the Soroptimists, looking back on a year that "took more—much more—than every spare minute" swore that she would never forget it, that "it helped me personally to 'get on my feet' and not be too frightened."[87]

Women's rights activity could give meaning to the lives of the women involved, as in the case of Rose Arnold Powell. Unhappy unless busy with her campaigns to make known the pioneers of the women's movement, she coped with discouragement because she enjoyed the struggle itself: "Oh, for ten years more of vigorous fighting!"[88] One woman, while considering whether to go to work for the Woman's Party, noted that "there is solace in being surrounded by people who are all dedicated to the point of fanaticism to the one goal. . . . It makes my twenty years in gov't seem sinfully wasted and boring."[89]

Lena Madesin Phillips, founder of the International BPW, admitted that something within her had asked many times whether it was worth the effort: "Always though I believe the answer has been immediately 'absolutely.' I have had my difficult times as you well know. But rewards have seemed to me rich beyond my desserts."[90] Marguerite Rawalt wrote that her years of service in the BPW "have enriched my life beyond measure."[91] Women in the BPW recognized, and expressed to Margaret Hickey, "how dear the Federation and its work is to your heart."[92] Wealthy social leader and diplomat Perle Mesta described in her autobiography how, as a newly widowed socialite, she found that work in the women's rights movement filled her life and gave it meaning.[93] A National Woman's Party officer wrote explicitly of the rewards of feminist activity in referring to a financial contribution from Fannie Ackley: "It seems such a pity to take money from her, yet I firmly believe that in that giving she is benefitted beyond what we can realize."[94]

The reward of feminist activism must sometimes have been clear even to outsiders. National Woman's Party members in particular were subject to accusations that they did not want the ERA to pass since it constituted their "life work." As Ethel Ernest Murrell warned Alice Paul in the midst of the conflict between them, the efforts of Woman's Party members could be interpreted as ends rather than means, as a "reason for being alive." Because feminist work *was* a "reason for being alive" for many women in the movement, they willingly devoted their time and money—in some cases their entire lives—to it. In exchange for their work, they enjoyed membership in a supportive group and experienced the satisfaction of finding meaningful "life work." As one woman expressed the joy of commitment, "It is as thrilling as a love affair, and lasts longer!!!!"[95]

Rewards of Participation: Women's Relationships

Women committed themselves to women's rights activity not only because they received satisfaction in working for a worthwhile cause, but also because they experienced friendship and "the joys of participation." It is impossible to read the correspondence of women's rights activists without finding evidence of close and loving friendships. Women addressed each other as "honey," "beloved," "my dear," "dearling," "dearest," and regularly sent their love to one another. That these were more than simple stylistic flourishes—that they expressed real love and affection—is attested to by more direct statements. Friendships often grew out of feminist work and in turn contributed to making that work possible.

Such relationships could bind together women within one organization.

For example, a longtime Woman's Party member wrote of her cousin and coworker who had just died: "We were 'best-friends' for seventy-five years and her nobility of character and love were a sustaining influence through all the years."[96] On the death of another member, her sister-in-law wrote: "I know what a sense of loss you feel in your organization, just as we feel here in the family circle. You, personally, loved her too, I know, just as she loved you."[97] Inez Haynes Irwin, the historian of the National Woman's Party, wrote of Maud Younger, an early suffragist who had died: "Outside of my two sisters, I have loved no woman as I loved her. . . . I think of her often and always with love, admiration and wonder."[98]

Other organizations fostered friendships among members in the same way that the Woman's Party did. A former AAUW member in Washington wrote to Mary Church Terrell to remind her "how much I love you and how much I miss you."[99] A woman involved in the AAUW at the local level reported in an interview that some of her best friends were women she had met in that organization.[100] Another woman, who became a national leader of the BPW, recalled that she first began to build a network of friends throughout the country through the National Association of Women Lawyers.[101] One woman in the BPW, having finished a year as president of the District of Columbia branch, wrote to Lucy Somerville Howorth that "[y]our friendship has meant so much to me.[102] Margaret Hickey and Lucy Somerville Howorth, both national leaders of the BPW, periodically lamented their too-infrequent opportunities to share ideas and talk over personal problems.

How vital friendship was is suggested by the post-schism history of the Woman's Party rebels. Two of the rebels, who had met at a Woman's Party convention in the 1930s and became close friends, continued to write each other twice a week into the 1980s, even after leaving the organization in 1947.[103] Although their ability to maintain their friendship over the years is unusual, the desire to hold onto friends made in the course of feminist work is not. A whole group of rebels maintained contact, gathering to celebrate the seventy-fifth birthday of one of their number in 1952. They continued to reminisce about "the delightful times we used to have together slinking around corners to avoid running into the ubiquitous AP [Alice Paul]" into the 1960s. For the majority of them, "most of the bitterness has disappeared and only the laughs and the priceless companionship remain alive."[104] Leaving the Woman's Party had raised the fear that friendships, too, would end: "Please," one woman wrote, "Never lose touch with me. I would consider it one of the big losses in my life. . . . You are one of my most treasured friends."[105] Having lost the sorority of the Woman's Party community, the rebels sought to maintain their friendships outside of the organizational structure that had originally fostered them.

These examples show that friendship could be important within individual organizations, but the experience of the Woman's Party rebels suggests that friendships could also exist outside the boundaries of a particular group. A number of examples of friendship circles show how work on particular goals might bring women with different organizational affiliations together and facilitate contact among organizations.

Connecticut Committee head Florence Kitchelt pursued her work on behalf of the ERA primarily through correspondence with a number of women, many of whom became her friends. Kitchelt corresponded regularly with Alice Paul and other major figures in the National Woman's Party but she also maintained close relations with the rebels, with whom she attempted to build a new coalition of state organizations devoted to passing the ERA. In the course of her ERA work, she built a warm friendship with Alma Lutz which, like other of her friendships, eventually led to social visits. Kitchelt attempted to win over her old friend Mary Beard to support of the ERA by inviting her to visit. Kitchelt met and built a close friendship with Jane Grant, who revived the Lucy Stone League in the 1950s. "Dear Jane, you don't know how much I prize your friendship!" Kitchelt wrote.[106] Both Grant and Kitchelt joined the Federation of Woman Shareholders, a group headed by Wilma Soss and dedicated to getting women on the boards of directors of every large corporation. Once Kitchelt and Soss had met, they began to correspond warmly about a number of issues affecting women. Kitchelt made friends as she made contacts with women's rights supporters, as is clear from a note she penned to Alma Lutz: "A grey, raw, day—a good time to have friends! You and Marguerite mean light and warmth to me!"[107] How important such friends were comes across in a letter to Caroline Babcock, one of the Woman's Party rebels. Kitchelt invited her for a visit, remarking that "Richard and I are alone, and I could talk three days straight if you would give me a chance. . . . Do you and your husband live alone as my husband and I do? I really want to talk to you dreadfully."[108]

A number of feminist women interested in politics became fast friends and worked together to improve political opportunities for women. Emma Guffey Miller, president of the National Woman's Party in the early 1960s and a longtime Democratic national committeewoman, exchanged warm and affectionate letters with Perle Mesta, also a Woman's Party member and Democratic political figure. Mesta responded to Miller's congratulations on her appointment as minister to Luxembourg by writing that the letter touched her deeply and that it helped "a great deal to *feel* kindness and confidence surrounding, & if I am successful it will be largely because of the faith of my friends—I have always been so glad that you are one of them."[109]

Unlike Perle Mesta, many of the Democratic women with whom Miller worked closely did not support the ERA, but New Jersey Congresswoman Mary Norton expressed well the sense that they were all working for the same cause, which she defined as "a chance for women to find their real place in our national life."[110] How the friendship circle could work toward this goal is illustrated in Perle Mesta's recounting of the origins of her appointment as minister to Luxembourg. She and India Edwards, the top woman in Truman's administration, had lunch, and the conversation turned to the need for more women in government. Edwards reported that her daughter thought Mesta would make a good ambassador, and that discussion led Edwards to broach the subject with Truman.[111] These Democratic women reaped personal as well as political benefits from working together. For example, one woman wrote to Miller: "I enjoy the Federation [of Democratic women] and the women very much, and cherish the friendships which I have been able to make, none, I might say, more than yours."[112]

One last example of a friendship circle outside the boundaries of a specific organization illustrates the ways in which women who knew each other only through correspondence might provide one another the support and friendship necessary for their feminist work. Rose Arnold Powell corresponded with women throughout the country who shared her interest in the pioneers of the suffrage movement. She began to correspond with Alma Lutz, who was working on a biography of Susan B. Anthony, the object of Powell's most intense devotion. Lutz regularly sent her "good cheer" and "love." Powell also maintained contact with Ethel Adamson, who headed the National Woman's Party Committee on Pioneers of the Woman Movement, and historian Mary Beard, drawing encouragement from Adamson's efforts and Beard's strongly held views on the power of women in history. Through correspondence she "met" Una Winter, director of the Susan B. Anthony Memorial Library of California, and Martha Howard Taylor, director of the Susan B. Anthony Memorial in Rochester. She also corresponded lovingly with Adelaide Johnson, the eccentric sculptor of the monument to the women's movement that stands in the Capitol, and Charl Ormond Williams, National Education Association leader and the woman to whom Powell entrusted her estate to be used for the women's movement.

Powell never met most of these women in person, but she gained sustenance from their letters. She had longed for "somebody to visit with—a woman who was a real friend—one whom I could talk to intimately about a lot of things" for a long time.[113] She wrote to Martha Howard that she wished she were near "women of vision—such as you! Letters are so limited."[114] She dreamed of a "heart-to-heart visit" with Una Winter.[115] Despite her desire for more direct personal contact, she did find solace in her correspondence. To Howard she wrote that she was "grateful for your loving, encouraging letters all along the way since our paths crossed."[116] Winter she thanked for keeping in touch. More than ever, she wrote in 1952, she appreciated "friends who 'buck me up,' as life seems so limited now."[117] It is clear from Powell's correspondence that friendship built entirely on the basis of shared interests among women who might never meet could provide support for otherwise isolated women and could facilitate cooperation among them on particular goals.

A second kind of relationship that women's rights workers formed was a leader-follower relationship characterized by intense devotion on the part of members of organizations toward the group's leader. Relationships between leaders and followers had complex consequences. Woman's Party members often seemed to join because they met Alice Paul and responded strongly to her commitment to the cause. Many expressions of love, admiration, and devotion came from awestruck followers. Like a distant leader, Paul maintained an odd formality in her correspondence that contrasted strangely with the loving sentiments she expressed. One woman who regularly wrote effusive letters commented, in a letter addressed "Dear lady:" "I always *wish* I could say 'dear Alice,' but your addresses to 'dear Mrs. Vickers' freezes me!!"[118] But this formality should not be taken to mean that Paul formed no close relationships. To Elizabeth Rogers, a wealthy New York woman with whom she stayed whenever she visited the city, she wrote: "I wanted so much to see you when I was in New York."[119] Rogers would send notes to Paul "just so you will know I love you dearly."[120]

Lavinia Dock, Paul's mentor from her settlement house days and a woman who lived in the Henry Street settlement house community for a long time in an intimate association with Lillian Wald, maintained a warm relationship with Paul throughout the years.[121] Dock and Paul—whose teacher-student relationship and twenty-seven-year age difference probably compounded Paul's usual formality—were highly supportive of one another. Paul' wrote in 1945: "You are a great inspiration to me when I feel too discouraged & weary to take another step."[122] Dock, who could not believe that Paul ever wearied, saw her as superhuman and addressed her as "My Beloved Deity."[123]

Paul does not seem to have formed an intimate friendship with any one woman, but she lived and worked within a close-knit female world. She lived, at least some of the time, at Belmont House; when she was away from Washington she lived either alone or with her sister, Helen Paul, in Vermont or later Connecticut. It is clear that Alice Paul's ties—whether to her sister or to close friends or admirers—served as a bond that knit the Woman's Party together. But as the history of conflict in the Woman's Party makes clear, the intense devotion Alice Paul inspired could also tear the organization asunder.

Alice Paul is not the only example of a leader who inspired love and devotion in her followers, although she is certainly the major figure in the movement in the 1940s and 1950s. Mary Church Terrell, for example, the founder of the National Association of Colored Women, received the same kind of accolades from her admirers. "You mean so much to me in a very personal way and will always be my inspiration," one woman wrote.[124] In the same vein, the president of the National Association of Negro Business and Professional Women's Clubs wrote to tell Terrell that she had been chosen the first honorary life member and added: "You have been an inspiration to me personally and I am certain you have touched many many lives during your life of service."[125] Letters to Terrell resounded with descriptions of her role as inspiration and beacon light to the women with whom she came in contact. "About a half century ago when I first met you," one woman wrote, "an awe overcame me, then I learned to know you were very approachable and your sincerity inspired love."[126] Members of the NACW addressed her as "My dear Precious One," "My dear Adorable friend," and "Dear Lady," and sent their love, best wishes, and celebratory poems to her as she grew older. Apparently Terrell helped women out personally whenever she could—one woman wrote to tell her how much her support had meant when she was having difficulty as a student at Howard University and another, who compared Terrell to Saint Francis, wrote to thank her for letting her drop in one evening to talk about her troubles.[127] Another woman wrote and identified herself as someone Terrell had complimented at a National Council of Negro Women meeting for her recording of minutes.[128]

It is obvious from Terrell's papers that this kind of sensitivity—just what Alice Paul was reputed to lack—endeared her to the women with whom she worked. As one woman summed it up, "It is a pleasure and privilege to work with, for and on behalf of the greatest woman of our times. Touching shoulders with you—calling you friend, begetting inspiration from your courage, tenacity and ability has enabled me to hold fast, to stand my ground and to try to help build a better world for all people."[129]

Terrell, like Alice Paul, was the "grand old lady" of her organization by the 1950s, and that enhanced her stature as a leader. The same was true of Lena Madesin Phillips, who inspired and called forth devotion from her followers. Women from national federations of business and professional women throughout Europe wrote to Phillips and looked forward to her frequent visits. This inspirational ability is suggested by the notes of Marjory Lacey-Baker, Phillips' partner, concerning Phillips' papers for one particular year: "There is the usual crop of letters to LMP following the Convention [of the BPW] from newly-met members in hero-worshipping mood—most of whom went on to be her good friends over the years."[130] One National Woman's Party member who had been associated with Phillips in international work for the BPW reinforced this impression when she reminisced about the first time she had met Phillips. She had followed her down the hall to watch her, feeling intuitively that she was right for the cause of women's rights, and had followed her ever since.[131] Another woman, who had the responsibility of recommending Phillips for a policy-making position in response to the BPW's push to get women in government, warned Phillips that: "Knowing how I feel about you, you know I'd have you sounding like a cross between Joan of Arc, the Statue of Liberty, the Virgin Mary . . . and the Almighty Himself."[132] Phillips' biographer concluded that her "magnetic personality drew to it women of vision and vigor similar to hers."[133]

As these three examples suggest, the phenomenon of leader-follower relationships was important. The various organizations were unapologetically hierarchical, and the centrality in them of women who had founded the groups or been associated with them in leadership positions for many years gave rise to a type of hero worship. The relationships between leaders and followers served primarily to strengthen the bonds within individual organizations.

We turn now to a third kind of relationship, couple relationships between women. These long-lasting, marriagelike partnerships between women must be considered in their historical context. Historians have pointed out that intense attachments between women were common and accepted among middle-class women in the nineteenth century.[134] As feminists, college graduates, and other independent women began to move such relationships outside the domestic sphere of marriage and family in the late nineteenth century, however, those relationships took on new meaning. By the dawn of the twentieth century, such relationships were no longer untouched by any hint of deviance. With increased recognition of women's sexuality and greater emphasis on heterosocial activity came the "discovery" of lesbianism and the application of that label to intense attachments between women.

Some women who remained active in women's rights lived together in long-term committed relationships, what were known in the nineteenth century as "Boston marriages." Yet the fact that lesbianism was socially recognized and that a lesbian culture, however marginal, did exist by the 1950s created a different context than that in which woman-committed women in the nineteenth century had lived. None of the women's rights activists would, as far as can be determined, have identified as lesbians. A great deal has been written on the problem of characterizing women's relationships in the past, but the important point here is that there *was* such a thing as a lesbian identity by the 1950s and these women did not claim it.[135] They lived together in couple relationships with apparent openness, security, and confidence. In part because of their age and class privilege, they seemed relatively immune to potential charges of deviance. In other words, they lived very much like nineteenth-century women in Boston marriages, but in a twentieth-century environment. They created a homosocial world within the boundaries of their organizations and maintained the kind of female public sphere that Estelle Freedman has argued was so essential to the survival of feminism.[136] An examination of several couples suggests how completely women's rights activists accepted couple relationships between women.

Jeannette Marks and Mary Woolley lived together for almost fifty years and belonged to a number of women's organizations, including the National Woman's Party. They met at Wellesley College in 1895 when Marks began her college education and Woolley arrived at the college as a history instructor. Less than five years later they made "a mutual declaration of ardent and exclusive love" and "exchanged tokens, a ring and a jeweled pin, with pledges of lifelong fidelity."[137] They spent the rest of their lives together, for many years at Mount Holyoke where Woolley served as president and Marks taught English. Jeannette Marks committed herself to suffrage and later, though the National Woman's Party, to the ERA. She pressured Woolley, always more cautious, to join the Woman's Party, which she did in 1942. Marks served as chairman of the New York state branch in the 1940s and played a powerful role as a rebel sympathizer in the schism. It is clear from Marks's correspondence that the two women's relationship was accepted as a primary commitment by their women's rights colleagues. Few letters to Marks in the 1940s fail to inquire about Woolley, who was seriously ill. One married woman, who found herself forced to withdraw from Woman's Party work because of her husband's health, acknowledged the centrality of Mark's and Woolley's commitment when she compared her own reasons for leaving to "those that have bound you to Westport."[138]

Lena Madesin Phillips lived for about thirty years with Marjory Lacey-Baker, an actress whom she first met in 1919. Phillips "lost her heart" to "[t]he most beautiful voice I ever heard" when she attended a pageant in which Lacey-Baker performed.[139] In the early twenties they moved in together, and they lived as a couple until Phillips' death in 1955. In 1924 the two women went different places at Easter, and the recollection, in the notes Lacey-Baker made for a biography of Phillips, caused her to quote from *The Prophet*: "Love knows not its own depth until the hour of separation."[140] Phillips described Lacey-Baker in her voluminous correspondence as "my best friend" or noted that she "shared a home with me."[141] Her friends and acquaintances regularly mentioned Lacey-Baker. Phillips happily described the tranquility of their life together at their farm, Apple Acres, to her many friends in the movement.

Madesin Phillips' papers seem to suggest that she and Marjory Lacey-Baker moved in a world of women friends. Phillips had devoted much of her energy to international work with women, and she kept in touch with European friends through her correspondence and through her and Lacey-Baker's regular trips to Europe. Gordon Holmes, for example, of the British Federation of Business and Professional Women, wrote regularly to "Madesin and Maggie." In a 1948 letter she teased Phillips by reporting that "two other of our oldest & closest Fed officers whom you know could get married but are refusing—as they are both more than middle-aged (never mind their looks) it suggests 50–60 is about the new dangerous age for women (look out for Maggie!)"[142] Phillips reported to Holmes on their social life: "With a new circle of friends around us here and a good many of our overseas members coming here for luncheon or tea with us the weeks slip by."[143] Lacey-Baker was always a part of Phillips' feminist world, and their relationship received acceptance and validation throughout the women's rights movement, on both the national and international level. When a colleague and friend from the International Federation published a biography of Phillips based on an unpublished autobiography and Lacey-Baker's notes, she described the two women as "family," although she downplayed the emotional commitment. But she did admit that Lacey-Baker "became increasingly concerned with 'causes' she had scarcely known existed before she met Madesin."[144]

The lifelong relationship between feminist biographer Alma Lutz and Marguerite Smith began when they roomed together at Vassar in the early years of the twentieth century. From 1918 to Smith's death in 1959, they shared a Boston apartment and a summer home, Highmeadow, in the Berkshires. Lutz worked both independently and with the National Woman's Party for the ERA and other causes. Smith, a librarian, also worked in the Woman's Party. Like Madesin Phillips, Lutz wrote to friends of their lives together: "We are very happy here in the country—each busy with her work and digging in the garden."[145] They traveled together, visiting Europe several times in the 1950s. Letters to one of them about feminist work invariably sent greetings or love to the other.

When Smith died in 1959, Lutz struggled with her grief. She wrote to her friend Florence Kitchelt, in response to condolences: "I am at Highmeadow trying to get my bearings . . . You will understand how hard it is. . . . It has been a very difficult anxious time for me."[146] She thanked another friend for her note and added: "It's a hard adjustment to make, but one we all have to face in one way or another and I am remembering that I have much to be grateful for."[147] Later she wrote to one of her regular correspondents to say that she was carrying on but it was very lonely.[148]

The fact that Lutz and Smith seemed to have had a number of friends who also lived in couple relationships with other women suggests that they had built a sort of community of women within the women's rights movement. Some of the sympathy directed to Lutz on the occasion of Smith's death came from women who had themselves suffered the loss of women partners. Perhaps most significant was the pattern of visiting among coupled women. Every year Mabel Vernon, a suffragist and worker for peace, and her friend and companion Consuelo Reyes, whom Vernon had met through her work with the Inter-American Commission on Women, spent the summer at Highmeadow. Vernon, one of Alice Paul's closest associates during the suffrage struggle, had met Reyes two weeks after her arrival in the United States from Costa Rica in 1942. They began to work together in Vernon's organization, People's Mandate, in 1943, and they shared a Washington apartment from 1951 until Vernon's death in 1975. Reyes received recognition in Vernon's obituaries as her "devoted companion" or "nurse-companion."[149]

Two other women in a long-term relationship, Alice Morgan Wright and Edith Goode, also kept in contact with Lutz, Smith, Vernon, and Reyes, sometimes visiting Highmeadow in the summer. Wright and Goode had met at Smith College and were described by one National Woman's Party member as "always together" although they did not live together.[150] They, like Lutz and Smith, worked together in the National Woman's Party, where they had also presumably met Vernon. Both Wright and Goode devoted themselves to two causes, women's rights and humane treatment for animals. Wright described herself as having "fallen between two stools—animals and wimmin."[151] The two women traveled together and looked after each other as age began to take its toll.

These examples suggest that an intimate relationship with a woman who shared a commitment to feminism could help facilitate work for women's rights. At the same time, the environment of the women's rights movement provided support for women who lived in relationships that the outside world might label deviant. Although none of the women identified as lesbians, that in itself would not have protected them from attack. Doris Stevens, who engaged in "red-baiting" after she left the National Woman's Party, had suspicions of another sort about some of her former friends. She recorded in her diary a conversation with a Woman's Party member about Jeannette Marks and Mary Woolley, noting that the member, who had attended Wellesley with Marks, "[d]iscreetly indicated that there was 'talk.' "[152] At another point she reported a conversation with a different Woman's Party member who had grown disillusioned about Alice Paul. Stevens noted that her informant related "weird goings on at Wash. hedquts. wherein it was clear she thought Paul a devotee of Lesbos & afflicted with Jeanne d'Arc identification."[153] Along the same lines, the daughter of a woman who had left the Woman's Party complained that Alice Paul and Anita Pollitzer sent her mother a telegram that "anybody with sense" would think "was from two people who were adolesant [sic] or from two who had imbied [sic] too much or else Lesbians to a Lesbian."[154]

Such conversations suggest that the intensity of women's relationships and the existence and acceptance of couple relationships in women's organizations had the potential, particularly during the McCarthy years, to attract denunciation. Doris Stevens herself wrote to the viciously right-wing and anti-Semitic columnist Westbrook Pegler to "thank you for knowing I'm not a queerie" despite the fact that she considered herself a feminist.[155] Steven's right-wing and homophobic views typified the connection made by McCarthyites between Communist subversion and homosexual "perversion." McCarthy himself attacked the "Communists and queers" in the State Department and Truman feared that the " 'Artists' with a capital A, the parlor pinks and the soprano voiced men" were banding together as a "sabotage front for Uncle Joe Stalin."[156] The association of "effeminate men" and "masculine women" with Communism makes clear the connection between sexuality and politics in the McCarthyite mind.[157] How real the threat was for women is suggested by three incidents.

A sketch of the first emerges from Doris Stevens' diary. Someone she knew reported that Anna Lord Strauss, the liberal leader of the League of Women Voters, was "not a bit interested in men" and thought Stevens "was kidding when asked who men friends were." Strauss apparently refused to be paired off with unattached men for dinner, leading Stevens to wonder if there was something "off color" going on. Stevens called a government official in 1953 to report what she had learned, apparently trying to discredit Strauss, far too liberal for her taste, with a charge of "unorthodox morals."[158]

The other two incidents, involving opposition to the appointment of women, are described in the memoirs of India Edwards, a top woman in the Truman administration.[159] In 1948, opposition to tax court judge Marion Harron's reappointment to the bench arose from Harron's fellow judges who cited her lack of judicial temperament and "unprovable charges of an ethical and moral nature." Harron had written letters to Lorena Hickok that even Hickok's biographer, who had denied the relationship between Hickok and Eleanor Roosevelt, had to admit were love letters.[160] The other case that Edwards described left no doubt about what ethical and moral charges were involved. When Truman appointed Kathryn McHale, longtime executive director of the American Association of University Women, to the Subversive Activities Control Board, Senator Pat McCarran advised Truman to withdraw her name and threatened to hold public hearings during which "information would be brought out that she was a lesbian." Truman sent Edwards to talk to McHale to give her the chance to decide what she wished to do, and McHale decided to stand for the appointment. "I will fight such a vicious lie with all the strength I have," Edwards quoted her as saying.

What these incidents suggest is that, at least in certain circles, participation in the women's rights movement was highly suspect and strong attachments between women were viewed as deviant. For woman-committed women, then, the acceptance and support of women's rights coworkers was all the more important. The women who lived in couple relationships managed to do so respectably, despite the emergence of a lesbian culture and the occasional charges of lesbianism that surfaced at the time, because they worked independently or in professional jobs, had the money to buy homes together, and enjoyed enough status to be beyond reproach in the world in which they moved. Women who later identified as lesbians but did not attach an identity to their emotions and behaviors in the 1950s describe that period as one in which women might live together without raising any eyebrows, but it is important to remember that even the class privilege that protected couple relationships would not necessarily protect women who sought to enter powerful male-dominated institutions.[161]

Friendships within organizations or among women active in different groups, leader-follower relationships, and couple relationships all helped in different ways to hold the women's rights movement together and provided rewards for the individual's commitment. Women who lived in couple relationships with other women found their commitments to each other validated. Leaders like Alice Paul, Mary Church Terrell, and Madesin Phillips inspired intense devotion from countless women who felt rewarded for their efforts by contact with their heroines. And friendships that grew up in feminist work brought satisfaction to women who saw little concrete progress as a result of their efforts.

Women's rights activists recognized how much the relationships they established meant to them. One National Woman's Party member accepted an invitation to attend the 1945 biennial convention with "a twinge of conscience that I go as a matter of duty when, in truth, I really do have such joy mixed in, as comes when one meets very good friends: and very good friends are not made easily but they grow into something very wonderful only when time has done its part."[162] A member of the National Association of Colored Women wrote to Mary Church Terrell to tell her that she had enjoyed every minute of the annual convention.[163] BPW president Minnie Maffett agreed that friendship was a major benefit of feminist work.[164] Marguerite Rawalt, a later Federation president, treasured "the joy and inspiration of friendships formed while working side by side with many of the great women of my country."[165] The women's rights movement, because of the importance of the relationships within it, could provide women support and foster the growth of feminist consciousness. Ethel Ernest Murrell, on the occasion of her resignation as national chairman, found her associations in the Woman's Party "happy ones, inspiring ones, among the most noble of my life." In the organization, she said, she had met women of integrity and dedication, "women whose beliefs, like mine, enshrine the ideal of emancipated womanhood."[166] Along the same lines, a BPW member found the annual convention "the one big highlight of my life so far. To meet and know so many fine women in the United States and the World makes me feel so proud to be of their sex."[167]

Without this kind of pride and satisfaction, it is hard to imagine how women could have continued to work for feminist causes throughout the 1940s and 1950s. But the importance of personal relationships also had significant consequences for recruitment.

Recruitment

Although the women's rights movement failed to attract a new following in the post-1945 years, personal relationships did bring some individuals into feminist work. One way that women came into women's rights work was through family connections. Scholars have noted the frequency with which women active in the nineteenth-century women's movement came from families committed to reform, suggesting that the reform impulse is often handed down from generation to generation.[168] In the same way, feminism in the post–1945 period seems sometimes to have been a kind of family legacy. Women with illustrious suffrage foremothers regularly used their ancestry as an important credential. Nora Stanton Barney, daughter of militant suffragist Harriot Stanton Blatch and granddaughter of pioneer suffragist Elizabeth Cady Stanton, proudly claimed her heritage and used it to support her work for the ERA. Her daughter, Rhoda Jenkins, carried the family commitment to feminism into the 1980s as vice president of the Greenwich, Connecticut, branch of the National Organization for Women.[169] But some in the movement criticized an over-reliance on ancestry. Susan B. Anthony II, grandniece of the famous suffragist, from the perspective of a National Woman Party member "trades on 'Aunt Susan's' name."[170] ERA supporters likewise complained that Anna Lord Strauss, League of Women Voters president and great-granddaughter of feminist leader Lucretia Mott, "play[s] ancestral connections constantly."[171] One League member, reporting on a panel discussion of the ERA in which Nora Stanton Barney had participated, lamented to Strauss: "I was very sorry you weren't there to boast about your grandmother [sic], Anna. I couldn't even dig up an ancestor who marched in a suffrage parade, much less a leader in the Woman Movement."[172]

Feminist foremothers were important as a means of recruitment as well as for status, but so were less historical and more immediate ties. Some

women had mothers who had been involved in feminist activity. Lucy Somerville Howorth's mother, Nellie Nugent Somerville, "battled for women's
rights at a time when it was universally unpopular."[173] When Howorth became the chairman of the Assembly of Women's Organizations for National
Security, a friend wrote to congratulate her and commented that the "women
in your family have never yet failed in taking responsibility."[174] A year later,
the AAUW named a grant after Howorth and her mother, prompting another friend to express her pleasure at the tribute to the mother-daughter
team.[175]

Other women, too, carried on their mothers' feminist work. Margaret
Hickey's mother had been involved in the suffrage movement. Mary H.
Kennedy and Mary C. Kennedy, mother and daughter, both devoted themselves to the National Woman's Party. A biography of the younger Kennedy
in the Woman's Party *Bulletin* mentioned proudly that the elder Kennedy
has passed on her enthusiasm to her daughter.[176] Lucy Gwynne Branham,
who opened a dining room at National Woman's Party headquarters during
the last days of the suffrage struggle, and her daughter Lucy Branham both
resided at Belmont House in 1959 and encouraged a younger member who
was considering serving meals there again.[177] The Woman's Party house
manager gratefully accepted a donation of furniture from Izetta Jewell
Brown Miller, whose family for three generations had worked for women,
happy not only to keep the interest of Miller but to gain the interest of her
daughter.[178] As one woman wrote on joining the National Woman's Party,
"Mother had fought so many battles for women, from suffrage on, that it
was time I took up the sword."[179]

The National Woman's Party seemed particularly likely to look to family
relationships as a form of recruitment. A number of relatives—not just
mother-daughter teams—worked together in the Woman's Party. Anita Pollitzer, whose two sisters and two aunts were "founders" of the Woman's
Party, joined while home on vacation from college. Pairs of sisters, in the
pattern of Alice and Helen Paul, worked together in the organization. Belief
in the strength of family ties is evident in the cases in which Woman's Party
officers asked relatives to take over the duties of a deceased member. For
example, one woman refused to take over a state chairmanship after her
mother-in-law died and vacated the position, although she felt that she
should and apologized for her unwillingness.[180] With more success, Alice
Paul asked Janet Griswold, sister-in-law of the deceased treasurer of the
Woman's Party, to take over her post. Griswold served for eight years,
commenting when she resigned to go to Europe on "how much it meant to
me eight years ago to follow in Mabel's interest."[181]

Woman's Party members sometimes looked to their own families when
they sought response from the younger generation of women. One member
wrote proudly that her sixteen-year-old granddaughter had joined the Junior
National Woman's Party on her own initiative.[182] Ernestine Powell, national
chairman of the Woman's Party in 1954, brought her two sons, her daughter,
and a friend of her daughter to Belmont House, where the two girls launched

a campaign to win the appointment of women as pages in Congress. Two years later, Powell's daughter went to Washington to work for the ERA and succeeded in winning the endorsement of another senator for the amendment. That such early training did not always lead to total commitment to a mother's cause is indicated by a letter Powell wrote twelve years later in which she reported that she was trying to get her daughter, now a university painting instructor, to give a talk on women in art for Susan B. Anthony Day. "As you can see," Powell noted, "I am trying to get her to become again interested in a world a little less Bohemian than that through which she traveled for a while."[183]

Family relationships might bring women into the women's rights movement, but so, too, could the other kinds of personal relationships we have explored here. Women might recruit their friends for feminist work. When a Woman's Party staff member suggested to retired linotype operator Fannie Ackley that she form a Washington state branch of the organization, Ackley responded by lining up a friend to head the new organization and recruited fellow members of her Typographical Union as members. Olive Hurlburt, a Woman's Party member involved in the 1947 schism, had joined the group when a fellow teacher suggested that she might enjoy it. Margaret Hickey attributed her interest in the women's movement to her contact with powerful women lawyers, including Florence Allen and Lena Madesin Phillips.

Women involved on the local level with such organizations as the AAUW emphasized in interviews the importance of friendship in recruitment. One woman joined AAUW as a student because a neighbor suggested that she might enjoy it, another had friends who belonged to AAUW and brought her in, and another noted that women in the 1960s would bring friends to meetings for support.[184] One National Woman's Party member found six women who wanted to join when she attended her local women's club meeting in Rhode Island.[185] The chairman of the Oklahoma branch noted in her resignation letter that she continued to do a great deal of speaking on the ERA "and am converting some of my friends."[186] Other Woman's Party members, too, reported the efforts they made to recruit their friends. In an article on the Woman's Party written in the 1970s, the author described Elizabeth Chittick, president of the group, as typical of the majority of members in that she came to the work "not out of a long standing interest in feminism but at the persuasion of friends."[187] In fact, Woman's Party leaders preferred recruitment through friends because it avoided the possibility of a takeover, which might result from a public membership drive.[188]

Recruitment through personal relationships—whether by family member, friend, leader, or life partner—ensured the stability of the women's rights movement, but it also perpetuated homogeneity and exclusiveness. Although women's rights supporters sometimes dreamed of a mass mobilization of women, they never attempted to reach outside their own personal circles. This is the cost associated with recruitment through personal relationships.

Commitment and community in the women's rights movement were important because they ensured survival in a climate of antifeminism. In the environment of the 1940s and 1950s, commitment to the women's movement could not be taken lightly. The intensity of commitment made feminist work possible, but it could also serve to circumscribe recruitment, not only by limiting the appeal of feminist activity, but also by creating structural barriers to participation by the vast majority of American women.

In the same way, the importance of women's relationships created close-knit worlds that gave women support and made feminist activism possible, but set up barriers to any substantial recruitment. The women's rights movement in this period was not, in the context of the Cold War and the conformity characteristic of the 1950s, likely to recruit a mass following in any case. But it is important nevertheless to recognize the tension between maintenance and recruitment that affected the movement's history. In the post-1945 period, the needs of women interested in feminist work were met at the cost of any outreach to groups not already represented in the women's rights movement. In this closed environment, feminist lives were possible.

6

Feminist Lives

Women's rights activists found support for their efforts in a community of like-minded women. But why and how did women become feminists? How did their involvement in women's movement activity fit into their lives? What did it really mean to be a feminist in the fifties? The lives of individual women suggest some answers to these questions. We explore here the lives of seven women involved in the women's rights movement. Although these women are neither the most important figures in the movement nor representative in any statistical sense, the range of experiences, personalities, activities, and ideas that they represent gives us a sense of different feminist lives.

Rose Arnold Powell

Rose Arnold Powell (1876–1961) devoted her life to the struggle to win recognition for the great women of the suffrage movement, especially Susan B. Anthony. Born in Minnesota, Powell attended a normal school and became a teacher, first in Minnesota and then in Seattle. She married in 1909 but her husband deserted her after seven years. After the end of the marriage, which had produced no children, she held a variety of jobs in different places. Years later, she recorded in her diary her thoughts about her marriage: "If my marriage had been a happy one, I might today have a home and security, but I am convinced that I never could have been able to make the contribution I have to the woman movement, to which so much of my life has been dedicated. I staked everything on a man's love, and it proved to be a delusion."[1]

In the 1920s, Powell read Anna Howard Shaw's *Story of a Pioneer*, and this led to her conversion to feminism. She described what she called her "awakening": "I grasped for the first time in my life, the deep import of the long struggle to secure for women a status of equality with men. I had been a teacher by profession and it awakened me to the fact that in all my experience American history taught only heroes—masculine greatness, cour-

112

age, bravery, achievements . . . I saw that for true democracy—democratic balance, there should be instruction in our public schools on the Woman Movement and its pioneers, especially Susan B. Anthony, who was 'the propulsive force' back of it during her life."[2] Powell worked for the Internal Revenue Service in Washington in the 1920s but resigned in 1930 in order to devote herself full-time to her new cause. The 1950s found her in Minneapolis, moving from place to place, usually living in a rented room in a private home but occasionally resorting to a nursing home. She continued to fight for recognition of the pioneers of the women's movement throughout the rest of her life, despite physical disability, difficulties with housing, and isolation.

Powell's major goal was to win a place for Susan B. Anthony on Mount Rushmore. She had finally persuaded the sculptor who created what she called the "Mount Rushmore National (Masculine) Memorial" that Anthony belonged on the mountain just before he died in 1941. She immediately started to work on his son, also a sculptor, and rejoiced when, in 1956, he agreed to sculpt Anthony if Powell raised the money and won Congressional approval—needless to say, no mean tasks. Undaunted, she set about trying to win support and cooperation from women's organizations, without much success.

This dream of Anthony ensconced in the Black Hills was Powell's fondest, but she also worked for recognition of Anthony in other, less dramatic ways, sometimes with more success. She supported a bill passed by the Minnesota legislature that proclaimed Anthony's birthday a state holiday and worked for the establishment of a national holiday. She also participated in a successful campaign to have Anthony elected to the National Hall of Fame, and she persuaded the Minneapolis school board to name a school after Anthony in 1956. Sensitive to discrimination wherever she encountered it, she regularly took to task those responsible for the injustices she saw around her.

Powell considered herself a feminist, although she did not often use the word, preferring instead simply to describe her anger at the male domination of society. For Powell, recognition of her beloved Susan B. Anthony would accomplish the goal of ending this "utter man-mindedness" that she saw all around her.[3] She looked to the women's rights movement to secure the "complete emancipation of women" and believed in the necessity of collective action: "As long as any woman is bound—we are not totally free, since all womanhood is *one*."[4]

Certainly all of this suggests a feminist vision of women uniting to fight the system that oppresses them, even if her point of attack seems inadequate. But Powell, despite her grand statements, often found women as "mired in masculinity" as the men and expressed disgust with the lack of support she found.[5] She often lapsed into self-pity for the "tired little mouse" meeting indifference, lack of gratitude, and fear at every turn. And in fact, few women, even among feminists, shared her enthusiasm for the Mount Rushmore project. Alma Lutz wrote that she respected Powell's interest but did not care to see "women's faces carved on the beauties of nature."[6] The head of the Susan B. Anthony Memorial in Rochester told Powell that Anthony, in response to a project to place her statue in a park during her lifetime, had commented that she did not like the idea of being out in the cold and the dark. Nevertheless, Powell remained committed to what she, like other feminists, described as her "life work." She was a deeply religious woman who saw her "mission" as a God-given one. In her less sacrificial moods, she recognized how beneficial her work was to her own state of mind. Perhaps most revealing of her own conception of her relationship to her work was her description of the Mount Rushmore project as her "child." As her comments on her marriage make clear, Powell saw this as more than a simple metaphor; she believed that the devotion she would have bestowed on a husband and children had instead been turned to the women's rights movement.[7]

Powell emerges from her diaries as a lonely and an occasionally troublesome woman—many of her moves, for example, resulted from her constant complaints about noise and other people's demands. She was essentially alone, despite the attempts of one woman, in whose house she lived the longest, to provide some of the comforts of family, such as parties on her birthday and celebrations on Anthony's birthday. But this should not be taken to suggest that feminists were simply women embittered by some personal loss. Powell did not seem to regret the turn her life had taken, and what seems sad about her life is not her failed marriage but her isolation. She longed for "a woman who was a real friend—one whom I could talk to intimately about a lot of things."[8] She relied on the support she received through the mail, although even that came, in most cases, from women she had never actually met. She expressed her relationships with women in the movement in terms of sisterhood, addressing some of her letters "Dear Sister," signing them "Yours in the Sisterhood of Woman," thanking correspondents for their sisterly regard, and noting that terms of endearment "warmed the cockles of my heart."[9]

Such sisterhood was clearly important to Powell, and she did not get enough of it in her correspondence, most of which was impersonal and distant. Why Powell chose this lonely crusade rather than involvement with one of the women's organizations is impossible to say. She belonged to the National Woman's Party, but her residence in Minneapolis made active participation impossible. She also belonged to the local branch of the General Federation of Women's Clubs, but the group's reluctance to cooperate with her Anthony schemes annoyed her.

Ultimately she put her faith in the next generation, believing that her work for recognition of Susan B. Anthony would inspire younger women. Perhaps her work did play some part in the later public recognition of Anthony. Powell died without seeing her "child" make it, but she never gave up her struggle. Although she could not leave us Susan B. Anthony on Mount Rushmore, she did leave a different legacy—a rich record of a feminist life. She worried about the fate of her papers and at the same time wondered who would ever read them, believing that it would be "an extraordinary person who would be interested enough to wade through" them, "but if anybody ever does, she—I can't imagine a *he* doing it—will get a partial idea of what my life has been like. The actual doing—speaking, writing, traveling; begging, failing, succeeding—the never ending impulsion that would not let me rest."[10]

Florence L. C. Kitchelt

Florence L. C. Kitchelt (1874–1961) lived and worked within the mainstream of American liberalism but, unlike the vast majority of American liberals in the 1950s, she worked actively for women's rights and, specifically, embraced the cause of the Equal Rights Amendment. Her life presents a dramatic contrast to the lonely efforts of Rose Arnold Powell.

Kitchelt's background followed the pattern for women reformers of her generation; she graduated from Wells College, a women's college, lived at home for a year, and then went into settlement house work. After a year abroad, she married artist and socialist Richard Kitchelt in 1911. She had no children; some biographical notes she wrote in 1951 stated simply, "ill, operation, no children."[11] She threw herself into suffrage work and, after 1920, into the League of Women Voters, progressive politics, and the League of Nations Association. In 1943, she announced her conversion to the ERA and devoted herself to it, through the Connecticut Committee for the ERA, until her death.

Kitchelt had met her husband when he arranged for Emmeline Pankhurst, the famous British suffragette, to speak in Rochester, where they both lived. Their wedding, an unorthodox ceremony in which they expressed their ideals of perfect union, began what was obviously a loving and companionable relationship. Kitchelt wrote in 1948: "My husband and I, married happily over 30 years, consider the home a cooperative institution, not to be maintained at the cost of development of one of the partners."[12] The Kitchelts moved to Ohio in 1956, where they lived with Kitchelt's sister, who taught at Central State College, a predominantly black school. Richard Kitchelt died in 1958 and, as Kitchelt and her sister put it in their annual holiday letter two years later, "F.K.'s pen went dry."[13] Although she missed him terribly and found that her physical condition increasingly prevented her from leaving the house, she continued to do what she could for the ERA, believing that her "elder years can be happy only as they are fruitful."[14] She died of a stroke in 1961 and her cremation and memorial service, as she had wished, were conducted according to the principles of a group that favored saner funerals. Friends paid tribute to her as "a genuine liberal and a fearless one," someone whose work for women was "ahead of her time."[15]

Most liberals in the 1950s opposed the ERA because they believed that it would harm working women by removing the protective labor laws for which reformers had fought so hard in the early twentieth century. Kitchelt accepted this argument until her "conversion" in 1943. For a leading figure of the anti-ERA League of Women Voters to change her mind and announce publicly that she supported the amendment was startling. Kitchelt immediately wrote to the National Woman's Party to announce that "life in general" had convinced her that no amount of special benefit could be enough to offset the damage done to human equality by accepting a second place for women in the social structure.[16] She never identified more precisely what led to her conversion on this issue, but held consistently to the "life in general" explanation that was most likely intended to assure her correspondents that she had not succumbed to the blandishments of the National Woman's Party, since she continued to belong to the League of Women Voters.

Once converted, Kitchelt joined the National Woman's Party, but she channeled her efforts on behalf of the ERA into the organization she helped to found, the Connecticut Committee. She became the driving force behind this group, which she envisaged as a "rallying-point for all persons and group[s], men and women" interested in endorsing the amendment.[17] The organization remained essentially a correspondence committee. Kitchelt sought prominent individuals, both men and women, for the advisory board; collected endorsements for the ERA from politicians, clergy, and others; urged members to write their representatives on behalf of the ERA; and penned countless letters to the editors of newspapers. Often the Committee functioned as no more than a name Kitchelt used in her own letter-writing campaigns; in 1949, for example, she noted that the work progressed, although it was a wonder, since "I work so much alone, with a one-finger movement on the typewriter!"[18]

Kitchelt dreamed of building a coalition of groups with broad membership to help push the ERA through Congress. She attempted to serve as a catalyst for the formation of groups like her Committee in other states, since she believed that grass-roots work at the state level would be essential for ratification, even if the national groups succeeded in getting the amendment through Congress. Kitchelt also used the Connecticut Committee in other campaigns. Like Rose Arnold Powell, she wanted to see Susan B. Anthony elected to the Hall of Fame and spent the summer of 1950 working toward this end. In 1955, she launched a campaign to persuade the American Civil Liberties Union, to which she belonged, to change its position on the ERA. Along with author Pearl Buck and Alma Lutz, she wrote to over one hundred ACLU members, succeeded in bringing the issue before the board, arranged testimony—but to no avail.

On a daily basis, Kitchelt, like Rose Arnold Powell, plugged away against attitudes she found demeaning to women whenever they came to her attention. She criticized the exploitation of women by the fashion industry, for example, which she found "sordid." She believed that the solution of women's problems would require "important social changes, like half-time jobs for women and much less attention to fashion and furnishings."[19]

She favored day nurseries as part of public education and believed that *f* were beginning to share in domestic responsibilities, a trend she heartily approved. But while she referred to women as a "subject class" and believed that the "whole world functions on women's subservience," she also characterized those who fought for identical rights as "extremists."

What this all boils down to is that Kitchelt had great faith that the system could be reformed and women could take their places as citizens and human beings first, with special consideration give to their responsibilities as women and especially as mothers. Most liberals believed that the ERA would eliminate such special consideration, but Kitchelt remained convinced that women could have equal, but not identical, rights.

Although Kitchelt put her work for women first, she also had contacts outside the women's rights movement in the American Civil Liberties Union, Americans for Democratic Action, and the National Association for the Advancement of Colored People, to which both she and her husband belonged. She tried to use these contacts for the good of her cause, and this involved, among other things, cooperation with men. Although she sometimes expressed disgust with male values—such as the time when informed that a young and attractive speaker for the ERA would make a better case to the ACLU—she still believed wholeheartedly in cooperation between women and men. The idea of a woman's party did not appeal to her, and she believed that it was especially important to have men write and speak for the ERA. No doubt her attitude grew in part out of her marriage, for she and Richard Kitchelt worked throughout their lives together for many of the same causes. Yet she sometimes expressed the desire for closer contact with other women involved in the women's rights movement.

The joy of working with women was part of what Kitchelt found exciting about her work for the ERA, and she remembered the suffrage movement as "thrilling," "uplifting," and "unifying." "To have shared in the comradeship of that movement of thousands of women devoted to a single goal— it carried a thrill the young women of today cannot guess," she wrote.[20] But however much the 1950s paled beside the grand old days of the suffrage struggle, Kitchelt maintained her commitment. "I am afraid it is in my blood to be a crusader," she commented, and several years later she described herself as "addicted" to the ERA.[21] Although she participated in many organizations, her priority remained the cause of winning equality, as she defined it, for women. "The life of reformers . . . is hard," she commented, but she obviously would not have had it any other way.[22]

Fannie Ackley

Critics of the Equal Rights Amendment liked to point out the wealth and privilege of many of the amendment's supporters in order to paint them as self-interested ladies willing to sacrifice working-class women to their own ambitions and needs. In response, ERA advocates, and especially the National Woman's Party, seized every opportunity to display proudly women like Fannie Ackley (1880–1955), a worker and union member who joined the National Woman's Party and devoted herself after retirement to the cause of equal rights for women.[23]

Ackley was a linotype operator in Spokane, Washington, when she first joined the Woman's Party in 1945. In explaining her support for the ERA and her opposition to protective labor legislation, she related her own experience in the labor force. Before World War I, she had worked at a poorly paying job, on a nine-hour shift with frequent overtime. After several years, she landed a job as a linotype operator on a weekly newspaper—a skilled job and a relatively well-paid one. She worked an eight-hour shift with one hour of overtime on press day. On the basis of protective legislation that limited women to eight hours of work daily, someone complained to the authorities about her weekly hour of overtime, and she lost her job to a man in 1914. For three years she went back to harder work for lower pay, until the entrance of the United States into World War I gave her a chance to return to linotyping. This experience convinced her that women needed to fight against protective laws. Although there is no record of her involvement in women's movement activity prior to 1945, it is clear that her feminist convictions date from her experiences during World War I. She wrote a booklet in 1919 about women's role as "molder of the race," and she later told a story about her individual activity on behalf of equal rights. In the 1920s she was working in St. Paul as a linotype operator on the night shift. Her union paper carried an article by a male linotype operator on the day shift arguing that women should not be allowed to work at night. The afternoon after the article appeared, she went to find the author as he left work and requested that he trade jobs with her since he opposed night work for women. Needless to say, he refused, confirming Ackley's belief in the selfishness and hypocrisy of union men who supported protective laws for women.

In 1945, Ackley was about to retire and wrote to the National Woman's Party for literature and to ask what women like herself, not at the "front," as she put it, could do for the ERA. Although she had to live on her union pension and social security and did not at first think she could afford the ten dollars for an active membership, she joined in December and continued to support the group financially as well as through active involvement in the ERA campaign. Although a skilled worker and relatively well-paid, she probably could not really afford to make the contributions that were so important to her. She apparently lived alone, eventually in a hotel, although her twin sister, who was married, also lived in Spokane. She seemed to be an educated woman, judging from her extraordinarily well-written letters, her ability to write what she called her "rimes"—doggerel about the ERA that she wrote, printed, and sent to senators and representatives—and her beautiful script that is difficult to distinguish from italic type. She belonged to the BPW, something that, despite the organization's rhetoric about bringing together the lowliest clerk with the top businesswoman, must have been unusual for a linotype operator and union member.

Ackley's working-class status affected how she participated in National Woman's Party activities and how she viewed the struggle for the ERA. Woman's Party officers, from their earliest contacts with Ackley, tried to make the most of her union membership. They suggested, for example, that she write to anti-ERA women's organizations such as the League of Women Voters, inform her senators and representatives that she and any other labor women she could round up supported the ERA, give radio speeches, and attend conventions. Ackley always tried to do her best and never seemed to get exasperated at suggestions that she travel across the country to attend meetings, something she could not afford and never did. Ackley never visited the Belmont House, nor did she ever meet most of the members with whom she corresponded. She did her work at home in Spokane, writing letters and her little booklets of doggerel, and eventually organized a short-lived Spokane branch of the Woman's Party. Her interest in eliminating protective legislation never wavered. "A law to 'protect' women out of the more desirable jobs and force them to take the menial, nerve-racking [*sic*] and status-lowering jobs as 'household servants,' or to walk the streets, is as undemocratic and un-Christian as Hitlerism," she wrote.[24]

Ackley believed that the ERA would mean new labor laws for women and men and could not understand how any man could oppose a measure of such simple justice. For she did blame men, and especially union men, for opposition to the ERA. Although she described herself as "strong for unionism" and would not, for example, buy ERA leaflets unless they carried the union label, she believed that union men supported protective legislation because it gave them an advantage in the labor market. Yet with some optimism she noted that no union had ever voted on the ERA, as far as she knew, which meant that the rank and file might not be well represented by their leaders.

Ackley insisted, however, that she was not a "man-hater"—she deplored the practice of dubbing women such because they believed in equal rights—but she did blame men for women's plight.[25] The first page of her booklet, "Woman—Molder of the Race," read "What men know about women" and the following page stated simply, "Nothing!" She opposed the usage of masculine pronouns to refer to both women and men. Yet she rejected the use of the word "feminist" because she believed that it had an unfavorable effect on so many people, especially men. She no doubt would have described herself—as she did in closing her letters—simply as "for equal rights."

Ackley's involvement in the formation of the Spokane branch reveals a great deal about the pattern of her participation. In 1947, she wrote to headquarters to report that no one in Spokane was doing anything for the ERA and to ask for some pointers in taking up the work. In response to suggestions, she drafted a letter to her congressional representative and gathered the signatures of nine other members of her typographical union. The national secretary of the Woman's Party then suggested that she form a Washington branch, no matter how small, since it would be helpful to the leadership, then embroiled in the lawsuit, at the next convention. Ackley did not feel that she could lead the branch herself, but she quickly found a friend, Mary Elizabeth Nye, who agreed to serve, and soon the branch had fifteen members. Nye found herself bogged down between work at the office and care of her aged mother, so Ackley continued to take care of correspondence for her. Despite all of her efforts in Spokane, she reported the demise of the branch in 1950. The cause is clear from a letter Nye wrote to the national chairman in which she explained that Washington was a one-woman chapter. She did not want Ackley to know that she had written so frankly, but she reported that Ackley did all the work and even sent in her own money for dues for the other members. In the face of advancing age, illness, and financial difficulties, Ackley continued to work for the ERA in her own way, but she could not maintain the branch through her own efforts.

Despite her experience in Spokane, Ackley believed that collective action was necessary in the ERA campaign. Although she sometimes found the indifference of the majority of women irritating, basically she believed that women simply needed to be awakened: "When women learn the truth about the Amendment, the women's equal-rights vote will be as strong as the labor vote."[26] Women would simply have to organize more thoroughly to defeat candidates who opposed full citizenship for women. Although she avoided the term "women movement" because, like "feminist," it sounded threatening to those who feared the "unsexing" of women, she put her faith in the work of the National Woman's Party: "I don't know what the Amendment would do if it wasn't for you women at Headquarters to keep up the fight."[27] She continued to do what she could as a small cog in the machinery of the Woman's Party, and she maintained confidence that the organization would achieve its goal.

Ackley was a remarkable woman but she consistently downplayed her contribution to the ERA fight. Her denigration of her verses, her insistence that she would not lead the local branch, represent the Woman's Party at national gatherings, or speak over the radio, as well as her unembarrassed appeals for direction and information about Congressional procedure, have a matter-of-fact tone lacking in false modesty or insecurity. "I'm not the right type to be chairman," she wrote. "I don't know anything about such work, and am like the fellow who couldn't sell peanuts to an elephant, and besides that disqualification, in health and years I'm getting like 'the old gray mare.' But I think I can start the ball to rolling."[28] She was always enormously pleased when she received compliments on her work.

And appreciation of her work did pour in from National Woman's Party members. Delighted with a dedicated, pro-ERA union member, coworkers praised her verses, if perhaps in a somewhat condescending manner. They told her that she was a "marvelous worker," a "miracle," and that they thought of her often. Feminist author Alma Lutz wrote to Florence Kitchelt and referred to Ackley affectionately as "quite a character."[29] Her approach to her ERA work does reveal a sort of endearing eccentricity that would not have been out of place in Woman's Party circles. For example, Ackley got "het-up" at Congress and threatened: "They are the ones who would need protection if I were within fist-reach of them!"[30] She wrote to relieve her pent-up indignation and longed to write letters so strong that they would have to be written on "asbestos paper." She felt no possessiveness about her verses and urged the Woman's Party to make use of them in any way that would help the cause. And she gave far too generously of her meager financial resources, in one case insisting on reimbursing the expenses of a wealthy Woman's Party member who had visited Spokane and, when the woman refused to accept the money, giving it to the Woman's Party as a contribution in the other woman's name.

Ackley herself admitted that she "almost went 'broke' physically and financially" trying to do her part for the ERA.[31] In November 1948, she underwent surgery to remove her appendix and gall bladder, and in 1955 she was operated on for cancer. She died in November of that year, a "brave spirit faithful to the cause with her last breath," as Mary Elizabeth Nye put it.[32] Her commitment inspired one member to call her a "blessed dove of a woman," "a rare soul—so sincere & inspired—so generous & expressive."[33]

Ackley's commitment to equal rights came out of her own experiences as a woman worker, and her devotion to the cause gave her life special meaning. As a working-class woman and union member, Ackley was highly unusual in the women's rights movement. Perhaps the fact that she participated almost solely through correspondence made it possible for her to maintain her belief in unionism while she fought with all of her resources for a cause at odds with the labor movement. Although Fannie Ackley was not a typical Woman's Party member, her commitment to the cause struck a chord in other members. Alice Paul, on learning of her last illness, wrote to express the sentiments of the membership: "We cannot begin to express the gratitude we all feel for the able, steadfast and courageous help you have given ever since you joined our ranks. We have been constantly encourage[d] by your unfailing devotion."[34] Fannie Ackley, like other more typical members, found reward for her wholehearted commitment in a sense of satisfaction with her "life work."

Lena Madesin Phillips

When Anna Lena Phillips (1881–1955) was about eleven years old, she dropped the "Anna" from her name and added "Madesin," her version of the French word for physician, in honor of her half brother's pursuit of a career in medicine.[35] From that point on, she used "Madesin" rather than "Lena." This early act of self-naming is typical of her life and career. Although the years from college to her mid-thirties were difficult, they seemed to be a period of troubled transition from a confident girlhood to a successful and dedicated feminist womanhood.

Madesin Phillips was born in Nicholasville, Kentucky, and grew up a tomboy in a comfortable and locally prominent family. She hated dolls and clothes as a child and longed to be a boy. She was, according to her own account, considered "queer" for a Kentucky girl. She hunted and fished with her father, a prominent county judge, rode horses bareback, and worked out with a punching bag. She attended the Jessamine Female Institute in her home town and, after graduation, enrolled in what later became Goucher College in Baltimore. Following in her musician mother's footsteps, she studied to become a concert pianist until an injury to her arm dashed her hopes for a musical career. She returned home and, in a pattern not unfamiliar in the lives of educated women in the nineteenth century, spent more than ten years trying to decide what to do with her life. Finally, after recovering from a nervous breakdown that landed her in a sanatorium, she set out on the course that led to her life work. She entered the University of Kentucky College of Law and two years later graduated with honors, the first woman to do so. She opened a law practice but soon came to the attention of the national leadership of the Young Women's Christian Association through her work with a state branch of that organization during World War I. Her war work with businesswomen, under the auspices of the YWCA, eventually culminated in the founding convention of the National Federation of Business and Professional Women's Clubs in 1919. Phillips saw the need for an organization devoted to the interests of business and professional women, and her election at the founding convention to the post of executive secretary launched her in what quickly became her real career.

Phillips wrote, but never completed, an autobiography that is included among her papers. In it, she analyzed in a straightforward manner her rebellious childhood. She recounted an incident in which she, as a very young girl, had bitten a male neighbor who had held her on his lap and pretended to be a bear about to gobble her up. This she identified as her "first move toward feminism." She was equally candid about her lack of interest in men and marriage. "I cared little for boys," she wrote, and she quoted a composition she had written for school at the age of seven: "There are so many little girls in the school and the thing i like about it is there are no boys in school. i like that about it." She noted that she never had any interest in marriage. "Only the first of the half dozen proposals of marriage which came my way had any sense of reality to me. They made no impression because I was wholly without desire or interest in the matter." She felt sorry for the first man she refused, but when he quickly married someone else, she realized that her sympathy had been wasted. She never felt sorry again; the other times, when the man "cried because I would not promise to marry him, I felt little more than for a luckless fly which had mistaken the fly-paper for a safe and pleasant place to stop."

Phillips was evidently oblivious to or unperturbed by possible Freudian explanations of her behavior. She herself explained unabashedly that she

envied boys their freedom. She pondered the relationship between her rebellious childhood and her feminist adulthood: "Has the subconscious made
me do life penance, because in childhood I was traitor to my sex? Have I
devoted so much of my life to the cause of women in guilt consciousness
for that disloyalty? I think not." She then explained her motivation as arising
from the realization that, contrary to the stories her father had told her,
there was no factory in Indiana that made girls into boys. This set her on
the way to feminism: "I set out to do my part in releasing my own sex from
senseless handicaps."

Phillips seemed equally impervious to the charges of lesbianism that
her memories might provoke. She mentioned in her autobiography the
"crushes" she had on girls at the Jessamine Female Institute—nothing out
of the ordinary for a young woman of her generation but perhaps revealing
for a woman who continued to devote her emotional energies to a woman.
In 1922, Phillips and Marjory Lacey-Baker, whom she had met in 1919,
rented an apartment together, the beginning, as Lacey-Baker noted, of
"thirty-three years of a shared home." Belying Phillips' occasional jests
about her "old maid" status, the relationship between the two women was
a deep and permanent one.

In the 1920s, while Phillips and Lacey-Baker established their relationship, Phillips practiced law in New York and served in various offices in the
new BPW. As president from 1926 to 1929, and later as a major, if unofficial,
force in the organization, Phillips supported feminist causes within the BPW.
She belonged to the Woman's Party, although she worked primarily through
the BPW and, after its founding in 1930, the International BPW. She traveled
widely in Europe prior to 1930 to set up the international group, and she
served as its president from 1930 to 1947. She and Lacey-Baker continued
to visit Europe frequently; in fact, Phillips died in 1955 on her way to attend
a conference on the organization of professional women in the Middle East.

It is clear from her analysis of her childhood that she identified as a
feminist from her earliest years. She acknowledged the influence of a friend
of her mother's who first drew her into her work with the YWCA; she
suspected that this woman's life and influence had something to do with her
"eventual trend toward feminism." As a result of her feminism, she gave,
as she saw it, "the best efforts of my life . . . to organized womanhood and
the causes which women espouse."[36]

Yet it is interesting that, despite her deep commitment to women and
the women's rights movement, she ultimately saw herself as working for
human, rather than women's, rights. She believed strongly that equality of
status for women would benefit both sexes. In 1950, she wrote to an old
friend from the National Woman's Party that she remained interested in the
ERA but that she could not take on any such cause at that point: "I have
not the energy of the former years and my thoughts are too deeply disturbed
about our world situation."[37] It was this concern that led her to Henry
Wallace's Progressive Party in 1948; despite the fact that it meant courting
charges of Communism, she ran for lieutenant governor of Connecticut on

the Progressive Party ticket. Her involvement in politics fit with what she identified in a 1944 interview in the *New York Times* as the goal of the feminist movement—a world of mixed societies that would see the dissolution of women's organizations.[38]

Her commitment to the organization she founded makes it hard to believe that she really longed for its dissolution in a future world. One friend wrote to Phillips on the occasion of the twenty-first anniversary of the International Federation and expressed her perception of the founder's relationship to the group: "it is your baby, only your baby. And now, that it comes of age, you can have the proud feeling of a mother who sees her son or daughter go down at commencement . . . the baby has grown up and you gave something lasting and wonderful to the world."[39] Although Phillips, who herself referred to the organization she had founded as her "international baby," did not invest this motherhood metaphor with the meaning that Rose Arnold Powell did, it is clear that she found that her feminist work gave her life meaning. By living in a supportive relationship and moving in a community that provided friendship and support, Madesin Phillips managed to hold a vision of a world in which gender would no longer serve as the basis for distributing the resources and rewards of a society.

Emma Guffey Miller

Emma Guffey Miller (1874–1970), "the matriarch of Pennsylvania Democrats," became National Chairman of the Woman's Party in 1960, at the age of 85.[40] In this way she brought her prestige in Democratic politics to bear on the struggle for passage of the ERA. Like Florence Kitchelt, Miller combined a commitment to liberal causes with her devotion to the equal rights struggle; in addition, she managed to maintain her influential role in partisan politics while heading the Woman's Party's nonpartisan fight on behalf of the ERA.

Miller came from an old Pennsylvania family marked by Revolutionary War ancestry, wealth, and a strong political tradition. Her father, uncle, and brother all made places for themselves in Democratic politics. She, too, brought up in a passionately political family, carved out a career as a Democratic national committeewoman from 1932 until her death in 1970, but she never ran for office.

Miller attended public and private schools in western Pennsylvania and then entered Bryn Mawr College, where she graduated in 1899. Her college experience seems to have led her toward feminism. She evinced great pride in Bryn Mawr and its feminist tradition under M. Carey Thomas, and she connected its tradition of excellence with its commitment to feminist issues. After college she taught, first in Pennsylvania and then in Japan, where she met and married Carroll Miller, an engineer. She bore four children, all sons, between 1905 and 1912; in 1907, the family returned to Pennsylvania, where she threw herself into political and volunteer activity, including work in the suffrage movement. After the passage of the Nineteenth Amendment in 1920, she joined the League of Women Voters as well as other anti-ERA organizations, such as the National Consumers League, but her enthusiastic involvement in Democratic politics fit poorly with the League of Women Voters' nonpartisan stand. She left her leadership position in the state League in the 1920s. By the 1930s, despite her participation in Democratic politics at the national level, she began to support the ERA. She and her husband moved to Washington in 1933 when he accepted an appointment on the Interstate Commerce Commission. Her involvement with the Women's Division of the Democratic National Committee in the 1930s solidified her feminism, but her position differed from that of other powerful women in the Democratic party, especially Eleanor Roosevelt and Molly Dewson, head of the Women's Division and a force to be reckoned with in Franklin Roosevelt's administration. Miller crossed swords with Roosevelt and Dewson over the issue of priorities—she argued for concentration on enlarging women's role in politics, while Roosevelt and Dewson elevated political concerns over women's issues. In the late 1930s, Miller began to work specifically for the ERA with the Woman's Party, although she never sacrificed her partisan politics to her feminism.

Although not the only Woman's Party member with stature in the political world—her close friend, Perle Mesta, for example, lent her prestige to the cause of the ERA—the level of her involvement with the Woman's Party and her willingness to chair the group in the 1960s singled her out. Her husband's death in 1949 incapacitated her temporarily; she wrote to the then-chairman of the Woman's Party of "the weeks of anxiety I went through only to be called upon to endure the deepest grief at the end." She continued: "and while I am ready to face life alone henceforth, I am still not in condition to assume any sort of real mental work."[41] It was not until her brother, a U.S. senator, died that she agreed to go to Washington to chair the Woman's Party. Although she suffered from disabling arthritis and broke her hip in 1966, her commitment never flagged.

Miller's position in the Democratic party enabled her to work in important ways for the ERA. She made use of her brother the senator in every possible way. She could, for example, get in to see the ERA's archenemy, Senator Carl Hayden, as "Senator Guffey's sister." Sometimes, reading the Woman's Party correspondence, it is difficult to remember that Joseph Guffey, rather than his sister, was the senator. Woman's Party members, for example, might write Miller asking what the senator thought of a particular issue and offering to send her information on it. In one particular case, Miller responded that it would not be necessary "to have my brother do anything about it."[42]

Miller also worked within the Democratic party itself, particularly on the platform committee to win the original endorsement of the ERA in 1944 and then to retain an ERA plank in later years. In addition, she used her influence with Democratic candidates and presidents to try to win support for the amendment. In 1945, Truman promised her that he supported the

ERA, but refused to make a public endorsement. Miller regularly sent the Trumans, as well as many other prominent figures, gifts of maple syrup and sausage from her farm in Slippery Rock, and from Truman she received personal thanks. She enthusiastically supported Adlai Stevenson but chided him after his loss in 1956 for his refusal to speak out forcefully for the ERA, remarking that "your refusal does not seem in character & I can never justify it to the great number of women[']s organization[s] who want 'Equality before the Law.' "[43] Miller, as chairman of the Woman's Party, sought to establish influence with Kennedy, despite her dislike of him, to no avail. She did not succeed in winning an appointment to Kennedy's Commission on the Status of Women.

Her leadership of the National Woman's Party certainly did her no good in political circles. In 1964, she had difficulty making an appointment with Lyndon Johnson to discuss the ERA. At first Johnson tried to arrange a meeting between the Woman's Party delegation and the First Lady; Alice Paul informed Elizabeth Carpenter, Lady Bird Johnson's press secretary, that they had the ERA, not a social visit, in mind. Delays and difficulties prompted Miller herself to wonder, in a letter to Carpenter, whether "there is some anti-feminist stalling such appointments, as there was previously," a reference to the influence of Esther Peterson in Kennedy's administration.[44] When Miller finally managed to see the president, she apparently offended him by bringing up the ERA; he had thought she was there simply to pay her respects. Miller afterwards commented that she was delighted to pay respects to a president she had so long admired and supported, but she asked, in typical fashion, "when I travel 350 miles from my farm to Washington and back can I be blamed for saying something more than 'Howdy?' "[45] What must have been her response, a few days later, when Johnson himself wrote to thank her for some clippings on the League of Women Voters and commented: "You and I both know there are only two things more changeable than a woman's mind, a baby's diaper and the weather."[46] Despite such discouragements and the National Woman's Party's failure to move Congress on the ERA issue before Miller's death, her efforts were undoubtedly significant in keeping the ERA alive as an issue in the Democratic party, if only in the platform committee deliberations every four years.

Miller also devoted herself to helping other women in politics. High-level women such as U.S. Tax Court Judge Marion Harron, Congresswoman Mary Norton, and DNC Women's Division head India Edwards appreciated Miller's efforts on behalf of women. Harron found her an "inspiration," and other women, too, thanked her for "all the extraordinary work you have done for all women."[47] Expressing the fear that politics and the women's rights movement did not mix, the president of the Federation of Democratic Women in Pennsylvania worried that Miller's election to the chairmanship of the Woman's Party might mean that she would "forget that you *belong* to the Federation," but Miller never did.[48] Woman's Party members and other ERA advocates believed that Miller's reputation would "help *mightily* in many ways and places for this uphill work."[49] They, like the women in Democratic politics, poured out their gratitude and appreciation of her work. In 1964, the Woman's Party held a reception in her honor and sent extravagant praise in a telegram on her ninetieth birthday. Miller's response is evident in a report on the reception by one member who found it a "wonderful thing to see someone *really* happy," and "a joy to watch her radiant face and hear her strong vibrant voice."[50]

Miller never ceased making use of her "strong vibrant voice," until her death from a heart attack in 1970. She attended the Democratic Convention in Chicago in 1968, where she spoke for the ERA but bitterly opposed Betty Friedan, president of the National Organization for Women, for supporting both the ERA and abortion and thus linking the amendment to an even more unpopular issue. This was typical, for Miller was above all a politician. That she chose to use her political skills and connections for the women's rights movement made her highly unusual for her time.

E. S. Pollock

In 1979, we interviewed Woman's Party member E. S. Pollock (born 1908) in Cincinnati, Ohio.[51] Although she was not any longer active in the women's rights movement, she still considered herself a feminist. At seventy-one, she recalled the good old days of the movement in the 1950s, when women of "integrity" were just, she believed, at the point of getting the ERA through Congress. She described certain aspects of the women's movement in the 1970s as a "disgrace" and regretted that feminists had turned men against them after her generation had spent years persuading men that they would still be good wives and mothers. She had to give up her ERA work in the National Woman's Party because of a battle with cancer, but she still believed fervently in women's rights, the ERA, and feminism as she defined it. She felt that the way to succeed was to make use of "genteel, sophisticated, nice, young, chic" women who speak well and believe, or at least appear to believe, in what they say. It is no coincidence that the approach she favored describes quite well the role that she herself played in the National Woman's Party during the years of her involvement.

Pollock was one of few relatively young women in the Woman's Party in the 1950s; she first made contact in the early 1940s when she was in her thirties. She saw her role as a "go-getter," bringing youth and fresh ideas to the organization. Pollock first heard about the Woman's Party while working for the Navy in Florida as a civilian assistant in the engineering department. She became angry when she was passed over for a promotion she felt she deserved in favor of a man she described as both incompetent and "macho" or, in the terms she would have used at the time, a "he-man." A woman friend who wrote for a local newspaper suggested that she contact the National Woman's Party for help. She did and, after joining, regularly sent reports of discrimination to headquarters, while her friend worked

stories on discrimination into local papers. Eventually Pollock was promoted
to head of the inspection department—in part, she believes, as a result of
the publicity—and later became the first woman in the Navy to hold the
title of Inspector of Supplies and Equipment. In 1948, she moved to Ohio,
where she worked for the Air Force and organized the Dayton branch of
the National Woman's Party.

Because Pollock was younger than most of the Woman's Party members, she felt that she had a unique role to play. She liked "what they had
to offer in so far as women's rights were concerned," which is why she joined
the organization, but she believed that a new approach would work better
than their straightforward and old-fashioned lobbying. She remembered that
Katharine St. George, the longtime sponsor of the ERA, told her that
members of Congress ran when they saw the old Woman's Party lobbyists
coming. Pollock and her close friends from Ohio—the three "honchos" of
their group—made use of their youth and attractiveness and charm in appealing to members of Congress. Pollock told how they made each other
laugh before they went into a Congressional office so that they would seem
cheerful and pleasant. She reported one incident in which they accidentally
walked in on a birthday party for the mayor of Des Moines in the private
office of a representative from Iowa and won a sponsor for the ERA by
drinking and joking with the men. Officers of the National Woman's Party
reported to the head of the Ohio branch that "the three attractive girls"
had been a great hit on Capitol Hill.

What lay behind this approach, for Pollock, was the conviction that
feminists had to work on men in order to get the ERA passed—that they
could not just work with women. "I advocated one thing," she said. "The
first thing I found when I got into the National Woman's Party was that
they seemed to cater to women. . . . And I said that we had to educate the
men, and let's bring men into this thing. It's the men we have to educate."
Since she believed in persuading men of the necessity for the ERA, she felt
it was important not to alienate them in any way.

Pollock had a lot of experience working with men—especially those
more than usually resistant to the idea of women's rights—because of her
nontraditional career. Before she went to work for the Navy, she had worked
as a model, owned a millinery shop, backed a play, operated a diaper service,
and played golf professionally. Once she decided on engineering drawing,
she worked mostly with men, although she always tried to get women into
the field and to help women in any way she could. She described numerous
battles with male supervisors and military officers over her treatment, battles
in which she would sometimes use what she called "devious" methods to
win. For example, she told of a discussion she had with a friend in the
personnel department about one of the officers who refused to accept women
in positions of authority, a discussion that, she suggested, may have been
related to the officer's transfer to an obscure post in Alaska shortly thereafter. Sometimes, however, Pollock could act directly, as she did in another
encounter in which an officer began to make sexual advances until she

confronted him. Despite the difficulties, she prided herself on her success in a man's world.

Pollock believed that it was essential to "live equal rights" and to work on male colleagues and family members. She urged women in the Woman's Party to begin their work for equal rights in their own homes by converting their husbands and sons. She herself was married briefly until her husband was killed in the war and reported that her husband had not much cared one way or the other about her work for the ERA, but was pleased that she brought home good money. Pollock later lived with a woman who said, during the interview, that she did not believe in equal rights.

As in the case of Fannie Ackley, Pollock's feminist views developed out of her experiences in the working world, although she was clearly a strong and spirited woman from the start. She also believed that she was influenced by her mother, a strong woman and a suffragist. At eighteen she ran away from home because her father would not allow her to go to college unless she became a teacher. With ambitions to study law, she secretly boarded a train for New York after three days at Kent Normal School (now Kent State University), carrying all her belongings in a hatbox. There she began her colorful career by boldly talking her way into a job as a model at Macy's. She also began her career as a feminist, though not yet actively, when she realized that only men advanced and that sexual objectification characterized the world of merchandising. Later, when her university refused to accept her in the engineering course because she was a woman, it only confirmed her views. Not surprising considering her experiences, she stated her feminist views in terms of women's rights to equal education, work opportunity, and pay.

Pollock considered herself to have been an extremist and a radical in her views. She proudly cited a newspaper interview in which she had discussed some of her ideas. She had said, for example, that it was time for women to begin selecting men instead of allowing men to do the choosing, and she had suggested that men begin to dress to please women and use perfume to make themselves more attractive. It seemed to us that Pollock liked to shock people. Some of her methods of working for the ERA showed real creativity and made clear why she was considered a "go-getter" within the Woman's Party. For example, she paid the women who cleaned office buildings at night in her city to leave ERA pamphlets on the bosses' desks. In the morning, no one knew how they had gotten there and discussion in the offices served to spread information about the ERA and the National Woman's Party. This kind of scheme no doubt won a lot of publicity for the ERA.

Pollock saw herself as a radical because she called herself a feminist, even though she was ridiculed for it. She described the importance to her of finding other women who were feminists: "And they were wonderful women. And I had this marvelous feeling that I was no longer alone. For so many years I had gone on alone. . . . But as soon as I became a member of the National Woman's Party, I felt this wonderful backing. . . . I was not quite as peculiar as I thought I was. . . . And then when I found that the women who were back of this were the most gorgeous wealthy women in the world, I said, 'Oh my God! . . . this is terrific, this is terrific.' " Although Pollock found the support of other feminists exhilarating, she also described some of the members as "old fuddy-duddies" and "hatchet-faced women," mostly wealthy women who did not have any experience in the world of work.

Pollock was younger than most of the active members and could only visit headquarters in Washington on occasion, but the support she found in the Woman's Party was extremely important to her. It is ironic that Pollock, the spirit of youth in the National Woman's Party, should denounce with such vigor the tactics of later feminists. Her scorn for the contemporary women's movement may have come in large part from her lack of contact with younger feminists. She herself wondered whether some of her views were based on media distortions of the women's movement. Her final words in the interview, appropriately enough given her role in the Woman's Party in the fifties, were a plea for cross-generational contact and understanding.

Alma Lutz

Of all the women discussed here, Alma Lutz (1890–1973) would be most comfortable in the women's movement of the 1980s. In fact, Lutz, a feminist biographer and longtime activist, alone of all these women, received recognition before her death in 1973. Ellen Goodman wrote about her in her column in the Boston *Globe* in 1970 and *Ms.* magazine in 1973 featured an article about suffragists that included a brief portrait of Lutz.[52] Lutz spent most of her long life dedicated to feminist causes, and she brought her feminism, nurtured in the suffrage movement, to the women's movement of the 1960s and 1970s.

She was born in North Dakota in 1890 and went east to the Emma Willard School and then Vassar College, where she graduated in 1912. Like Florence Kitchelt and Emma Guffey Miller, she moved toward feminism in the environment of a women's college. After five years of suffrage work in her native North Dakota, she moved to Boston and made a home with Marguerite Smith, her Vassar roommate, with whom she lived for forty-one years. Throughout the years, she published biographies of Emma Willard, Elizabeth Cady Stanton, Harriot Stanton Blatch, and Susan B. Anthony, as well as a book about women in World War II and a study of women in the abolition movement. She belonged to the National Woman's Party, the BPW, and the American Association of University Women, and she joined the National Organization for Women when it took over leadership of the ERA struggle. She remained hopeful and involved in the women's movement until her death at the age of eighty-three.

Lutz had joined the Woman's Party in 1928, and she contributed to the organization's journal, including a column entitled, "A Feminist Thinks It

Over." She stopped paying dues and resigned from the National Council in 1946 at the time of the schism but kept informed about the group's activities, in part through her friend Mabel Vernon, who had left the Woman's Party earlier but kept in touch with Alice Paul. Lutz never ceased criticizing the weaknesses she saw in the organization, but she maintained respect for its hard work and dedication and she always hoped for a "new lease on life" for it. She admired its constancy, early deciding that there were advantages in its "very one track" program, and she admitted in 1955 that it at least kept "hammering away at the Amendment." In 1963, after she had rejoined the National Council, she wrote to an officer that "the cause means a great deal to me and only the National Woman's Party keeps the Equal Rights Amendment before the people."[53]

Despite Lutz's breach with the Woman's Party, however, she did not work in isolation in the post-1945 years. One of her contacts was Florence Kitchelt, whom she first met in 1945. Lutz disagreed with Kitchelt's position on protective legislation, but believed that they could "argue and speak frankly without ill feeling."[54] Lutz hoped that Kitchelt's Connecticut Committee (on whose National Advisory Committee she served) would act as the nucleus for a national coalition for the ERA made up of state groups, and she urged Kitchelt throughout the years to expand her committee to all of New England. Lutz worked to set up a state committee in Massachusetts and finally persuaded an acquaintance to head it in 1955. She reported in 1956 that the committee was not doing well enough to suit her but put up a "good front" so that Massachusetts senators and representatives would "know some women in Mass. want the Amendment."[55] The Massachusetts Committee still existed, however minimally, in 1960 when Lutz, in its name, solicited pro-ERA letters to candidates for Congress and the presidency, and in 1964, when she called for support for the ERA in the context of passage of the Civil Rights Act.

In addition to her efforts to prod a Massachusetts Committee for the ERA into existence and her cooperation with Kitchelt on Connecticut Committee work, Lutz corresponded with Rose Arnold Powell about her work on behalf of Susan B. Anthony, joined with Kitchelt in the campaign to change the American Civil Liberties Union's position on the ERA, regularly submitted articles on the ERA to national magazines and newspapers, taught a Radcliffe graduate seminar on American women's history, worked with the Radcliffe Archives (now the Schlesinger Library) to collect and preserve the papers of feminists, and continued to write her feminist biographies.

Lutz saw all of her efforts as part of the same struggle. She believed that teaching and writing women's history was one of the best ways to reach the younger generation, which she considered extremely important. She identified strongly as a feminist, and she used the term regularly in her correspondence, in a complimentary fashion, to describe other women. She recognized the societal hostility to feminism and deplored publicly in 1948 the fact that it was the "fashion today as in 1848 to ridicule feminism." She wondered to her friends why feminism "bores everyone but us" and why women "are so afraid of the word feminism."[56] She lost her patience, she claimed, over women's apathy and once declared that women were their own worst enemies. But most of the time she seemed able to maintain her optimism: "Without doubt the feminism of the old days is dead, the fervor of the last years of the suffrage campaign is lacking, but I do believe feminism is taking new forms in keeping with the age," she wrote in 1951. "After all, the feminists have always been a minority."[57] She agreed with Florence Kitchelt that child care and housekeeping arrangements would have to change and deplored the assumption that women, even if they took up professions, would have to take responsibility for the care of children. She worked almost entirely with women but she believed that women and men should work together for the ERA and for other progressive measures. Above all, she believed that feminists had to educate women about the ERA, since Congress would not listen until large numbers of women demanded the amendment.

Lutz, like Florence Kitchelt, Lena Madesin Phillips, and Emma Guffey Miller, held liberal political views. She denounced the "red hysteria" of the McCarthy era, anti-Semitism in the Woman's Party, and racism in the women's rights movement. Lutz continued her involvement as the years wore on. She took great interest in the President's Commission on the Status of Women in the early 1960s and appeared before it to speak for the ERA, and she followed the course of the Civil Rights Act of 1964, enthusiastic about Title VII's inclusion of discrimination in employment on the basis of sex. Her commitment as well as her optimism comes through in a letter she wrote in 1950: "A new woman has been added to our family. On New Year's Day my niece's little girl was born. . . . I hope she won't still be asking for women's rights in 1970."[58] In contrast to E. S. Pollock, Alma Lutz maintained her contact with, and involvement in, the women's movement. Despite her wishes for her new grandniece, Lutz was still working for women's rights in 1970.

Patterns

These biographical sketches offer some details about the lives of seven feminists in order to give a sense of their views and activities in the post-1945 period. The feminist lives of Rose Arnold Powell, Florence Kitchelt, Fannie Ackley, Lena Madesin Phillips, Emma Guffey Miller, E. S. Pollock, and Alma Lutz illuminate the human dimension of several aspects of the women's rights movement. Perhaps most significantly, all of the women belonged at one time or another to the National Woman's Party, although only Ackley, Miller, and Pollock put their major efforts into it. This illustrates well the central role that the Woman's Party played. In addition, all of the women except Pollock belonged also to other women's organizations. Five of the women—all except Ackley and Kitchelt—explicitly identified as feminists, and the two who did not held world views that others easily identified as feminist. None of the women was young; in 1950, Pollock, who stood out for her youth, was forty-two, Phillips was fifty-nine, Lutz was sixty, Ackley was seventy, Powell was seventy-four, and Kitchelt and Miller were both seventy-six. Miller, alone of these women, had children, and she and Kitchelt had the only lasting marriages. Thus family responsibilities did not stand in the way of their commitment. All of the women found rewards in the friendships they established in the course of their work for women's rights.

But these sketches also provide insight into the development and maintenance of feminist consciousness in the post-1945 period. Although it is not possible to draw definitive conclusions, these lives suggest patterns in the process of becoming a feminist. Two experiences emerge as central—contact in some form with a woman who was a feminist or with a women's institution, and recognition of obstacles to equal opportunity. Six of the women were influenced by other women. Rose Arnold Powell, in a pattern that continued in her feminist activity, made contact through the written word. Her reading of feminist Anna Howard Shaw's biography propelled her into her lifelong involvement with feminism. Madesin Phillips credited a friend of her mother's with drawing her into YWCA work and serving as a model for her own feminist commitment. E. S. Pollock cited her mother, a strong woman and a suffragist, as a major influence in her life, and found the Woman's Party through the help of a woman friend. Florence Kitchelt, Emma Guffey Miller, and Alma Lutz attended women's colleges, which seemed to have introduced them to feminism. Miller remembered the influence of M. Carey Thomas, the strong feminist who left her mark on Bryn Mawr College even after her presidency. Upon graduation from Vassar, Lutz went to work for the suffrage movement. Kitchelt graduated from Wells College and moved into settlement house work and the suffrage movement. In each case, the experience of a women's college seems to have been decisive. Thus the pattern emerges of a committed woman or a women's institution influencing a developing feminist. Many other women, in their correspondence, mention such an influence.

The other experience, recognition of obstacles to equal opportunity, emerges in fewer cases but is nevertheless suggestive. Phillips related her feminism to her childhood realization that her choices in life were limited by gender. She believed that her commitment to the women's rights movement grew out of a sense of injustice. Ackley based her feminism on the experience of having been "protected" out of a job. Pollock, too, recognized in her first job that being a woman was a handicap in the labor market, and in her nontraditional work for the military she experienced further direct discrimination. Other women who worked for the military found the discrimination there conducive to feminist activism. One who worked as a civilian for the Air Force during World War II, for example, became interested in employment discrimination and equal pay as a result of resistance she met and joined the AAUW in the early 1950s.[59] Another woman, educator Helen Schleman, dated her feminist consciousness to the experience of serving in the Coast Guard for four years.[60]

Women did not have to experience the military, however, to learn about employment discrimination. Margaret Hickey, a BPW leader, found in her first job as a reporter that she was resented because she was a woman, and when she went to law school she was shocked that she could not, as a woman, join the debate team.[61] Another feminist lawyer graduated from law school in 1939 only to find that no one would hire her.[62] Black lawyer and activist Pauli Murray graduated at the top of her class at Howard University in the 1940s, an accomplishment traditionally accompanied by a fellowship to Harvard, only to find that Harvard would not accept her because she was a woman.[63] Such stories abound in interviews with women who participated in the women's rights movement in the 1950s. If such a route to feminism were a common one for women, it would suggest a concrete link between the changes in women's material conditions—increasing labor force participation, so often cited as a precondition for feminism—and the growth of the movement in the 1960s.

Of course, not all women who knew feminists or attended women's colleges or experienced discrimination became feminists, so these patterns do not provide us with a formula for producing activists. But they are suggestive. What is striking about these seven women is that all of them except Pollock became feminists in the early years of the twentieth century, when the women's movement was large and visible. Kitchelt and Lutz had been part of the suffrage movement. Phillips developed her feminist beliefs in childhood and became involved in women's organizations during World War I. Miller encountered feminism at college and began to participate in movement activity in the 1930s. Both Powell and Ackley dated the origins of their feminist beliefs in the 1920s. This suggests that the women's rights movement in the post-1945 period was made up primarily of women who had developed their feminist convictions earlier. What is perhaps most remarkable about them is that they maintained their commitment through the years, even when little "new blood" was flowing into the movement.

The women's rights movement functioned well in allowing previously committed women to maintain their involvement in feminist activities in a hostile environment. But feminists did not, in this atmosphere and period, make positive connections with other social movements. We turn now to the interactions of feminists with McCarthyism, the labor movement, and the civil rights movement.

7

The "Isolationism" of the Women's Rights Movement: Relations with Other Social Movements

In the years after World War II, the women's rights movement brought together those who shared a commitment to improving the status of women in American society and created a world within which personal relationships flourished, but it did not mobilize large numbers of American women or make positive connections with other activists working for social change. The fact that women's rights supporters were a homogeneous group, with individuals holding widely divergent views on all other political and social issues other than that of equality for women, affected the movement's relations with the other social movements in this period.

Russell Curtis and Louis Zurcher contend that social movements and their environments are best understood as multi-organizational fields in which various social movements and organizations with different interests establish linkages through which they compete and cooperate in pursuit of their goals.[1] Exchanges between groups can be friendly or unfriendly, frequent or rare, but whatever the nature of the relationship, all social movements exist within the context of other social movements that compete for the time, commitment, loyalty, and money of potential sympathizers as well as for legitimacy and "symbolic dominance" among the larger social movement sector.[2] The period between the end of World War II and the mid-1960s was not, as we have seen, characterized by widespread social movement activity. Yet social movements and debate over social and political issues did exist, and these other movements and political debates touched the women's movement in various ways, affecting the growth, membership characteristics, structure, and strategies of the movement.

In this chapter we explore the relationship of the women's rights movement with other important social issues, focusing on its relations with McCarthyism, the labor movement, and the civil rights movement. Because the ERA was the most controversial of feminist goals, it served as the focus of much of the interaction with other social movements. Advocates of women's rights in this period never forged true alliances with activists in other movements, but rather pursued a policy of isolationism. Nevertheless, feminists attempted to use the goals and activities of other movements for their own ends.

Communism, Anti-Communism, and the Co-optation of McCarthyism

Since feminism of any variety represents a challenge to the traditional social order, it is perhaps not surprising that the women's rights movement appeared to some observers in the Cold War atmosphere of the postwar years to stand with Communism as an enemy of the American way of life. A Knights of Columbus member, for example, addressed the District of Columbia BPW in 1949 and denounced the ERA as a Communist measure.[3] One Woman's Party member reported that a prominent woman in her home town of Madison, Wisconsin, whose son worked for the FBI, believed that the Woman's Party was a Communist organization or at least one with Communist leanings.[4] It is clear that such absurd charges were a consequence not of any radical ideology associated with women's rights activism, but rather of the lingering suspicion that feminists were somehow linked to Communists. During the height of anti-Communist feeling, any kind of deviance could be labeled Communist. As one National Woman's Party member remembered, it was a simple thing to hurl that epithet: "Anyone that disagreed with anyone, you know, was a Communist."[5]

Despite charges of Communism, it is clear that the vast majority of women's rights supporters, like the population at large, were anti-Communists to one degree or another. Yet within that general consensus lay a great deal of diversity on the issue of Communist witch-hunting and the more extreme manifestations of anti-Communism associated with Senator Joseph McCarthy. Just as the women's rights movement brought together Republicans and Democrats, conservatives and liberals, segregationists and integrationists, so too it included both supporters and opponents of McCarthy, all of whom managed to work together for women's rights.

Some feminists explicitly supported McCarthy. Alice Paul, in reference to the Woman's Party, told an interviewer that "so many, almost all our members seemed to be of the conservative school."[6] Ethel Ernest Murrell, chairman of the group in the mid-1950s, always stressed "the fact that we are ultra-conservative, and that the very groups which sponsor socialism are our strongest opposition."[7] Virginia Starr Freedom was a National Woman's Party member and outspoken supporter of McCarthy. She was "horrified at the villainy presently going on in Washington in the Administration's attempts to silence Senator McCarthy."[8] She believed in an armed conspiracy against all non-Communist governments and felt that it was time to "stop this whinnying about misunderstood 'liberals.' " She wanted to give accused individuals sixty days to resign from the Communist Party and its front organizations and, if they refused, seize them as "Stalinist guerilla [*sic*] fighters here on our soil." One woman active in the leadership of the Woman's Party in the 1950s could not say whether the majority of members were pro- or anti-McCarthy, although she did state that "many of our women applauded what he was doing, much to my dismay. I feel that a lot of them thought that he was a courageous and helpful person."[9]

Perhaps the most virulent opponent of Communism in the women's rights movement was Doris Stevens, who left the National Woman's Party after the 1947 lawsuit. In 1949, she addressed the Women's National Republican Club and accused the Democratic Party of being controlled by Communists who incited minorities to make demands on the government.[10] The FBI relied on Stevens and her husband, Jonathan Mitchell, who wrote for William Buckley's *National Review* in the 1950s, for information. In 1949, FBI agents came to Stevens' home in Croton-on-Hudson, New York, a community of artists and intellectuals known for leftist sympathies, to inquire about the Hiss case, and in 1950 they arrived to serve a subpoena on a neighbor and raked leaves on the Stevens-Mitchell lawn in an attempt to disguise their purpose for being in the area.[11] Stevens described in her diary numerous encounters with anti-McCarthy sentiments at social events. At one party peopled by, in Stevens' words, "socialist jews," a woman lawyer called Stevens a fascist; at another gathering, Mitchell got into an argument with some "crypto-commie types"; when old friends from the National Woman's Party came to visit, Stevens found two of them "very, very pro C.P." on everything, "[t]erribly anti-MacArthur, anti-McCarthy—not a patriot word or tho't."[12] Stevens believed that Hiss had gone about his Communist machinations in government for so long without being detected because both Roosevelt and Truman were "captive of the Communist conspiracy." Although she herself engaged in "queer baiting" as well as "red baiting," she doubted that Hiss and Chambers, in spite of their fondness for one another, were "queeries," and marveled at the "character smears that communists can think up."[13] Not surprisingly, Stevens gave her full support to McCarthy.

Stevens had less to do with the women's rights movement in the 1950s than she had in the 1920s and 1930s, before her views swung to the right, but she maintained her strong feminist convictions and remained active in the Lucy Stone League. Other women continued, unlike Stevens, to devote most of their energy to women's rights but incorporated an anti-Communist outlook into their work. Lucy Somerville Howorth, for example, spoke about Communism in a radio interview for Voice of America and advised women to shun groups, especially peace groups, that included Communists.[14] Burnita Shelton Matthews, a National Woman's Party member appointed to a federal judgeship in 1949, served as the president of her local unit of the American Legion Auxiliary and received an invitation to speak on Communism for a program on Americanism.[15] One woman wrote to the National Woman's Party for patriotic literature because she wanted to do her part in the fight against Communism.[16] Clearly, anti-Communist sentiment in the women's rights movement, as in the population at large, was strong.

Not all women, however, responded to McCarthyism by incorporating anti-Communism into their repertoire. Although even many liberals feared Communism, they criticized McCarthy's tactics and deplored his excesses. Thus, some were denounced as Communists or unwitting dupes. Woman's Party member Nora Stanton Barney, for example, merited a special section in a report of the House Un-American Activities Committee on the Congress of American Women, to which Barney belonged.[17] The HUAC material reported her "participation" in the Paul Robeson concert in Peekskill, New York, her support for the National Council of American-Soviet Friendship, her attendance at the Women's International Democratic Federation convention in Budapest, and her statement on her return that non-Communists in Hungary enjoyed greater freedom than Communists in the United States. Yet Barney was critical of the Communist Party's position on women's rights. She attempted to engage Bella Dodd, a member of the Communist Party, in a debate on the ERA, insisting that the Communists' "vitriolic opposition" resulted from a mistrust of the individuality of women, and she informed Emanuel Celler, the representative who kept the Judiciary Committee from reporting on the ERA to the House, that women were waking up to the fact that "the Communist Party and the C.I.O. are opposing the granting of civil rights to women."[18] Barney worked in liberal organizations for peace and for political prisoners and refugees, but she was not a Communist and criticized the United States for seeming "to copy all the Russians' faults while inveighing against her."[19] One National Woman's Party member saw Barney's name in a left-wing publication and wrote to the Woman's Party chairman that "[i]f it is true that Mrs. B. is a member of their group and leans so far left, you all at headquarters probably know it."[20] Another Woman's Party member wondered about Barney when she remembered that her mother, militant suffragist Harriot Stanton Blatch, had announced on her eightieth birthday that she had been converted to Communism.[21]

Other women connected in a greater or lesser degree to the women's rights movement also came under suspicion of Communist leanings. Lena Madesin Phillips, who ran for office in Connecticut on the Progressive Party ticket, was attacked for her association with the party of Henry Wallace. An officer of the Soroptimist Club, for example, reported to the National Woman's Party that she had heard a rumor that Phillips boasted of her Communist affiliation and wondered if it were true. "If it be true, we surely should not include her in correspondence or plans; and if it be not true, the rumor should be scotched."[22] Wilma Soss, the founder of the Federation of Women Shareholders, inquired of the Connecticut Committee for the ERA, of which Phillips was Honorary Chairman, whether Phillips' name appeared on the Attorney General's list of members of subversive groups.[23] Yet Phillips received a warm response at the 1948 BPW convention despite her fears of disapproval.[24] Sarah Hughes of the BPW faced charges in her bid for reelection as a district judge in Texas that she was a Communist and controlled by the CIO.[25] Susan B. Anthony II, who made appearances at a number of women's rights events, was denounced, like Nora Stanton Barney, by HUAC in its reports on the Congress of American Women. Mary Church Terrell, billed as a "non-Communist Negro leader," took an active part in defending the rights of Communist Party leaders and in the defense of the Rosenbergs, thus opening herself up to charges of Communist subversion.

Some women spoke out against the excesses of McCarthyism even though they fought Communism. Florence Kitchelt, for example, was "shocked beyond expression at the way that leading citizens who have the soul and courage to be concerned about the welfare of other people . . . have mud thrown at them by bigots."[26] She believed that the way to fight Communism was to reform American society, not "throw mud at some of the best minds in our country." Kitchelt took this stand publicly, writing to congratulate members of Congress who stood up to McCarthy, expressing her appreciation of journalists who dared to criticize Communist witch-hunting, and corresponding with victims of the purges. Alma Lutz believed that it was important to preserve American liberties in "this red hysteria which is being whipped up" and reported that Mabel Vernon was disgusted with Alice Paul's "anti-Communist fury."[27] An officer of the National Council of Women defended Susan B. Anthony II as a brilliant woman devoted to progressive and liberal causes and deplored the fact that many organizations that gave lip service to liberal ideas feared to implement them because left-wing women supported them as well.[28] One National Woman's Party member even feared that the anti-Communist hysteria would eventually turn against the ERA itself: "If that amendment don't get through soon," she wrote, "the red-herring hunters will tell us, we must not have Equal rights for women. . . . [ellipses in original] as that idea originated in Russia."[29]

Despite the diversity of women's rights supporters' views, McCarthyism directly affected the women's rights movement, as it did all institutions within American society, by creating an atmosphere of fear and suspicion. The hostile environment reinforced organizations' tendencies to withdraw into themselves, distrust outsiders, including potential new members, and shun coalitions. For example, the right-wing magazine *American Mercury* in 1958 criticized the BPW as an organization with a left-leaning leadership, despite the fact that its board had taken a firm stand against Communism in 1953 by issuing a Declaration of Principles that urged members "to remain alert and to organize their programs and efforts to combat communism." In response, the BPW president and legal counsel trotted out the record of anti-Communist activities, winning a refutation in 1959 and the admission that the Federation's strong anti-Communist stand was well-known. But even success in such a showdown took its toll—the leadership decided to use greater discretion in making contacts with other organizations.[30] One anti-McCarthy women's rights activist remembered how cautious groups became about their membership. "It was a very sick time," she recalled.[31]

The atmosphere of anti-Communism in society at large could also make women's organizations suspicious of internal conflict. This was clearly the case in the 1947 schism in the National Woman's Party. Early on, some who sympathized with Alice Paul compared the tactics of the rebels to those of Communists.[33] Alice Paul requested that a close associate who knew J. Edgar Hoover personally ask the FBI to investigate the conflict in the National Woman's Party.[33] Another Woman's Party member close to Alice Paul remembered that the FBI accused the Communists of creating chaos in women's organizations and remarked that the bulletins Doris Stevens issued attacking Alice Paul seemed to follow the Communist strategy. She hoped that an attack in the *Daily Worker*, the Communist Party newspaper, on the Woman's Party would prove useful in the trial, since the level of anti-Communist sentiment in the country might win the established faction sympathy.[34] Yet another Woman's Party member consulted an ex-Communist professor who assured her that the trouble in the Woman's Party resulted from Communist infiltration. Some members continued to believe that the trouble had originated with Communist infiltrators long after the hysteria of McCarthyism had passed.[35]

The impact of McCarthyism on women's organizations is clear in another, more minor, conflict within the National Woman's Party. In 1949, Alice Katchadourian, a member of the National Council and past head of the Maryland branch, made a number of radio broadcasts on the issues of civil rights, vivisection, socialized medicine, and the trials of Communists. Some of the right-wing members of the group criticized Katchadourian for using her Woman's Party titles while speaking about issues other than equal rights. Other members defended her right to speak on any issues whatever. Those with right-wing leanings not only disagreed with Katchadourian's progressive perspective, they also claimed to fear the publicity. After Katchadourian's criticism of Judge Medina, who had presided over the trial of Communist Party leaders in 1949, the Daughters of the American Revolution, the Committee on Constitutional Government, the American Legion, and the Knights of Columbus complained to the FBI that the National Woman's Party sponsored subversive broadcasts. Katchadourian apparently agreed to be careful in her broadcasts, but she bitterly resented the fact that a rumor had circulated to the effect that the Woman's Party headquarters had asked for an injunction to prevent her from using the name of the group. One of the Woman's Party leaders had in fact suggested just such a measure, but apparently it had not proven necessary.[36] In any case, the "Katchadourian affair" indicates how fearful and suspicious the anti-Communist atmosphere could make an organization. When a woman in Portland, Oregon, wanted to start a branch of the Woman's Party, the executive secretary took pains to obtain some information on her: "ordinarily we would not raise such questions, but in these days, when we hear so much about 'subversive elements' we are a little cautious about giving approval without asking questions."[37]

The fear, distrust, and suspicion created by anti-Communism made women's rights groups more hesitant to bring in large numbers of new members, created fear when conflicts erupted, and dampened enthusiasm for coalitions. While it was McCarthyism that brought about these consequences, some women believed it was a result of Communism. Commenting on difficulties within organizations as diverse as the National Woman's Party, the League of Women Voters, the YWCA, and the DAR, one woman concluded that "[i]t must be true that communists are trying to destroy the women's organizations in this country."[38] The House Un-American Activities Committee certainly believed that Communists were attempting to dupe the members of women's organizations. Its report on the Congress of American Women claimed that the Congress served only the interests of international Communism but paraded as a legitimate American women's organization in order to "ensnare idealistically minded but politically gullible women." HUAC specifically stated that it did not wish to leave the impression that it was critical of women's organizations sincerely interested in social reform or world peace, and it disassociated the Congress of American Women from established women's groups such as the AAUW, BPW, League of Women Voters, and National Councils of Catholic, Jewish, and Negro Women. But, despite the disclaimers, it seems that one effect of this kind of report was to create suspicion among women's organizations; the "rumors" about Nora Stanton Barney's Communist affiliations, for example, may have originated in, or at least drawn strength from, the HUAC report.[39] It is no wonder, then, that as one working-class National Woman's Party member said, "People are afraid to open their mouths for fear of being accused as Communists."[40]

Despite the negative consequences of McCarthyism for the women's rights movement, the National Woman's Party attempted to co-opt McCarthyism for its own ends. This strategy grew out of the Communist Party's position on the woman question in the 1950s. Its platform called for "an end to any and all political, social and economic inequalities practiced against women" and demanded the maintenance and extension of existing protective legislation.[41] That put Communists on the side of the liberals who advocated equality but opposed the Equal Rights Amendment as the means to that end. According to the orthodox Marxist line, full emancipation for women would follow inevitably from the establishment of the socialist state. Thus women in the Communist Party did not form autonomous women's organizations and did not place much emphasis on any analysis of gender-based oppression.[42] The women's rights movement they denounced as bourgeois.

In 1950, Elizabeth Gurley Flynn, the Communist Party leader who spent two years in prison under the anti-Communist Smith Act, attacked the women's rights movement and the Equal Rights Amendment in an article in the Communist newspaper, the *Daily Worker*: "Platitudes 27 years old are repeated by reactionaries posing as progressives, and Dixiecrats exposing their insulting 'Jimcrow' 'chivalry.' " Flynn based her accusations on the fact that the ERA had won the support of conservative lawmakers. She denounced the National Woman's Party as an "extreme feminist group, which believes even rape laws should apply equally to both sexes" and dismissed the BPW and other supporters as "middle and upper class groups."[43]

It is not at all surprising that the Communist Party in the 1950s opposed the ERA, not just as a useless bourgeois reform but as a weapon of the employers. This was, after all, the position that the labor movement took on the measure. But the Communist Party's public statements on the ERA had important consequences simply because, during the McCarthy witch-hunts, the principle of guilt by association assured that many groups wanted at all costs to avoid siding with the Communists on any issue.

The National Woman's Party leadership ruthlessly exploited this atmosphere to try to win support for the ERA. Woman's Party members reminded ERA opponents that the Communist Party opposed the amendment and questioned the legitimacy of a position that attracted Communist support. This tactic seemed to be particularly popular in dealings with the Catholic Church, but Woman's Party members used it with especially vulnerable labor union officials and others as well.

In 1949, for example, Alice Paul suggested that a Woman's Party member win an endorsement from Cardinal Spellman by making use of the Communist position. She hoped in this way to win the backing of the National Council of Catholic Women.[44] In 1960, another Woman's Party member went to see Cardinal Cushing, armed with Elizabeth Gurley Flynn's *Daily Worker* article, and succeeded in convincing him of the need for the ERA.[45] This same member visited a top AFL-CIO official and a Harvard Law professor, whose anti-ERA statements had been used for years by the opposition, and threatened both men with the fact that their statements had been used by Lee Pressman, general counsel to the CIO and a member of the Communist Party. Another Woman's Party member sent a *Daily Worker* article attacking the ERA to the chairman of the House of Representatives subcommittee considering the amendment in order to explain "the source of much of the opposition that we are having to meet."[46]

Woman's Party members even stooped to the McCarthyite tactic of comparing a legislator's votes to those of Vito Marcantonio, an American Labor Party representative from New York. This tactic, used successfully against liberal Helen Gahagan Douglas by Richard Nixon in his 1950 Senate campaign, appealed to one Woman's Party member who read a newspaper article commenting on the fact that Emanuel Celler, liberal representative and arch-foe of the ERA, voted with Marcantonio against bills designed to regulate Communist activity. "There ought to be some way that, as legislation against Communists and feeling against them develops, we could do something about Celler preventing women of the United States from getting their rights," she wrote.[47] As late as 1962—long after the excesses of anti-Communism associated with McCarthy were over—liberal Emma Guffey Miller, then president of the National Woman's Party, used the threat of guilt by association. She wrote to Senator Joseph Clark, enclosing a copy of the old *Daily Worker* attack on the ERA, and pointed out that Elizabeth

Gurley Flynn was chairman of the Communist Party: "although according to the old saying, 'Politics makes strange bed-fellows,' I hate to think of your sleeping with Elizabeth."[48]

These tactics seemed to have some success, although they fell short of attaining the final goal. The National Woman's Party leadership believed that the opposition of the Communists and the left wing of the labor movement had increased the support among Catholics for the ERA.[49]

What is perhaps most shocking is the fact that even anti-McCarthy individuals within the Woman's Party used McCarthyism to try to win support for the ERA. One Woman's Party member who described herself as "so vocal in opposition" was the woman who visited Cardinal Cushing, the Harvard law professor, and the AFL-CIO official to inquire how they felt about siding with the Communists. When the Harvard professor denied knowing that his statement had been used by Communists, she warned him to be careful: "That's all I'm saying, you better be careful. *We* intend to make an issue out of it," she remembered telling him.[50] She apparently saw no conflict between her opposition to McCarthy and her use of anti-Communism on behalf of the ERA.

This single-minded focus on the ERA could work the other way as well. Before the end of the war, Alice Paul's mentor from her settlement house days, Lavinia Dock, wrote to her "Communist friend" Elizabeth Gurley Flynn to try to persuade her to support the amendment. Flynn, in a letter addressed to "My Dear Friend," insisted that "fundamentally we do not disagree on equality for women" but argued that a better amendment could be drawn up.[51] In 1945, the National Woman's Party received secondhand information to the effect that Flynn had changed her mind on the ERA. Although Ella Sherwin, the head of the Industrial Women's League for Equality, the group of women workers organized under Woman's Party auspices, feared the consequences of such a switch, Alice Paul immediately tried to get an endorsement.[52] The rumor turned out to be untrue, but what is particularly revealing is Paul's willingness, even eagerness, to accept the endorsements of anyone, even Communists. Paul herself, as already indicated, was extremely conservative—one dissident Woman's Party member labeled her a "reactionary"—but all that mattered to her was a person's position on the ERA.[53] It was this single-tissue intensity that allowed members to use McCarthyite tactics that they deplored in the service of "the Cause." It was just this kind of sacrifice of means for ends—or sacrifice of all other issues to the cause of women—that bred distrust of and dislike for the Woman's Party among other women's organizations.

Perhaps the most important key to understanding the attitude of the National Woman's Party toward Communism lies in labor's opposition to the ERA. By the 1950s, many woman's rights supporters had correctly identified the labor movement, and the CIO in particular, as a major source of opposition to the amendment. In this situation, it is not surprising that the National Woman's Party in particular gained a reputation as a reactionary antilabor organization. It *was* antilabor because labor opposed the amendment, and in the 1950s, McCarthyism proved to be a very effective weapon for fighting unions. Just as antiunionism was a basis for McCarthyism among conservative businessmen, so too might hostility to labor have contributed to McCarthyite sympathies and the willingness to use McCarthyite tactics in the Woman's Party.[54]

The Labor Movement: Opposition and Neutralization

From the perspective of the women's rights movement, the labor movement in the period after 1945 was an increasingly powerful opposition force that had to be dealt with in some way. Labor's direct opposition to the Equal Rights Amendment came to seem the major obstacle in the path of progress toward legal equality, thus pitting the labor movement against supporters of the ERA, particularly the National Woman's Party.

ERA advocates believed that union men opposed the ERA for selfish reasons. The secretary of the Woman's Party, for example, informed a correspondent that a CIO leader had admitted that opposition to the amendment stemmed from a desire to hold the best paid positions for men.[55] To another member she explained that labor was afraid that women would withdraw from existing unions, form their own, and gain control of the industry. "It always boils down to fear of women—opposition to the E.R.A.," she added.[56] If this suggests an unrealistic analysis of women's power in the labor market, it at least makes clear the sense of conflict between women and men that pervaded discussions of the labor movement in the Woman's Party.

Some National Woman's Party members expressed themselves more unrestrainedly. An Arizona attorney, for example, blasted the president of the Phoenix Central Labor Council for his opposition to the ERA. She would not be so disgusted with his opposition, she wrote, if he came out frankly and admitted that it grew from the "fierce male desire to keep women in a subservient position." What especially annoyed her was his "Damnable hypocrisy . . . in trying to cloud the issue and divert attention from the real facts, through a red herring trail cloaked in piously expressed falsehood, ranting of 'booby traps' and a desire to 'protect' women."[57] In an allusion to the consequences of burgeoning anti-Communism in American society, she warned that women would remember labor's stance when the antilabor sentiment already growing caused labor to cast about for friends.

In 1945, the secretary of the Woman's Party reported that her group had met with a CIO attorney in the office of the chief sponsor of the amendment in the Senate, where the CIO representative had claimed credit for preventing a vote on the ERA in both the House and Senate.[58] Although Alice Paul preferred to devote her energies to combatting anti-ERA sentiment among women's organizations, most ERA supporters probably would have agreed with Florence Kitchelt, who insisted in 1950 that "[o]ur big enemy is organized labor."[59] Even Paul admitted in 1958 that the only active opposition to the ERA came from the AFL-CIO lobby.[60] She even suspected the CIO of sending spies to Woman's Party headquarters in an attempt to make trouble.[61]

If Paul's suspicions sound a bit paranoid, it is important to remember that the Woman's Party had worked long and hard for the ERA by 1958 and believed that the labor unions stood in the way of passage. Although with hindsight it seems as if belief in passage of the amendment was hopelessly unrealistic in the 1950s, the ERA had gained a large sponsorship and had come up for a vote in the Senate in that decade. That the women's rights movement was not entirely ineffective is indicated by the fact that, in 1960, an AFL-CIO official sent a memo to the union lobbyists reminding them that the supporters of the ERA "keep up a constant barrage of communications to the members of Congress," so that it was important to get at least one opposing statement on each desk.[62]

Pro-ERA union women, like other ERA supporters, explained union opposition by insisting that only the leaders, not the rank and file of union members, opposed the amendment. Florence Kitchelt believed that union women were "carefully protected from any information about the Amendment by their male officers."[63] One ex-CIO member reported that practically everyone in the CIO favored the ERA but did not have the chance to express themselves on the issue.[64] Fannie Ackley argued that no union had ever voted on the ERA and that opposition came from the top only.[65] Although there is no evidence to support such claims and it is unlikely that most union members had even heard of the ERA, it is true that the opposition of the leadership would have prevented union support. In 1954, retired AFL official Florence Thorne met with Alice Paul to discuss ways to persuade AFL head George Meany of the necessity of supporting the ERA; but even though Thorne had retired, she did not feel that she could make her position on the amendment public.[66] In the mid-1960s, when women from the United Auto Workers helped to form the National Organization for Women, opposition to the ERA from the AFL-CIO forced them to withdraw from active participation when the new organization endorsed the ERA, even though they themselves did not oppose the amendment.[67] Had grass-roots support for the ERA existed in the ranks of labor during the 1950s, it is unlikely that it would have found expression.

Although the antilabor sentiment of some National Woman's Party members makes suspect some of the charges hurled at the unions, it is difficult to avoid the conclusion that labor leaders acted hypocritically in their opposition to the ERA. The unions claimed that protective legislation was necessary to win for women protection that men could gain by collective bargaining. It is true, of course, that only a very small percentage of the female labor force belonged to a union, but it is equally true that the labor movement itself had to take a large share of the blame for that situation. Regardless of the complex reasons why women's union membership lagged behind that of men, however, the position of the AFL on equal pay gave the lie to their anti-ERA arguments. The AFL did not support equal pay legislation because it believed that this was a matter for collective bargaining, not government control.[68] That union officials could see equal pay, but not, for example, night work, as the proper province of collective bargaining is revealing.

Women's rights supporters suspected that union leaders opposed the ERA out of general opposition to women in the labor force, but they could not explain the opposition of women's labor groups so easily. The split in the 1920s between women who focused their efforts on the amelioration of working conditions for women—particularly the National Women's Trade Union League (which disbanded in 1950) and the Women's Bureau of the Department of Labor—and those who continued to champion legal equality for women had created, by 1945, a situation of deeply rooted antagonism on both sides. Women's labor groups, like the other anti-ERA women's organizations, continued to identify women's rights advocates, and especially the National Woman's Party, as antilabor and reactionary, a tool of the employers or perhaps simply a collection of selfish wealthy or professional women.

In 1951, for example, two members of the Woman's Party visited Elisabeth Christman, who had headed the Women's Trade Union League in its last years, in order to ascertain her objections to the ERA. They did not tell her of their affiliation with the Woman's Party, but they reported that she became violent when they mentioned the ERA. "Her hatred of N.W.P. is great," they noted. "She said the N.W.P. sent speakers in ermine coats (or collars) to the Trade Women's League in N.Y.C. 'We hated them. We will always work against them.' "[69] Women on both sides of this conflict mistrusted and even detested their opponents, despite their common concern for improving women's lives. Florence Kitchelt was at least partly right in her belief that the anti-ERA convictions of the League of Women Voters, which supported the Women's Trade Union League and the Women's Bureau, stemmed from personal animosity toward the National Woman's Party, animosity that harked back to the dispute over the Woman's Party's use of militant tactics during the final days of the suffrage struggle. Certainly the long history of conflict did little to build trust between the two opposing camps.

Although organized union women and their supporters did not favor the ERA in the 1950s and 1960s, that does not mean that they were unconcerned about women's issues. The Women's Bureau of the Department of Labor, for example, staffed mostly by educated professional women supportive of unionism, continued to lead the fight against the ERA but worked for other measures, particularly equal pay, that it believed would benefit working women.[70] The Women's Bureau maintained contact with the major women's organizations, except for the National Woman's Party. Sarah Hughes, while president of the pro-ERA BPW, served on the Advisory Committee of the Women's Bureau, and Margaret Hickey, an ex-president of the BPW, also gave her support to the agency. The reform-oriented women's organizations served as a base of support for the Bureau, which sponsored a number of conferences on women's issues. In 1945, for example, the agency held a conference of Women Representatives of Labor Unions on War and Postwar Adjustment of Women Workers. At another conference the next year with female union leaders, a discussion of the political and civil status of women focused on the identification of areas in which women's legal status was weak and resulted in support for a state-by-state approach to removing discriminatory laws. The Bureau fought for women but disassociated itself from the fight for women's rights.

The Women's Bureau played a key role in orchestrating opposition to the ERA from its introduction in 1923 until the 1970s. When Frieda Miller was sworn in as director in 1944 she made clear her opposition to the amendment.[71] The Woman's Party hoped for a change of course under Alice Leopold, who served from 1953 to 1960, since she belonged to the Connecticut Committee for the ERA. Alice Paul, a Connecticut neighbor of Leopold, warned the new director that the Bureau's staff would bring strong pressure to bear on the ERA issue, and Leopold reportedly responded that she had already experienced such pressure.[72] A small group of feminists, including Florence Kitchelt and Alice Paul, visited Leopold to discuss the ERA and rejoiced when she announced that the Women's Bureau had no position on the amendment.[73] This, in addition to Leopold's background in merchandising and state politics, did not endear her to labor.[74] Perhaps as a result, in 1955, Leopold asked that her name be deleted from the advisory committee to the Connecticut Committee and reminded Florence Kitchelt that the Connecticut Committee distorted the truth when it implied that she had removed the Women's Bureau opposition to the ERA, since in fact she had simply stated that the Bureau had no position on the issue.[75]

Esther Peterson, appointed to replace Leopold in 1961, renewed the agency's strong anti-ERA stance. As a key woman in Kennedy's administration, she made sure that the President's Commission on the Status of Women did not recommend passage of the amendment and coordinated much of the opposition. She worked with unions to ensure that opponents of the measure would meet the onslaught of pro-ERA letters, and she worked with ERA foes in Congress who were anxious to "put on strong opposition witnesses and not allow 'the circus' to take over."[76]

It is perhaps not surprising, given the long history of conflict between pro-ERA feminists and anti-ERA labor women, that hostility between the two camps continued even after the protective legislation issue lost its salience. As one National Woman's Party member commented: "[a]ll those who most violently oppose this amendment concede that it is 'right in principle.' "[77] For example, two Woman's Party members who attended a convention of the UAW Women's Bureau in 1956 reported that all of the women there believed in equality for women.[78] As Pauline Newman, an International Ladies Garment Workers Union official associated with the U.S. Women's Bureau, remarked, "I think we were probably advocating equality long before the equal rights people came on board."[79] But even women who, like the union members at the 1953 National Conference on the Problems of Working Women, protested employers' use of women to lower wages for all workers, denounced the ERA as a "fraud on the working women of the country."[80] Nancy Gabin has concluded that the UAW Women's Bureau activists would have balked at the label "feminist": "in the 1940s and 1950s, feminists were regarded by unionists as professional or leisured middle-to-upper class women who lacked an understanding of the needs and interests of working class women."[81]

The suspicions of the ERA opposition about feminists' lack of concern for working-class women and their antilabor bias were occasionally confirmed. For example, Mary Anderson, ex-director of the Women's Bureau, reported to the California branch of the Woman's Party that Alice Paul had said that she did not care what kind of legislation, if any, would be passed in the aftermath of the ERA.[82] A woman who sent a contribution to the Woman's Party criticized a publication that denounced labor for the level of its political contributions. "If labor opposes the amendment," she wrote, "the answer is to convince labor it is mistaken, not to criticize it, and it alone, for political expenditures. . . . For years, the Woman's Party has been considered anti-labor, and has done nothing to correct this opinion among labor people."[83] Alma Lutz agreed that the Woman's Party had not made sufficient effort to win friends among labor leaders.[84]

In response to the opposition to the ERA of the labor movement as a whole, including the women's labor groups, the National Woman's Party employed three strategies. The first strategy attempted to eliminate the opposition. The National Woman's Party directly attacked the Women's Bureau of the Department of Labor. One NWP member denounced the use of "nefarious ways" by the Women's Bureau to defeat the ERA and insisted that the agency's staff of women lawyers simply wanted to keep their jobs by preserving protective legislation for women. Like other Woman's Party members, she saw the Women's Bureau as all-powerful, directing the actions of government agencies, senators, and representatives on the ERA issue. "[N]o matter what we tried to do, they would axe it," she commented.[85] In fact, the agency was poorly funded and unable to exercise much influence over either industry or the unions. But because the Woman's Party saw the Bureau as such a threat to its program, it attempted to destroy it. Edna Capewell, a CIO member who agitated for the Woman's Party, wrote to Truman in 1945 to urge him to eliminate the Women's Bureau or at least appoint a friend of women to head it.[86] In 1947, a bill to eliminate or greatly reduce the budget of the agency worried Women's Bureau supporters, and the League of Women Voters rallied to the cause of saving it.[87] At the 1951 National Woman's Party convention, a resolution to ask the president to replace the Women's Bureau with an "independent" agency generated debate over the propriety of "unladylike" behavior; one member considered the resolution too acrimonious, while others, recalling the grand old days of the suffrage movement, argued the need for direct action. The resolution eventually passed unanimously and the call for a new agency made the Washington papers.[88]

Alice Paul, photograph inscribed to Helen Hunt West. By Harris & Ewing, Washington. Courtesy of the Schlesinger Library, Radcliffe College.

National Woman's Party members celebrating Susan B. Anthony's birthday at the Suffrage Monument, United States Capitol Crypt, February 14, 1945. Courtesy of the National Woman's Party.

The Alva Belmont House, headquarters of the National Woman's Party. Courtesy of the National Woman's Party.

The National Council of the National Woman's Party, January 12, 1947. *l–r, seated:* Miriam Y. Holden, Mabel Griswold, Alice Paul, Marie Moore Forrest, Emma Guffey Miller, Anita Pollitzer, Clara Snell Wolfe, Margaret C. Williams, Gladys Greiner, Lucia Hanna Hadley. *l–r, standing:* Florence E. Kennard, Elizabeth Forbes, Mary E. Owens, Meta Grace Keebler, Helen Vanderberg, Cecil Norton Broy, Mary A. Murray. Courtesy of the National Woman's Party.

National Woman's Party insurgents sitting outside Anna Kelton Wiley's home, Washington, D.C., ca. 1947. Courtesy of the National Woman's Party.

Ethel Ernest Murrell, controversial chairman of the National Woman's Party, 1951–1953. Courtesy of the National Woman's Party.

Ernestine Hale Bellamy, one of the few young members of the National Woman's Party in the 1950s. Courtesy of the National Woman's Party.

Rose Arnold Powell, with photographs of her beloved Susan B. Anthony and Elizabeth Cady Stanton. Courtesy of the Schlesinger Library, Radcliffe College.

Lucy Somerville Howorth, AAUW and BPW leader, 1955. By Miller of Washington. Courtesy of the Schlesinger Library, Radcliffe College.

A tribute to Emma Guffey Miller (center), Democratic committeewoman and National Woman's Party member. Courtesy of the Schlesinger Library, Radcliffe College.

Perle Mesta, Washington socialite and hostess, United States minister to Luxembourg, and National Woman's Party member. Courtesy of the National Woman's Party.

Meeting of the Women's Joint Legislative Committee, Belmont House, 1956. Courtesy of the National Woman's Party.

Lena Madesin Phillips, founder of the International Federation of Business and Professional Women's Clubs, 1954. By G. Maillard Tesslere. Courtesy of the Schlesinger Library, Radcliffe College.

Agnes Wells *(left)*, chairman of the National Woman's Party, 1949–1951, with an unidentified friend. Courtesy of the Schlesinger Library, Radcliffe College.

Florence Kitchelt, Connecticut Committee for the ERA leader, with her husband Richard, 1953. Courtesy of the Schlesinger Library, Radcliffe College.

Ruth Crane's television show featuring "The Equal Rights Review" enacted by National Woman's Party members, May 24, 1951. *l–r:* Mildred Palmer as Abigail Adams, Dora Ogle as Lucretia Mott, Ernestine Bellamy as herself, Queen Walker Boardman as the Woman of the West, Ruth Crane, and Amelia Himes Walker as the Suffragette. Courtesy of the National Woman's Party.

The National Woman's Party leadership believed that the Women's Bureau was returning the attack years later when two reporters came to headquarters to inquire whether it were true that the group would lose its property when the ERA passed. Woman's Party head Emma Guffey Miller saw this as a further attempt of Esther Peterson, Women's Bureau director under Kennedy, to defeat the Woman's Party and the ERA.[89] One Woman's Party member who remained active in the 1980s insisted that the Women's Bureau was "the most archly anti-feminist group that ever was."[90]

A second strategy involved attempting to win endorsements, especially from the leadership of the labor movement. At the end of the war, for example, Nora Stanton Barney tried to make use of postwar uncertainties to win support from the AFL and the CIO.[91] Emma Guffey Miller, too, pointed out to Phillip Murray, head of the CIO, that women CIO members had lost their jobs and seniority at the end of the war, that there had been no "protection" for them.[92] At the end of 1945, Alice Paul reported that John Lewis, head of the United Mine Workers, supported the ERA. "He tells us that the Senators & Congressmen thoroughly understand that he is for the Amendment. He still does not want to make a public statement as he says it might add to his internal union difficulties to do so at present," she noted.[93] Lewis' daughter Kathryn belonged to the National Woman's Party and agreed to try to help with her father.[94]

Edna Capewell, hiding her Woman's Party membership and advertising her UAW affiliation, visited Walter and Victor Reuther in Detroit in 1945 to discuss the ERA. She devised a plan that she believed would win over the union. Walter Reuther, who she claimed favored the ERA, was running for president of the UAW, and if he won, she believed she could obtain the names of UAW women who had worked during the war who would surely support the ERA. In order to help Reuther win the election, she proposed to volunteer her services for the convention, but she needed money to do so. She considered asking the Woman's Party for support, but eventually dropped the idea out of hesitation to approach the leadership with this request.[95]

In 1954, the Woman's Party made a concerted effort to win an endorsement of the ERA from the AFL. With some difficulty, the leadership induced a union official—the international president of the American Federation of State, County, and Municipal Employees—to introduce a pro-ERA resolution at the AFL convention. The Woman's Party hoped for withdrawal of support for the Hayden rider and a position of neutrality on the amendment. What it got was a recommendation that the executive council undertake a thorough study of the issue and then take whatever action seemed warranted. Shortly after the convention, Alice Paul spoke with AFL head George Meany, described by Ella Sherwin as an "old brute," and reported that he seemed friendly and interested.[96] The CIO convention three months later proved more difficult, and the Woman's Party made no progress. When the two federations merged in 1955, the new AFL-CIO, headed by Meany, passed a resolution opposing the ERA. Alice Paul wrote to Meany to express her hope that the new federation would continue the study of the ERA begun in the AFL, adding that the Fair Labor Standards Act "makes us feel that there is no reason for conflict between the Labor Movement and the Woman Movement."[97] Another Woman's Party member echoed this plea: "We know from our own struggles to win friends, how much every reform movement, struggling for its rights needs friends." Would it not be advantageous, she asked, if they could revive the friendship that had existed during the suffrage struggle?[98] Meany denied the existence of hostility between labor and the women's movement, and the AFL-CIO continued to oppose the ERA until after its passage by Congress in 1972.

A third strategy used by the National Woman's Party was to attempt to create the impression that labor was not opposed to the ERA. It did this in two ways: by using working-class, and especially union, women who supported the ERA as token members; and by establishing a pro-ERA front organization of working-class and union women.

The few working-class members of the National Woman's Party often seemed to have joined after experiencing job discrimination as a result of protective legislation. Because there were so few women in the Woman's Party who could counter the opposition's charges of indifference to the plight of working-class women, NWP leaders encouraged the working-class members to speak out. The executive secretary, for example, asked Fannie Ackley to have her group of union women send a telegram to the convention of the League of Women Voters stating their views on protective legislation.[99] Four years later, another officer asked Ackley to write out the history of her work and union experience, including the details of losing her job due to protective legislation.[100] The Woman's Party often paid the travel expenses of working-class women in order to get them to Washington to testify on the ERA. As one member cynically described the dealings of the leadership with a woman who sold tickets in the New York subway, "they paid her to leave her job and paid her more so that she could go around and work, to show that they weren't all patricians."[101] Mary Markajani, an electrical worker and CIO member, lobbied for the ERA and made public appearances on behalf of the Woman's Party, despite her realization that she was too "rough" for most members. An ERA opponent described Markajani, who participated in a panel discussion on the ERA, as "Quite proud of her lack of education."[102] How the National Woman's Party leadership felt about such pride is hard to know; although they may have been pleased to be able to show off a member so obviously not a "patrician," they almost certainly patronized such women. One member, for example, reported that she had found a woman taxi driver who "talks well" and would be able to represent the group publicly.[103] Members of the AFL-CIO were especially sought after. One woman, for example, joined the Woman's Party in 1955, telling of the discrimination she had suffered as a worker. Alice Paul inquired immediately whether she belonged to the AFL or CIO in hopes that she could make her views known there. Unfortunately, she turned out to be violently antiunion, and thus not useful in the struggle to win union endorsement.[104]

The National Woman's Party recognized both its lack of working-class membership and the need for its support, but it never really attempted to recruit working-class women. Instead, it tried to prove by whatever means possible that some working-class women did support the ERA. Especially useful for this purpose was a front organization, the Industrial Women's League for Equality, created in 1944 by Alice Paul and Nora Stanton Barney out of an old New York labor ally of the Woman's Party. Ella Sherwin, a proofreader, linotype operator, and member of the International Typographical Union, became president. She had once lost a job through the impact of protective legislation and had become convinced of the need for the ERA.[105]

The Industrial Women's League for Equality never grew into a large organization and it quietly faded away sometime in the 1950s. The group did its only recruiting by sending out personal letters to women it thought might like to join and, not surprisingly, it grew slowly.[106] But it, like the National Woman's Party that financed it, did not aim at large membership and mass mobilization. Instead, it served largely as a proof of the support for the ERA. When a speaker stated that workers opposed the ERA, a National Woman's Party member reported to a friend that she "corrected this misstatement good-naturedly by telling about the League of Industrial Women."[107] The national chairman of the Woman's Party wrote to Ella Sherwin in 1945 to tell her how important it was to have the organization backing the amendment.[108] For these purposes, it was enough that the group simply existed. This use of the League makes clear the NWP's attitude toward labor. While the Party felt it had to convince other women's organizations and legislators that labor was not opposed to the amendment, it put little or no effort into persuading working-class women of the merits of a constitutional amendment granting equality.

Although the Woman's Party in reality set up and financed the Industrial Women's League, both groups pretended to be independent in order to strengthen the League's claims to represent working-class women. One California attorney who criticized the domination of her local branch of the Woman's Party by teachers who organized expensive luncheons and dinners only of interest to women with money told Alice Paul that "[y]our Industrial League for Equality will not function until the same group of teachers take their hands off it and let the working women bring up their own child."[109] Ella Sherwin was horrified when someone put her on a write-in slate for vice president of the Woman's Party, for she especially wanted to keep the League distinct from the Woman's Party.[110] And Josephine Casey, who lobbied on behalf of the Industrial Women's League, denied that she had been hired by the Woman's Party.[111]

Ella Sherwin's relationship with the leadership of the National Woman's Party gives insight into the attitude of the officers toward working-class women. Sherwin was often deferential, thanking Alice Paul for making possible a visit to Washington and for making her so comfortable, deprecating her abilities as a lobbyist and a leader, and apologizing for her "impudence" when giving advice.[112] On the other hand, she knew the kind of approach that would work with union officials and could and did assert herself. When Florence Kitchelt began to publicize her argument that the ERA would not eliminate protective legislation, Sherwin took her on directly. She curtly dismissed one of the legal opinions that Kitchelt had solicited: "Professor Bakke writes an excellent opinion—for a professor. So many people in his position find it difficult to see the real world, they are so familiar with the world of books."[113] And she also wrote Kitchelt that "[i]f you can't 'take it' from union spokesmen, we working women can."[114] Alice Paul praised Sherwin for the amount of work she had been able to do in time away from her job, recognizing, at least, the difficulties of participation for women who worked for an employer on a regular schedule.[114]

Women like Ella Sherwin had a difficult balance to maintain since they were, as Fannie Ackley put it, "strong for unionism" yet critical of union attitudes toward women.[116] They also had to confront elitist and antiunion attitudes from women's rights supporters. One woman whose husband was a shop steward for the Screen Cartoonists Guild in California rejoiced that the writers and cartoonists had organized, bringing "some enlightened and cultured people" with "intelligence and thinking ability above the ordinary longshoreman level" to the labor movement.[117] Another woman believed that the CIO represented "lower labor groups whose education and outlook are somewhat meager."[118] Many women believed that female unionists blindly followed the lead of their male leaders, unlike "intelligent & independent women such as the assoc. of women lawyers, doctors, business & prof. women."[119] Working-class women were not unaware of the class differences and elitism that separated them from their coworkers in the equal rights movement. One woman, for example, commented to the secretary of the National Woman's Party that "I bet I sound very crude to you" and wrote to Alice Paul that "I believe and think that the girls in ERA Woman's Party feel that I am to [sic] rough—to [sic] outspoken, but this is the right way that the Union members understand."[120]

A common trait of all three strategies for dealing with labor opposition to the ERA is the absence of any actual attempt to mobilize working-class and union women. Looking back, it is possible to see that the postwar era had potential for a cross-class movement of women determined to eliminate gender-based discrimination in American society. At the end of the war, women who lost their factory jobs to returning veterans protested not the policy of allowing men who had fought for their country to return to their jobs, but rather the denial of their seniority rights. At least one National Woman's Party member, CIO member Edna Capewell, connected the protests of women workers to the ERA, suggesting that the amendment was the only way to protect women against the refusal of employers to rehire women in accordance with their seniority.[121] Another ERA advocate, a wealthy and flamboyant manufacturer who belonged to the Connecticut Committee, threatened, as if she were herself an industrial worker, that if an attempt were made to reinstate the curfew law for women in her state "we will pick up our hammers, our welding torches, our wrenches, and all of the other hand tools which you men have taught us to use so expertly, and we won't use them to make munitions, either."[122] Yet as Susan Hartmann has argued, class and race kept women's organizations of various kinds from uniting during the war.[123]

Union women continued to protest the discrimination they experienced after the war, and several women active in the United Auto Workers union in the post-1945 period emerged as founders of the National Organization for Women in the 1960s.[124] That connection is an important one. If feminists devoted to winning women equal rights had been interested in making connections with those who were angry about losing their jobs and seniority at the end of the war, the history of the women's movement might have taken quite a different course. But instead, the UAW Women's Bureau and the strong women unionists associated with it opposed the ERA and complained about having the National Woman's Party on their necks at every UAW convention.[125]

In 1970, the UAW endorsed the ERA, although the AFL-CIO did not follow suit until 1973, having opposed passage in Congress and supplied opposition in the early state campaigns for ratification. But the unions swung around to support of the ERA not only after federal legislation such as the Civil Rights Act of 1964 changed the context of the ERA, but also after new organizations coalesced as part of the resurgence of the women's movement. The National Organization for Women brought together union women and BPW members, government officials and longtime activists, and paved the way for a real alliance between the women's movement and the labor movement. When NOW sponsored a massive ERA march in the summer of 1978, the unions turned out in force. That show of solidarity and strength could not have come to pass in the atmosphere of the 1950s, when the women's rights movement, from a class-bound perspective, saw labor as an enemy whose opposition had to be neutralized.

Civil Rights and the Opportunism of the Women's Rights Movement

The issue of race was central in the women's rights movement from its very inception, emerging as it did out of the abolition movement in the 1840s. In its early years, just after the Civil War, the women's movement split apart over the issue of whether or not to support the Fifteenth Amendment, which enfranchised black men but not women, and the reunited movement later made use of racist arguments—that women's suffrage would maintain white supremacy in the South—to support the demand for the vote.[126]

Before 1920, the National Woman's Party stood, at least theoretically, on the side of black women by supporting a federal commitment to universal suffrage. This was in contrast to the National American Woman Suffrage Association, which took a states' rights position. After suffrage, however, the National Woman's Party refused to work explicitly to ensure black women the right to vote, dismissing that as a race issue.[127] But even the pre-1920 theoretical commitment to universal suffrage did not ensure a nonracist approach to suffrage work. A controversy arose over the participation of black women in the suffrage parade of 1913 and, although it is unclear who asked that black women not march, it is clear that fear of offending white Southern women made the Woman's Party hesitant, at best, to accept the participation of black women.[128] In 1921, the same year that Mary Church Terrell sought Woman's Party support for black women's voting rights in the South, Freda Kirchwey denounced the Woman's Party for racism in the *Nation*, although NWP officers denied her charges that Alice Paul had sought to deny the black delegates the floor and that black women had been forbidden use of the elevator at the annual convention.[129] In 1922, Alice Paul dealt with the information that Southern members would not tolerate admission of black women by assuring her organizers that the Woman's Party would not actively recruit blacks. Two years later, Paul excluded black speakers at a ceremony at the grave of Inez Millholland, a National Woman's Party member who had also been active in civil rights work, again in order to appease the Southern delegates. This incident received a great deal of press coverage and resulted in the withdrawal of some members from the Woman's Party.[130]

In response to publicity about the racism of the Woman's Party, Mary Church Terrell, apparently at the prompting of another member, rose to the organization's defense. In a press release sent to NAACP official Walter White, Terrell was quoted as saying that the National Woman's Party, "of all women's organizations throughout the country . . . has been the most broadminded and friendly in its attitude to the colored people." Black women, she reported, picketed with white women during the suffrage struggle and were included in parades, conventions, banquets, meetings, and social events on the same basis as other members. Although she admitted that the group's refusal to fight for the voting rights of black women came as a grievous disappointment, she understood, she said, that it was committed to work only on discrimination that affected all women. "The Woman's Party has been very close to my heart in spite of that one disappointment about the resolution, and I cannot bear to let such an unjust accusation against them stand unrefuted," she concluded.[131] This may have been a statement intended to keep conflict in the women's rights movement private, since Terrell knew that white suffrage leaders had been willing to accept a suffrage amendment that excluded black women.[132]

The women's rights movement, throughout its history, defined its priorities with reference to white middle- or upper-class women. Thus "discrimination that affected all women" included the right of owning property but not black women's voting rights. Black women had formed their own organizations, including the National Association of Colored Women and the National Council of Negro Women, to fight racial discrimination and foster solidarity among black women. Both organizations recognized, as Mary Church Terrell put it, that black women were "double-crossed," with "two heavy loads to carry through an unfriendly world, the burden of race as well as that of sex."[133] They sought contact with white women's organizations but were often overlooked if they did not insist on inclusion in coalitions. As two National Council of Negro Women leaders explained, their organization endeavored "to integrate Negro women into national and international programs of women to unite the spirit and energies of women of all races, creeds and colors, thus building finer and deeper understanding."[134]

The National Association of Colored Women fought for the ERA, among its other activities, and the National Council of Negro Women, although opposed to the ERA, supported the appointment of women to high-level posts. Mary McLeod Bethune, one of the most influential black women in the United States, saw her work in supporting the appointment of women like Judge Burnita Shelton Matthews, a National Woman's Party member, as all of a piece with her endless struggle to raise money for her own Bethune-Cookman College.[135] Black women in the period after 1945 continued to work on issues of central concern to them, including segregation, lynching, and discrimination. In the late 1940s and early 1950s, for example, Woman's Party member Mary Church Terrell led a drive to desegregate restaurants in Washington.

As the efforts of black organizations began to make progress in the area of civil rights, the women's rights movement was forced to respond. The movement adopted four strategies: the maintenance of largely segregated organizations, despite a diversity of views within the movement on integration; the use of token members and groups; competition with the growing civil rights movement; and an attempt to "piggy-back" on the successes of the civil rights movement.

Women's organizations in this period, like their nineteenth-century predecessors, remained largely segregated. The Alva Belmont House, headquarters of the National Woman's Party, was a mansion staffed by black servants in a segregated city. Alice Paul noted that "we didn't have very many Negroes in our group."[136] When painter Betsy Reyneau, who did portraits for the Woman's Party, tried to find a photograph of herself for an unveiling ceremony of one of her works, she reported that she could not find one without the portrait of a black person, which "unfortunately might irritate some of our members." Although she herself worked on the "race question," she believed that it was not a good idea to "mix issues."[137] When the rebel group met at the Mayflower Hotel in Washington in 1947 for their "Rump Convention," they only gained access to the hotel facilities by assuring the management that "we would not have 'niggers' at our convention," much to the distress of some of the rebels who learned of the promise later.[138]

As far as can be determined, the issue of integration arose only once in the National Woman's Party. A black woman asked to join the Maryland branch in 1950, and the member who reported this to headquarters as a serious problem wondered if her request had been motivated by sincere desire or if it had been prompted by the group's enemies. She reported that several members of the Maryland board were "vehement" in not wanting black members. The national leadership had never discussed the issue of integration within the organization, according to the secretary, who suggested that each branch should vote on its own membership policy.[139]

Despite the publicity accorded the racism of the Woman's Party by historians and 1920s commentators, black feminist and civil rights activist Pauli Murray suggested that the Woman's Party was no different from other women's organizations at that time. The BPW, like the Woman's Party, was a virtually all-white group. The International Federation left decisions about membership policy to national organizations and Americans, according to founder Lena Madesin Phillips, "feel more violently about this than any other of our groups." She knew of no more than a very few women of color in the BPW, although the national Federation had nothing in its constitution preventing such membership.[140] But since individuals could become members of the BPW only when recommended by a member in good standing, it is perhaps not surprising that the ranks remained overwhelmingly white. Phillips herself had supported the membership of the first black woman in a BPW club in 1940, and her support helped keep the black woman in the club and the club in the Federation, despite the minor crisis this engendered.[141] A Norwegian woman from the International Federation was discouraged by the "prejudiced minds" of some of the BPW women she met in Atlanta and surprised that the northern clubs were not interracial.[142] When Florence Kitchelt read an article in *Ebony* about black women lawyers reporting that they found gender more of a barrier than race, she wrote to the BPW to urge support for a second article on black women and the ERA, and she inquired about the "BPW Negro Club," which she hoped could serve as the basis for the article. The BPW office gave her the address of the president of the National Association of Negro Business and Professional Women and informed her that the two organizations were not affiliated in any way, since the black group "does not want us encroaching on their territory."[143]

Integration never became a public issue in either of these groups, but a controversy over membership policy erupted in the Washington, D.C., branch of the American Association of University Women in the late 1940s. The fight over integration illustrates the lack of consensus on racial issues in the women's rights movement, makes clear the barriers to black women's participation in women's organizations, and demonstrates the impact of civil rights activity on the women's rights movement.[144]

In 1946, Mary Church Terrell, who had joined the AAUW in the last years of the nineteenth century, applied for readmission to the District of Columbia branch. She was eligible by all criteria, and a member of AAUW sponsored her application. The Executive Committee voted not to approve her application since it involved a larger issue—integration—but decided instead to poll the members to see if they wanted to admit "members of the Negro race."[145] This poll made plain what the issue was, despite later attempts to focus attention on the branch's right to screen members in order to keep out "Communists and undesireables."[146] About sixty percent of the members who answered opposed the admittance of black women. The national headquarters of the AAUW responded to the branch's decision to reject Terrell with a strong statement against discrimination on the basis of race, religion, or politics, but the branch refused to change its decision on the grounds that it needed to retain the right to control membership in order to protect against subversive elements and to avert racial friction.

The conflict erupted into a court battle between national headquarters and the branch, with the majority of the D.C. members opposed to integration and national control. Three hundred D.C. members joined the court battle on the national side, declaring their "reputations as 'liberal women' " were at stake.[147] The court ruled in favor of the D.C. branch, however, because the national constitution did not include a statement to the effect that members would be accepted regardless of race. So a convention of the national organization amended the constitution by a huge majority, recognized the pro-integration minority group in the District of Columbia as the official branch, and reinstated three black women, including Terrell, in the new branch. A number of AAUW members who resided in the District of Columbia seceded to form their own group, which one critic described as including "those hampered by race prejudice, the bridge and dance devotees, and those who on general principles look upon and wish the Branch to be as they have named it, the University Women's Club, with social aims and restricted membership."[148]

A number of women's rights supporters belonged to the D.C. branch of the AAUW, and a debate between one of them, who chose to secede rather than submit to integration, and a coworker in favor of integration reveals the diversity of views on the issue. Anna Kelton Wiley, a longtime National Woman's Party member, admitted that Mary Church Terrell was a "very fine woman" whom "anyone would be delighted to have," but opposed her as a "figurehead to force us to do away with seggregation [sic]."[149] She asserted that the time was not ripe to end segregation in Washington, since it was a Southern city and so heavily black. "Many of them are fine people but many are near to the brute," she insisted. Such crude racism must have offended Florence Kitchelt, to whom Wiley addressed her comments. Kitchelt responded, obviously trying to be gentle with an old coworker, by suggesting that admitting qualified black women to the AAUW did not mean abolishing segregation throughout the District. In addition, she asserted, "the same standard of culture makes people akin—regardless of skin!"[150] In reply, Wiley became cruder: segregation was the only possible protection against intermarriage. Would Kitchelt like a Negro grandchild? (Kitchelt penciled on the letter: "It's all right with me!!") White people have a different inheritance, background, standard of conduct, and code of behavior, Wiley asserted. "We have been produced after centuries of effort and self-sacrifice. The N. on the other hand were an inferior race brought over against their will." And she signed off, "Yours for race purity."[151]

Wiley's views, like those of other women's rights advocates involved in the AAUW secession, were offensive to Kitchelt and also to Alma Lutz. Lutz found Wiley's letters "hard to take" and she was distressed that Wiley expressed herself on the stationery of the Connecticut Committee. She could not understand how "such a sweet person" could hold such views.[152]

The fight over the D.C. branch had ramifications elsewhere in the AAUW. The St. Louis branch suffered a lawsuit over its clubhouse when those opposed to integration withdrew. In Richmond, someone offered the branch a clubhouse but withdrew the offer after the national convention amended the constitution. At lease one black woman who wanted to join another branch wrote Terrell seeking advice. The new D.C. branch began inviting black members to join slowly, because "some of our people lack experience, even though their belief is firm," but soon grew to include about fifty black women out of a total membership of about four hundred.[153] One black woman active in the AAUW on the local level described the organization as not particularly open to black women until the late 1960s. She was one of the first to join, in the 1970s, in Columbus, Ohio, and recruited about ten friends for the group. The branch did not reject black women, she said, but it operated as a kind of social clique and did not seek out black members. Her friends joined but rarely participated.[154]

Terrell had applied for membership, just as her opponents feared, as part of her struggle against racism. In writing to AAUW president Althea Hottel, Terrell emphasized that this was not an individual issue but a matter of justice. She referred to her membership in the National Woman's Party when she pointed out that "a group of your women—a National Organization—has honored a colored woman by having her portrait painted which will be hung in Headquarters here with the portraits of three women of the dominant race."[155] Letters poured in congratulating Terrell for the victory "for our women," praising her for her willingness to serve as a "test case," lauding her as "the symbol of our cause."[156] Terrell's challenge to the exclusionary policies of the District of Columbia branch of the AAUW certainly served as a symbolic victory by winning national attention and wresting from a women's organization a constitutional commitment to racial equality. It did not, however, change the fact that women's organizations remained largely segregated.

Although women's rights organizations made no large-scale efforts to open to black women, they did use the same strategy of token representation of individuals and groups that the National Woman's Party pursued in the case of working-class and union women. The Woman's Party sought to take advantage of Mary Church Terrell's—and her daughter's—membership. Alice Paul, for example, asked Terrell, whom she considered "one of our ablest and best members," to serve on the national board and to obtain endorsements of the ERA "by other leading colored people."[157] Mabel

Vernon, a friend of Paul's since their days together at Swarthmore, told an interviewer that Paul was very nice to individual black women but admitted that Paul had "prejudices." Vernon also noted Paul's "antagonism for Jews," despite her close friendship with Anita Pollitzer, who was Jewish. Vernon reported that if, in response to Paul's anti-Semitism, she pointed out that Pollitzer was Jewish, Paul would answer, "Yes, but she's different."[158] Paul's anti-Semitism was apparently no secret in the Woman's Party. One member reported, for example, that a woman who did not get a job with the group asked if Paul had objected because she was Jewish.[159]

The National Woman's Party not only used token members to try to advance the cause of the ERA, but also sought to use Terrell's National Association of Colored Women in the same way. This group had participated in the suffrage campaign and, according to Alice Paul, took up support of the ERA at the very start.[160] It regularly sent representatives to the Women's Joint Legislative Committee for Equal Rights, joined the World Woman's Party in 1946, and sent Terrell to testify for the ERA before Congressional subcommittees. Alice Paul welcomed such support of the ERA, but reported in 1943 that a Woman's Party officer had asked her if a way could not be found to put the group off the Joint Legislative Committee.[161] Clearly even interest in token representation was not unanimous.

The same kind of tokenism characterized a coalition of women's groups organized in 1950. When the BPW called together the leaders of women's organizations to discuss women's role in civil defense and military mobilization, a representative from the National Council of Negro women pointed out that her group had been left out of the planning for the meeting. She reported, in a spirit of constructive criticism, that when she had taken up this issue with a woman from another organization, she had been told that some of the members were not as "broad" as they might be. She warned that civil defense meant a concern with intercultural relations in the community as well as a concern with the atomic bomb.[162] As a result of her comments, the group decided to add the National Council of Negro Women, as well as groups representing other "interests," such as patriotic, labor, and service organizations, to the steering committee of the group that eventually became the Assembly of Women's Organizations for National Security.

Not only did the women's rights organizations show little real interest in connecting with black women, but throughout this period women's rights leaders tended to perceive the burgeoning civil rights movement as a competitor likely to win rights for black women and men that were still denied to white women. Just as the disappointment of feminists at the failure of the Fifteenth Amendment to include women's rights caused some who had fought bravely for abolition to lash out in racist terms, so the frustration of feminists in the 1940s and 1950s sometimes led even those who were committed to racial equality to adopt positions that set black and Jewish women and men in competition with white Christian women.

In 1944, for example, Florence Kitchelt, herself a member of the

NAACP, wrote a letter to the *New York Times* publicizing a case in which black teachers in South Carolina won equal pay with white teachers under the Fourteenth Amendment in order to point out that the Constitution protected black women and men but not white women. The Assistant Special Counsel of the NAACP Legal Defense and Educational Fund wrote to express his disagreement with her assertion that the Fourteenth Amendment did not protect white women. The National Woman's Party sent a copy of Kitchelt's letter to Mary Church Terrell "to get the reaction from the colored viewpoint."[163] Terrell strenuously objected to the Kitchelt letter. She found it misleading, since black women received the lowest wages of any group in society. The Woman's Party member through whom the letter had been transmitted explained to Kitchelt that "when it comes to the question of racial equality, the negroes have no sense of humor whatever."[164] What Terrell should have found amusing about Kitchelt's implied statement that black women and men received the same wages as white men while white women earned less is a mystery, but Kitchelt continued to mull over the idea of using this approach on behalf of the ERA. Four years later she drafted a leaflet, entitled "The Constitution protects Negroes but not Women in demands for equal pay in public schools," and sent it to Constance Baker Motley, then a legal research assistant at the NAACP, for comment. Motley found the title "generally not offensive" but suggested that Kitchelt make clear that she approved of the Supreme Court's upholding of equal pay for black people.[165] Alma Lutz was less approving. She pointed out that, despite the law, black people suffered worse discrimination than white women, and she warned that "we will be regarded in some quarters as anti-negro—which we are not."[166] Even a well-intentioned liberal such as Kitchelt, who considered adding race to the ERA and sought NAACP representation on her Connecticut Committee, could place white women in competition with blacks in the struggle for equality.

Kitchelt was by no means the only women's rights supporter to take this approach. National Woman's Party members in California, seeking to add sex to a state anti-discrimination bill and threatening to work against it if women were excluded, believed that "colored and Jewish men" stood against them.[167] Rose Arnold Powell deplored the fact that white men legislated for the benefit of black people, "apparently undisturbed over the degradation of their own mothers, wives and daughters by this action."[168] The National Woman's Party sent out a letter in 1957 that asked: "Doesn't all the 'hullabaloo' about the civil rights of certain racial, color, religious and national groups sound fantastic in face of the fact that these groups have already received the guarantees of the Constitution, while the women of the nation have no Constitutional rights except the right to vote?"[169] Alice Paul tried to win the support of Virginia senators by emphasizing the discrimination against white native-born American women. Another Woman's Party member found it "passing strange that women are actually as far as the Constitution is concerned, inferrior [*sic*] to the Negro."[170]

Another incident further suggests that the women's rights movement saw itself in competition with the civil rights movement. In 1945, Nora Stanton Barney, feminist granddaughter of Elizabeth Cady Stanton, wrote a letter to the editor of the New York *Herald Tribune* that sparked great controversy among women's rights supporters. Although Barney prided herself as a "champion of equality and justice for all races, creeds and colors," she attacked a proposed bill to give the Fair Employment Practices Commission, the government agency set up in response to demands from organized black groups, the right to set up courts to hear complaints of discrimination on the basis of race and religion. "Other elements in our population, notably the Hebrews and the Negroes, are asking to have teeth put in the civic rights that they are fully guaranteed by the Constitution, well knowing that women have no civic rights," she complained. Black, Jewish, labor, and Communist groups turned a deaf ear to women's pleas for equality, she accused, and she feared that this situation would "fan racial animosity."[171]

Response to Barney's letter was immediate and vehement. Rebecca Hourwich Reyher, a Jewish militant suffragist and ex-Woman's Party member, fired off a telegram to Alice Paul that denounced Barney's "appeal to Protestant white women to regard themselves as a special class apart from other women" as a reflection of the totalitarian age.[172] She accused Barney of pandering to race and class hatred in order to gain support for the ERA, something she found shocking and disgraceful. Another Woman's Party member, also so horrified that she sent a telegram to Alice Paul, believed that the Nazis could not have written a more anti-American letter. Barney's response to her critics took a more moderate tone, and she emphasized her desire to include sex-based discrimination in the bill rather than her resentment of the ostensible privileges of blacks and Jews. But her hostility was clear in other letters. She wrote to ask the governor of New York why he bowed down to the rabble and signed a state anti-discrimination bill when the Republican Party could not get the votes of the CIO or the "Colored" anyway, since they were "almost all definitely to the left." She warned that "the resentment of women is mounting day by day as they find themselves being pushed around and placed on the lowest economic level, below the Negroes and the Jews."[173] To a sister Woman's Party member, Barney explained that black women were demanding white women's jobs.

Some women who responded to Barney saw nothing offensive about her letter or accused her at most of a lack of tact, but Alma Lutz warned that, especially in light of the "strong anti-Jewish feeling" at Woman's Party headquarters, the group should take care to keep free of the anti-black and anti-Jewish groups fighting the FEPC bill. That Lutz's fears were not groundless was borne out by the positive response Barney received. One man praised her letter and indicated his desire to send her a copy of his book, *The Lost White Race.* A woman wrote to express her agreement that it was "frightful that negro and jewish women have protection that our real American women, native born, with our Anglo-Saxon heritage, have not."[174] Although friends defended Barney as a well-known liberal "entirely free of any race or religious prejudice," Barney clearly used racist and anti-Semitic arguments to try to win support for the ERA. Her view of rights won by any other group as a loss for white Christian women was typical of the women's rights movement's response to other liberation movements.

This kind of competition over equality was a consequence of the renewed activity and early successes of the civil rights movement in the period after 1945. In addition to competing with the efforts of civil rights activists women's rights supporters attempted to "piggy-back" on the successes of the civil rights movement. Edith Goode, for example, a national Woman's Party member and a close friend of black leader Mary McLeod Bethune, represented the Woman's Party at the San Francisco conference that established the United Nations. A member of the NAACP almost from its beginning, she met with NAACP official Walter White and asked him to be sure that wherever a document outlawed discrimination on the basis of race, it also forbade discrimination on the basis of sex.[175] This was the same tactic that would be used successfully in the case of the Civil Rights Act of 1964.

Sometimes, however, the backlash to the civil rights movement was used. Alice Paul, for example, believed that the Woman's Party benefitted from the support of those opposed to the civil rights movement. In 1956, she and two other Woman's Party members had lunch with several Southern senators to discuss equal rights for women. Paul's sister Helen reported that the Southerners seemed "to be galvanized into action probably due to the fact that the Republicans have been basing their appeal for votes very largely on Civil Rights for negroes. This seems to have made the southern group sympathetic with our cause of civil rights for women."[176] Although women's rights supporters sometimes realized that the general interest in civil rights would help their cause, their racism could make them afraid. A friend of Alice Paul's, for example, advised her to be careful during the 1963 march on Washington: "I shudder at the thought of any 100,000 persons' 'march on the Capitol' let alone Civil Rights negroes."[177] A Woman's Party convention scheduled in Washington two days before that march had to be canceled due to the exodus of local members and the reluctance of members who lived elsewhere to come into the city. Yet, at the same time, the Woman's Party, as an affiliate of the National Council of Women, endorsed the march and women's representation in it under a banner reading "Equal Rights for Women Now."[178]

What is clear about the relationship between the women's rights movement and the civil rights movement in this period is that, despite a diversity of views on the issues, the women's rights movement often sought to make whatever alliance might further its cause. A controversy within the National Woman's Party during World War II over the propriety of associating with an anti-Semitic group suggests that for that organization, once again, the acid test was simply support for the amendment.

In 1943, the Woman's Party list of organizations that had endorsed the ERA included a group called We, the Mothers, Mobilize for America, Inc., one of several "mothers' groups" that sprang up at the time of the war to oppose American involvement in the fighting. These groups of largely working-class women based their opposition to the war on their desire as mothers to protect their sons, but they grew out of and maintained contact with more traditional right-wing isolationist groups.[179] When Florence Kitchelt read about the right-wing, to her mind subversive, politics of We, the Mothers, she suggested to Alice Paul that the organization be eliminated from the Woman's Party list of endorsing organizations. Paul took the position, however, that the campaign for the ERA benefitted from the support of any organization, regardless of its politics, and pointed out to Kitchelt that some Woman's Party members would like to eliminate the National Association of Colored Women from the list. Elizabeth Forbes, a Woman's Party member who belonged to We, the Mothers and represented it on the Women's Joint Legislative Committee, defended her organization in a letter to Kitchelt. She found the phenomenon of the mothers' organizations encouraging to a feminist because they mobilized previously nonpolitical women, mostly housewives, from the lower income brackets. She considered it quite a coup to have gotten their cooperation on the ERA "because they are apt to consider us radical."[180]

Two years after this, the issue resurfaced when another National Woman's Party member objected to a different mothers' group that also belonged to the Women's Joint Legislative Committee, where it too was represented by a Woman's Party member. Katharine Norris, the convenor of the Women's Joint Legislative Committee, consulted with several women in an attempt to decide what to do about the two groups. Alice Paul's response to the information that We, the Mothers was anti-Semitic was that everyone was entitled to his or her own opinions and that only the group's endorsement of the ERA need concern the Woman's Party. Norris herself agreed that "we should be tolerant even of intolerance," since her only concern was the effect on the Joint Legislative Committee, the Woman's Party, and the ERA.[181] Even Anita Pollitzer, chairman of the Woman's Party and herself Jewish, tended to defend the groups since she valued the lobbying of their representatives. Eventually Norris asked the chairman of the Joint Legislative Committee to appoint a committee to consider the problem.

Not all of the women involved, however, believed that an endorsement was worth any price. Dorothy Shipley Granger, the founder of the St. Joan Society, a group of Catholic women for the ERA, believed that association with the mothers' groups would spell doom for the amendment: "[W]hen leaders of a cause are so bankrupt for followers that they allow groups which are known to be subversive affiliate with them, and flaunt their disregard for public opinion—especially in the period of a war—it is foolhardy to continue association with them," she argued.[182] Clearly more concerned with their isolationist than their anti-Semitic views, she pointed out the close connection of Woman's Party members in Maryland with America First, the right-wing isolationist group, and asserted that the open statement of reactionary politics had hurt the Woman's Party.

Emma Guffey Miller, who had opposed association with these groups from the beginning, argued that the fact that the representatives of the two groups admitted they were anti-Semitic was sufficient reason to drop both from the Joint Legislative Committee. She, unlike so many other women's rights advocates, insisted that one could not be anti-Semitic and in favor of equality under the law at the same time. When Katharine Norris finally read the publication of We, the Mothers, its anti-Semitism shocked her into two sleepless nights, and she realized that the group did not even believe in equal rights for women. But still the Joint Legislative Committee made no decision. Elizabeth Forbes, who represented We, the Mothers, asked the House Un-American Activities Committee to hold hearings to determine who was responsible for spreading lies about the members of the board of another isolationist group that, like We, the Mothers, she claimed was anti-Communist but not anti-Semitic. She announced that she had many good Jewish friends, but she worried that even though Pollitzer had not objected to We, the Mothers "the fact that our chairman is now a Jewess would not be overlooked" in the event that the Committee asked the group to resign. Finally, in 1947, We, the Mothers took the problem out of the hands of the wavering Joint Legislative Committee by withdrawing, having been informed that the supporters of the ERA were "advocating communistic doctrines."[183]

This incident is important because it illustrates the opportunism of the Woman's Party and the Joint Legislative Committee. Alice Paul and others believed strongly that any individual or group that supported the ERA, regardless of views on other issues, should work with the Woman's Party and would contribute to the struggle. Other members likewise considered devotion to the ERA the only issue but believed that groups with poor reputations would harm rather than advance the cause. Only a few members seemed to believe, like Emma Guffey Miller, that a commitment to equality meant a belief in the equality of all people.

As in the case of the labor movement, the women's rights movement made no real attempt to connect in a positive way with the burgeoning civil rights movement. Although individual women supported organizations devoted to winning racial equality—and although women like the executive secretary of the Woman's Party wrote to another member in 1951 that "[w]e need all women, white, black, yellow, red. We need all women Protestants, Catholics, Jews, Mohamedans"—the women's rights movement as a whole did little to forge such alliances.[184] When Susan B. Anthony II criticized the leaders of women's organizations for failing to "tie up the status of women with the oppressed, because of color or class," she held up a very modern feminist vision of a world free of both racial and sexual inequality.[185] She was, as we have seen, denounced by the House Un-American Activities Committee for such views. What is significant is not so much the attitude of women's rights supporters, for they were probably neither more nor less racist than other Americans, but the ways in which the women's rights movement responded to the challenge of the civil rights movement. Not until the founding of the National Organization for Women in 1966—which itself received impetus from and to some extent grew out of the civil rights movement—and the emergence of the radical branch of the women's movement in the late 1960s were organizational connections between the women's rights movement and the civil rights movement forged. In the 1950s, for the most part, segregation, tokenism. competition, and opportunism prevailed.

Conclusion

In the three cases examined here, the women's rights movement attempted to use the issues and activities associated with other social movements for its own purposes, but did not attempt to mobilize groups of women or forge alliances with other movements. The National Woman's Party, in particular, emerges as an organization that evaluated any individual or group solely on the basis of support of, or opposition to, the ERA. The Woman's Party sought to co-opt McCarthyism to support the amendment but also would have welcomed an endorsement from the Communist Party. Women's rights supporters saw labor as an enemy and tried to neutralize its opposition, and tried to compete with or piggy-back on the successes of the civil rights movement. The consequences for the women's rights movement was a kind of isolation from other groups. Both the hostile social environment and the nature of the women's rights movement made direct mobilization or true coalition-building unlikely.

Not until the evolution of a new stage of the women's movement, with a broader and more diverse base and access to a range of resources, could the movement attempt to mobilize large numbers of women and build coalitions with the labor and civil rights movements. Although the women's movement in the 1980s remains in practice far from its theoretical goal of uniting all women into a powerful egalitarian movement, it has at least begun to grapple with the issues of class and race, and it has to some extent adopted the strategy of coalition with other progressive movements for social change. How this happened is a story that carries us into the 1960s.

8

Making Connections:
The Resurgence
of the Women's Movement

By the end of the 1960s, a resurgent women's movement had taken up the fight against gender-based inequality in American society. The founding of the National Organization for Women in 1966—the beginning, according to many accounts, of the contemporary women's movement and the end of our exploration of the women's rights movement of the post–1945 period—came about through the convergence of women from different backgrounds, all of whom were concerned about women in American society. The establishment of NOW thus serves as a useful signpost in mapping the course of the transition from the women's rights movement of the 1940s and 1950s to the resurgence of a broader women's movement in the 1960s and 1970s. NOW brought together labor union activists, government employees, and longtime feminist activists and quickly took leadership of what has come to be called the liberal branch of the women's movement. At the same time, some of the women involved in the civil rights and New Left movements began to apply their more radical critique of American society to their own lives and experiences in and outside of their movements, thus launching the more amorphous and locally organized radical branch of the movement. Analyses of the women's movement in the 1960s have seldom recognized the connections the already existing women's rights movement made with the newly resurgent movement.

The President's Commission on the Status of Women

When President John F. Kennedy announced the formation of his Commission on the Status of Women in 1961, he implicitly recognized the existence of gender-based discrimination in American society. The mere establishment of the body is probably the best evidence that the women's rights movement in the 1950s made some impact on American society. For the Commission was to a large extent a response to the challenge of the existing women's rights movement and, at the same time, an important factor in the resurgence of a broader women's movement.

The concept of a commission on women's status in American society was not new in the 1960s. Opponents of the ERA had supported Congressional legislation to establish such a commission in the 1940s and 1950s, hoping that a body to review the status of women and declare a national policy against discrimination would serve as an alternative to the amendment that they believed would harm women by eliminating protective legislation.[1] Katherine Ellickson, who became the executive secretary of the President's Commission, in her 1976 unpublished history explained the formation of the Commission as a reaction to the increasing political activity of women both for and against the ERA, as well as to the growth of the civil rights movement.[2]

ERA opponent Esther Peterson, head of the Women's Bureau and a top woman in Kennedy's new administration, revived the idea of a commission when the activity of ERA supporters raised the issue of the amendment in the 1960 election. The National Woman's Party had been especially active in this regard, but the national legislation chairman of the BPW had also reminded the presidential candidates and all candidates for Congress of both parties' support for equal rights.[3] The National Woman's Party, of course, regularly fought for inclusion of the ERA as a plank in both major party platforms and solicited support from presidential candidates. In 1960, the Democratic platform, for the first time since 1944, carefully avoided a specific commitment to the ERA. Despite the efforts of ERA supporters, the 1960 platform plank supported "legislation which will guarantee to women equality of rights under the law including equal pay for equal work." Although National Woman's Party president and Democratic committeewoman Emma Guffey Miller insisted that the plank favored the ERA and represented an improvement over the 1956 plank because it included the important word "guarantee," in fact the plank deliberately avoided mentioning the amendment, a consequence of strong opposition from high-level Democratic women, especially Peterson.

When Kennedy received the nomination at the convention, Miller forged ahead in an attempt to gain an endorsement from him, despite the fact that he had long refused to commit himself on the question, making it quite clear to other Woman's Party members that he was not a supporter.[4] Nevertheless, Miller triumphantly proclaimed his support, quoting a letter over his signature that assured her he would interpret the platform, "as I know it is intended, to bring about, through concrete actions including the adoption of the Equal Rights for Women Amendment, the full equality for women which advocates of the equal rights amendment have always sought."[5]

This puzzling announcement of support was not, however, genuine. Esther Peterson, who worked closely with Kennedy in his campaign, had drafted his response, which did not include the phrase "including the adoption of the Equal Rights for Women Amendment" in the portion of the letter quoted above. Her files include two versions of the letter, one with and one without the offending phrase. Peterson later privately denounced the revised letter as a "forgery" and blamed it on a "mishap of the headquarters office" that she had not wanted to discuss openly because it reflected badly on the operation of the campaign. Notes in Peterson's papers from the 1970s make clear her conviction that a National Woman's Party member in the campaign office had added the phrase and taken the letter to the press.[6] According to Cynthia Harrison, Miller herself had amended the letter and used her connections at the Democratic National Committee to have it typed on Kennedy letterhead and signed, presumably by machine.[7] To Peterson's fury, the Woman's Party took full advantage of the altered letter. Emma Guffey Miller, after the election, continued to act as if Kennedy might indeed support the amendment. She assured Alice Paul that Kennedy would do his best to carry out his promises and took heart from what she interpreted as Lyndon Johnson's hearty support of the Woman's Party.[8] She wrote to Kennedy to call for the appointment of a Secretary of Labor in favor of legal equality for women and fired off a telegram to object when Esther Peterson was named head of the Women's Bureau.[9] Miller, like other women's rights supporters, soon complained bitterly about Kennedy's poor record of appointing women to high-level offices. But her wrath, like that of other ERA supporters, continued to be directed primarily against Esther Peterson.

In response to this situation, Peterson, with a great deal of effort, persuaded Kennedy to appoint the Commission on the Status of Women.[10] Scholars who have written about the President's Commission are agreed that it represented in part an attempt to sidetrack the ERA, and as one pointed out, its existence suggests that there was enough pressure for the ERA to force its opponents to propose countermeasures.[11] Peterson met with a group of union women to discuss the possibility of establishing a commission to study discrimination against women. Katherine Ellickson, an AFL-CIO official who drafted the proposal and who then became the commission's executive secretary, urged that such a commission focus on the status of women rather than on discrimination. According to Peterson's notes, Ellickson argued that women would be in a better position to advance their cause if they stayed away from the feminist view. She feared that, if the commission included "equal righters," they would insist on issuing minority reports.[12] Peterson used the threat of the ERA, and the issue of the Kennedy letter, to win support from Kennedy for the idea of a commission. She complained that "[a]lmost no day passes but what I have to draft some letter" about the ERA. "I will continue to do this," she wrote to one of Kennedy's aides, "but I do think you should be aware that the Equal Righters, since they have nothing else to do, are intensifying their campaign in every State." She hoped that the commission, which she explicitly labeled a counterproposal, would solve her problem.[13]

In her proposal for a Commission on Women, presented to her immediate superior, Secretary of Labor Arthur Goldberg, she stated that it would "substitute constructive recommendations for the present troublesome and futile agitation about the 'equal rights amendment.' "[14] She named the Business and Professional Women as the key proponent of the amendment but, having identified a strong anti-ERA sentiment in the BPW, indicated that she believed that the establishment of a commission would get the organization off the hook and thereby help to turn the tide against the amendment. National Woman's Party members wrote to Goldberg to try to win an endorsement of the ERA, and Emma Guffey Miller, repeating the strategy she had used with Kennedy during the election, chose to interpret his vague statement in favor of equality as support for the ERA. She encouraged her coworkers to send letters of thanks to Goldberg, taking delight in the fact that "[s]uch letters have certainly gotten under the skin of some of the anti women in his department."[15] Goldberg recommended the establishment of a commission as proposed by Peterson at the end of 1961 and, echoing Peterson's words, described it as an alternative to the ERA.[16] In December, Kennedy announced the establishment of the Commission.

Kennedy named "the first lady of the world," Eleanor Roosevelt, to head the Commission and Esther Peterson to serve as executive vice chairman. Peterson herself chose the Commission members.[17] Prominent men from a variety of fields joined women from Congress, higher education, the labor movement, and women's organizations. Members included official representatives of the National Councils of Negro, Jewish, and Catholic Women, but noticeably none from the pro-ERA women's groups. ERA supporters, especially from the Woman's Party, complained of their exclusion, despite the appointment of one ERA advocate, Marguerite Rawalt, a former president of the BPW and, at least on paper, a National Woman's Party member. Esther Peterson knew of Rawalt's position on the ERA but considered her "open to reasonable argument."[18] But the Woman's Party was not impressed; one member accused Rawalt of having done "nothing for our cause" as president of the National Association of Women Lawyers and dismissed her as an opportunist and "no friend of ours."[19]

Pro-ERA women tended to see the Commission as an attempt to placate them and avoid passage of the amendment.[20] Emma Guffey Miller denounced the Commission as "anti-equal rights," a plot of Esther Peterson to defeat the ERA.[21] She planned to contact some of the Commission members whom she believed to be sympathetic to the ERA to "start some trouble."[22] Another Woman's Party member described the Commission as "a 'loaded dice and packed deck' commission to see that ERA is not passed."[23]

The National Woman's Party tried valiantly to get Emma Guffey Miller appointed to the Commission throughout its term. David Lawrence, the Democratic governor of Pennsylvania who had swung his state's delegation behind Kennedy at the 1960 convention, requested Miller's appointment, taking the tack that her failure to be named must have been an "inadvertent but unfortunate" omission, since "it would seem to me natural to include the woman who heads the only national women's organization exclusively concerned with advancing the status of women."[24] Kennedy promised Lawrence that Miller would be appointed to the committee that would grapple with the ERA, but she never was. Miller suspected Peterson's opposition, but Peterson claimed that Edith Green, the Congresswoman who headed the committee, refused to have Miller.[25] The Woman's Party, without success, continued to push Miller's appointment into the last year of the Commission, hoping she might at least participate in the writing of the final report. When Kennedy invited Miller to serve on the President's Committee on the Employment of the Handicapped in 1963, she declined, explaining that she had devoted her life to the advancement of women and was best qualified to sit on the Status of Women Commission.[26]

    Supporters of the ERA in the women's rights movement were cognizant of the anti-ERA complexion of the President's Commission and as a result dismissed the possibility that the Commission might help women in other ways. Alice Paul predicted, accurately as it turned out, that the establishment of the body would be used to delay action on the ERA in Congress until the report appeared in 1963.[27] Just two days after Paul made this prediction in a letter to a sister Woman's Party member, Esther Peterson urged the deputy attorney general to avoid any official pronouncements on the ERA until the Commission had made its recommendations.[28] Woman's Party members continued to lobby for the ERA throughout the tenure of the Commission, but they frequently encountered legislators unwilling to anticipate the report by taking a stand.

    The Commission met for the first time in February 1962. Kennedy welcomed the members with the remark that he had established the Commission for two reasons: one was that the question of women's status had taken on national and international importance since women had come to make up one-third of the labor force, and the other was to protect himself from reporter May Craig, who asked him every two or three weeks what he was doing for women.[29] Craig, famous for her flowered hats and tart questions, was a longtime feminist activist and National Woman's Party member who bombarded presidents from Eisenhower to Johnson with questions about equality for women, winning encouragement and gratitude from her colleagues in the Woman's Party. Although Kennedy got the expected laughter from his reference to Craig's insistent questions, the national attention that Craig helped to focus on women's status was taken quite seriously by women's rights advocates.

    The Commission tackled a wide range of issues concerning women although, interestingly enough, it, like the women's rights movement, avoided the issue of birth control as too controversial.[30] The ERA was, in this context, controversial enough. At the first meeting, after a good deal of talk about protective legislation as one of six designated areas for study, the representative of the Commerce Department, with apologies for rushing in like the proverbial fool, mentioned the ERA for the first time. Pointing out that it was the issue that had divided the country for so many years, he expressed the hope that the Commission could reach some accommodation on it. If this suggests a surprising familiarity with the ERA, the comment of the president of the National Council of Jewish Women confirms the impression that the members of the Commission understood the relevance of the ERA, if not always correctly. She reported that she had heard rumors that the Commission had been organized to put through the ERA, a statement which Eleanor Roosevelt, Esther Peterson, and Marguerite Rawalt all found strange since the opposite bias was evident to informed women. But all hoped that the Commission would help to heal existing divisions.[31]

Despite such hopes, the ERA remained divisive throughout the history of the Commission. One member believed that the ERA was "really the heart of our work" and the basic item on the agenda.[32] The Committee on Civil and Political Rights had the responsibility for hearing arguments on both sides and making some recommendation. Representatives of women's organizations and labor unions sent statements or appeared in person before the Committee. The BPW presented a brief in favor of the ERA, and women's historian Miriam Holden drafted an argument in support of the amendment.[33] Although the Woman's Party never succeeded in winning a place for Emma Guffey Miller on either the Commission or the Committee, it, like other women's organizations, sent representatives to present pro-ERA arguments. That the Woman's Party's well-known support for the amendment entered into the deliberations is clear from the attitude of Pauli Murray, a black lawyer who served on the Civil and Political Rights Committee and who favored an attempt to use the Fourteenth Amendment to win women a guarantee of equality through the courts. Murray hoped that "we could steer a complete path between not opposing the equal rights amendment and not bringing down the wr[a]th of the very influential Women's Party . . . upon our fresh green heads."[34] But the Committee could not get around the basic issue. As one member stated, taking no position would be "like a book about elephants that doesn't talk about elephants," since "some feel that this is one of the reasons why this Commission was constituted."[35] Eventually the Committee, which included ERA supporter Marguerite Rawalt, opponents of the amendment, and advocates of an alternative method such as Pauli Murray, agreed to the statement that appeared in the Commission's final report: "Since the Commission is convinced that the U.S. Constitution now embodies equality of rights for men and women, we conclude that a constitutional amendment need not now be sought in order to establish this principle."[36] Marguerite Rawalt was responsible for the word "now" in the phrase "need not now be sought," and that represented the most that she could do as an ERA supporter on the Committee.[37]

To no one's surprise, then, the Commission report tried to dampen the ardor of ERA advocates, but the impact of the Commission was far more complex. Its recommendations put a new weapon in the hands of ERA opponents, who after 1963 could cite the Commission's findings in support of their position on the amendment. Chief Senate opponent Carl Hayden, for example, replied to an inquiry from an ERA supporter that "[i]n view of the recommendations made in the Commission's report, I doubt that it would be possible to obtain a two-thirds majority" in Congress.[38] Two years after the issuance of the report, President Lyndon Johnson responded to a letter from Emma Guffey Miller by citing the President's Commission recommendation.[39] National Woman's Party members recognized the significance of the report's recommendation (or lack of one) on the ERA and expressed their chagrin. "What a disappointing document the Commission on the Status of Women produced!" exclaimed one member.[40] Emma Guffey Miller labeled the report "nothing but a compromise," a consequence of the fact that Esther Peterson "loaded" the Commission.[41] On the cover of her own copy of the report, Miller summed up her verdict: "A contradictory report from a packed committee."

But ERA supporters did not give up because of the Commission's work. One member suggested that the Woman's Party put out a counterreport.[42] Jane Grant, a Lucy Stone League member, praised the Woman's Party for a newly revised pamphlet and hoped that it would help counteract the effects of the Commission report.[43] Alice Paul believed that the best way to meet any new opposition or any weakening of existing support would be to show an immediate increase of strength for the amendment, and she worked to add sponsors in the House.[44]

The ERA, of course, was not the only issue with which the Commission grappled, but ironically the Commission did give a boost to the ERA by serving as a stimulus for the growth of the women's movement. Commission member Margaret Hickey believed that the report sparked a "woman hunt" in Washington, the objective of which was "not to give women special privilege but to open up new opportunities . . . so that women will be given a chance to work as equal partners with men throughout society."[45] Katherine Ellickson, executive secretary of the Commission, commented in her unpublished history that it "furnished stimulus, institutional frameworks, and content for much of the subsequent growth of the women's movement," and other commentators have agreed.[46] It did this by raising the issue of women's status to the level of national discussion and by creating an institutional structure that brought together women not previously involved in feminist activity.

Katherine Ellickson reported that the work of the Commission was widely publicized throughout the country, although not so much through the media as by distribution of copies of the final report, which became a bestseller. It is not surprising that the media often failed to accord the Commission serious consideration, given the media's traditional approach to women's issues. Ellickson lamented the humorous, condescending, or sexually tinged reports in the press and complained that the stories that were printed often appeared only in the women's sections of newspapers. But a presidential commission could not be completely ignored, and another knowledgeable observer reported that "the work of the Commission is having a very worthwhile effect" in pointing out the existence of discrimination and the need for change.[47] Throughout the existence of the Commission, women in a variety of situations wrote to the president or the Commission itself in praise of the national attention focused on the status of women. A woman worker in the Springfield Armory in Massachusetts, for example, congratulated Kennedy for giving women a fair chance; a woman newspaper columnist in Illinois added her recommendations to the "flood of letters from latter-day Susan B. Anthonys" that had descended on the Commission; a thirteen-year-old girl who wanted to be a Senate page requested Kennedy to refer her problem to the Commission; and a woman with a full-time job offered to volunteer on nights or weekends to help the Commission in any way she could.[48] While clearly no representative sample of public opinion, these letters do suggest that some women saw the Commission as a body able to bring change to women's lives.

What changes those would be concerned both opponents and supporters of the ERA. Esther Peterson worried that the Commission's deliberations would seem to support the feminist position. She hoped that Kennedy would familiarize himself with the report, since he had earlier referred to the "Commission on Equal Rights," a blunder that disturbed Peterson and other women leaders opposed to the ERA.[49] Peterson sponsored a meeting of women's groups, after the report appeared, to discuss implementation of the Commission's recommendation with regard to winning a judicial reinterpretation of the Fifth and Fourteenth Amendments as applicable to women. The president of the BPW invited the National Woman's Party to attend, and Alice Paul hoped that the group could be used to support the ERA, or at least to prevent organized opposition to it, if pro-ERA organizations stayed in it.[50] Esther Peterson made sure to send a copy of the Commission report to Carl Hayden, thanking him for his support of protective legislation and hoping that the ERA could be deferred while the approach recommended by the Commission was tried out. If all else failed, she urged Hayden to reintroduce his nullifying rider.[51]

Perhaps even more significant for the growth of the women's movement was the continuing institutional structure that the Commission created. Participation in the work of the Commission played a crucial role in the development of feminist consciousness for several key figures. Catherine East and Mary Eastwood, both government employees assigned to assist the Commission, played important roles in the resurgence of the women's movement. Pauli Murray described the experience as "high level intensive consciousness-raising." She had had personal experiences with sex discrimination, but she described the Commission as "kind of heaven for me, this was like throwing brer rabbit in the briar patch, because it was the first time in my life that I had really sat down and researched the status of women."[52] Historian Miriam Holden heard Murray speak at a 1962 All-Women Conference sponsored by the National Council of Women and reported to Alice Paul that Murray was the most impressive speaker at the conference and an "ardent advocate for equal rights."[53] Clearly the Commission brought together women interested in women's status, raised their consciousness in some cases, and created a structure through which they could continue to meet.

In its final recommendation, the Commission asked Kennedy to establish both a permanent commission of executive department and agency heads and an advisory committee of private citizens. These took form, respectively, as the Interdepartmental Committee on the Status of Women and the Citizens' Advisory Council on the Status of Women. In addition, the BPW, with the approval of the Commission, urged its local branches to set up state commissions on the status of women throughout the country in conjunction with other women's organizations. The National Woman's Party leadership could not understand why the pro-ERA BPW would want to perpetuate the anti-ERA Commission, but eventually the Woman's Party president, Emma Guffey Miller, welcomed an appointment to the Pennsylvania Citizens' Advisory Council on the Status of Women and used it to propagandize for the ERA.[54] Even opposition to the ERA from the state commissions seemed to one Woman's Party member a hopeful sign that the anti-ERA forces were feeling the pressure that ERA supporters put on politicians and lawmakers.[55]

At the same time that the BPW worked to set up state commissions, Kennedy appointed members to the new national-level Citizen's Advisory Council. Among them was one Woman's Party member, New Jersey banker Mary Roebling. One women's rights supporter found this group a "curious assortment of persons," with Roebling "strangest of all."[56] Emma Guffey Miller commented that Roebling "has never been at all militant."[57] The National Woman's Party continued, throughout the years and administrations, to inform this group on the ERA and rejoiced when it endorsed the amendment in 1970.

Even at the time, members of the President's Commission recognized that their work had both grown out of and in turn stimulated activity among existing women's organizations. During Committee discussions, Pauli Murray asserted that everyone associated with the Commission in any way would act as a catalytic agent.[58] It is not surprising, then, that the Commission has been identified as a key institution in the growth of the liberal branch of the women's movement in the 1960s. Although, as Katherine Ellickson noted, she and Esther Peterson "were not primarily concerned with feminist efforts," the Commission that they shaped had important feminist results.[59] Commission members whom Ellickson contacted in 1976 in the process of writing her history of the body agreed that "our Commission did much to shape the emerging women's movement."[60] Historian Caroline Ware identified the Commission as "the turn-around point from the anti-feminist swing of the post-war years, first to stock-taking and then to new dimensions for action."[61]

The Equal Pay Bill

The President's Commission also had consequences for another landmark in the history of women's struggles in the twentieth century: the Equal Pay

Bill of 1963. The idea of equal pay for equal work was by no means new—the Women's Bureau and its allies had been fighting for such a bill since 1945—but victory for such a simple and self-evidently just principle both grew out of and contributed to the growing awareness in American society of gender-based discrimination.[62]

During World War II, the government supported the principle of equal pay for equal work in industry, reversing the precedent set by the National Recovery Administration's Depression-era codes, which institutionalized existing wage differentials. Throughout the postwar years, the Women's Bureau spearheaded the attack on unequal pay, with little success. The Federation of Business and Professional Women, too, worked for an equal pay bill, the earliest objective of the organization and one that had received continuous support since the group's founding in 1919. While no women's rights supporter would have opposed the principle of equal pay for equal work, the conflict over the ERA created an odd situation. While the BPW supported equal pay, the National Woman's Party distrusted those who worked for the issue, believing that it was less important and a distraction from the real issue, the ERA. A California member of the Woman's Party, for example, reported in 1945 that the state BPW insisted on introducing an equal pay bill, despite her attempts to talk them out of it. While she did not oppose the bill, she planned to let the BPW do most of the work and spend its money on it.[63]

When the Women's Bureau sponsored a National Equal Pay Conference in 1952, the National Woman's Party was not invited to participate. Nevertheless, one member represented the group as an observer and reported that Women's Bureau director Frieda Miller treated her warmly until she learned of her Woman's Party affiliation. When the member inquired about the ERA, she was ruled out of order with what she interpreted as clear hatred from Miller.[64] After the conference, Miller received a report that another Woman's Party member, denied a badge, ' more or less crashed the conference.'"[65] Without Woman's Party participation, a committee of representatives of women's organizations concerned with the issue of equal pay began to meet regularly. Only the BPW, which offered secretarial services to the new committee, combined active work for equal pay with support for the ERA. Although the executive secretary of the Woman's Party suggested to Alice Paul that the organization publicly support equal pay, a move that she believed might help win the support of the AFL and CIO for the ERA, the Woman's Party continued to maintain its distance from the issue.[66]

When the Equal Pay Bill, which the President's Commission had strongly recommended, passed Congress in 1963, Emma Guffey Miller insisted on referring to it as the "so called Equal pay bill."[67] Another member complained that equal pay "was certainly a smart dodge for the people who want to spike the movement for real equality."[68] No doubt the real reason for the hostility of the Woman's Party was the long-standing feud with the Women's Bureau. Although the NWP's one-issue stand would have kept it from working actively for equal pay in any case, the years of distrust of the Women's Bureau made the Woman's Party unlikely to cooperate. Likewise, Esther Peterson remarked in a 1979 interview that both she and Eleanor Roosevelt had continued to oppose the ERA for years because the Woman's Party fought them on equal pay.[69]

Despite the conflict over the Equal Pay Bill and its limited effect on women's wages, it represented a significant first step toward winning the government's commitment to eliminating gender-based discrimination. As a result, the passage of the bill, like the establishment of the President's Commission, focused attention on women's issues and brought together women from government, labor, and women's organizations.

Title VII of the Civil Rights Act

Perhaps the most peculiar episode in the history of the women's movement occurred one year after the passage of the Equal Pay Bill. In 1964, Congress passed a Civil Rights Act that included a prohibition against discrimination in employment on the basis of sex. Title VII, the only section of the bill to include sex as well as race discrimination, is a cornerstone of the women's movement; it continues to serve as the major piece of legislation safeguarding women's employment rights, and the issue of its nonenforcement by the government became an early rallying cry for the liberal branch of the movement in the mid-1960s. Yet most interpretations of the process by which Title VII came to include gender-based as well as racial discrimination emphasize the fact that the Civil Rights bill's opponents introduced and supported this amendment, including sex discrimination in an attempt to ridicule and defeat the Civil Rights Act by dividing its supporters. But the story, as several commentators have recognized, is much more complex and shows that Title VII was not just an unexpected gift to the women's movement.[70]

The idea of linking sex and race in the Civil Rights bill did not suddenly occur to Howard Smith, the conservative Virginian who introduced the amendment on the floor of the House of Representatives. The National Woman's Party, as we have seen, had already tried to benefit from civil rights activity in the 1950s. For example, in 1955 the Woman's Party protested an executive order that prohibited discrimination on the basis of race, religion, and national origin—but not sex—by those who held government contracts.[71] Eight years before the famous debate on Title VII, the House of Representatives debated a similar amendment to a civil rights bill, at Alice Paul's instigation. Howard Smith even spoke in support of the amendment.[72] So in late 1963, in anticipation of the debate on the new civil rights bill, the National Council of the Woman's Party passed a resolution calling for an amendment of the proposed bill to include discrimination on the basis of sex. Two members then wrote to their representative, Howard Smith, to suggest that a sex amendment would divert some of the pressure for passage of the bill. Presumably the Woman's Pary was trying to persuade Smith that a sex amendment would serve his interests, but the members who wrote were themselves both sincere about wanting to include sex discrimination and by inclination opposed to the civil rights movement.[73]

Smith responded to the letters from the Woman's Party members with an indication that he expected someone to make an amendment from the floor during the debate, but he did not offer to make it himself. This in itself indicates that the idea of including sex discrimination was not such a far-fetched idea in 1964. In January, he mentioned the sex amendment in a committee hearing, commenting that the "National Woman's Party were serious about it."[74] Late in January, Smith appeared on "Meet the Press," where the persistent May Craig asked him about the amendment. He said that he might introduce it. And, when the House took up debate of the bill, he did. Smith's presentation and the subject itself provoked a great deal of laughter, contributing to the legend of what came to be known as "Ladies' Day in the House."

Smith's motives in all of this are not totally clear. While it is certainly true that he wanted to defeat the Civil Rights Act, it is also true that he had served as a sponsor of the ERA since 1943. The tone of his remarks in the House debate suggested that he intended his amendment only as a weapon against the entire bill, and Martha Griffiths, the representative who fought hard for the amendment, reported that Smith explicitly told her that he meant the amendment as a joke. On the other hand, Smith also denied that he was insincere and his legislative assistants reported that he would not have voted for something in which he did not believe.[75] Women involved in the fight for Title VII tend to believe that Smith may have acted in part out of a kind of Southern "chivalry" toward white women.[76] In any case, Smith's support for the ERA and his opposition to civil rights was not an unusual combination for the time; probably he hoped that the amendment would help defeat the bill but believed that, if it did not, it would be better at least to give white women the same protection accorded to black women and men.

Once Smith introduced the amendment, forces other than those hostile to civil rights jumped into the fray, and this support from feminists has been ignored in the popular version of the story. Martha Griffiths, a member— if only on paper—of the National Woman's Party since 1955, had planned to introduce a similar amendment but had allowed Smith to do it since he would bring in the votes of the Southern Democrats.[77] All of the other women members of the House, with the exception of Edith Green, who followed the lead of the Women's Bureau, supported the amendment also. The National Woman's Party lobbied, sometimes using the argument that the bill would discriminate against white, native-born Christian women without the amendment. Although Martha Griffiths considered the Woman's Party ineffectual, members did write and contact Congressional representatives.[78] Caruthers Berger, a lawyer for the government and a Woman's Party member, wrote a defense of Title VII but could not sign her name to it since she worked for the Department of Labor, which opposed the amendment.[79]

Once the bill passed the House—with all but one of the men who had voted for the amendment voting against it—both President Johnson and the Women's Bureau dropped opposition, since they wanted to see the same version of the Civil Rights Act voted on in the Senate in order to expedite passage. The National Woman's Party set up a special emergency committee to work on keeping the sex amendment in the Senate version of the bill. Pauli Murray wrote a memorandum in support of the amendment, which the BPW distributed in the Senate. With support from the administration, the bill passed in the Senate. Emma Guffey Miller, president of the Woman's Party, reported that women's support had proven crucial.[80]

Afterwards, the Woman's Party claimed credit. Alice Paul sought publicity for her group, insisting that "[w]e bore the entire burden of this battle with absolutely no help of any kind from other women's organizations."[81] Despite Paul's refusal to recognize the work of other women, it is important historically that women's rights supporters helped to instigate Smith's amendment and lobbied for it. Marguerite Rawalt, a government lawyer and a founder of NOW, recognized the hard work that the Woman's Party put into the Title VII fight.[82] Women involved in the struggle recognized that the small number of women in the feminist "underground" had an impact out of all proportion to their numbers. Said one feminist: "I seriously doubt that women could have gotten Title VII without just a very small number of us. As a matter of fact I think it was a tremendous political coup, if you want to know the truth."[83] And another: "our guess is that the Senators did not know how few there were of us and got the feeling that this . . . would really touch off a woman's revolt and that they better not fool with it and so they went along."[84]

The National Woman's Party saw Title VII as a "vital wedge" in the law that might lead to passage of the ERA, so it also took an interest in the enforcement of Title VII by the Equal Employment Opportunity Commission.[85] In 1965, the National Council of Women met to discuss Title VII and the EEOC. At this meeting, the significance of the legend of the passage of Title VII began to emerge. Franklin Roosevelt, Jr., head of the EEOC, suggested that the inclusion of sex discrimination in Title VII presented a problem, since the intent was to ridicule the concept of equal employment opportunity. Miriam Holden, who attended the meeting, reported to Alice Paul that he seemed to be indicating that the EEOC would not enforce the sex provision.[86] This, of course, was just what happened, and the public image of the sex amendment as "a mischievous joke" and "cumbersome irrelevancy" supported the government's decision to downplay sex discrimination.[87] As Ernestine Powell, former national chairman of the Woman's Party, commented in a Columbus, Ohio, newspaper: "It was a great achievement. Oh, I know most men call it a fluke, but it wasn't anything of the sort. It was a long hard fight and we won."[88] That victory—which probably could not have been won by the women's rights movement alone—shows how women's rights supporters could link up with diverse interests, not all of them progressive, to make significant progress for women. After the founding of the National Organization for Women in 1966, the enforcement of Title VII became a primary goal of the liberal branch of the movement.

The Founding of NOW and the Resurgence of the Women's Movement

As a result of the President's Commission, the Equal Pay Bill, and the inclusion of gender-based employment discrimination in Title VII, the issue of women's status in American society began to edge into the public consciousness. One National Woman's party member believed that "Equality for Women is in politicians' consciousness."[89] Another member put out a call, under the name WOMEN (BIPARTISAN), for a protest march at Madison Square Garden against the lack of attention paid women by the government.[90] The BPW national board issued a resolution protesting the newly formed EEOC's ruling that Title VII permitted sex-segregated want ads.[91] After the passage of the Civil Rights Act, Pauli Murray wrote to Marguerite Rawalt to suggest the formation of a national ad hoc committee of women to fight for change.[92] The flamboyant head of the Federation of Women Shareholders, Wilma Soss, promoted Title VII and the ERA in her radio broadcasts.[93]

The publication in 1963 of Betty Friedan's *The Feminine Mystique* had also alerted women throughout the country to the collective character of what had once seemed personal dissatisfaction. The "unprecedented passion" of the letters Friedan received from some of the three million or so individuals who bought her book led to a request from *McCall's* for an article on the responses to her work.[94] One woman who belonged to both the League of Women Voters and the AAUW wrote to Alice Paul of her joy at finally finding the "vestiges of the movements begun by the early crusaders" and commented that Friedan had done women a big favor by writing *The Feminine Mystique*.[95] Caruthers Berger, a government attorney who had joined the Woman's Party around 1958, learned of Friedan's book when a secretary at work told her that Friedan sounded so much like Berger that she thought they had been in contact. Berger immediately bought the book, happy to find a kindred soul.[96] Helen Schleman, a leader in the National Association of Women Deans and a member of a committee of the President's Commission, believed that Friedan "really . . . galvanized things." She managed with some difficulty to persuade her organization to invite Friedan to address its convention.[97]

Meanwhile, women continued to meet in the status of women organizations sparked by the President's Commission. Patricia Zelman argues that these women were reluctant to work specifically for women, trained as they were in the tradition of self-sacrifice, but that the blatant ignoring of women in Johnson's War on Poverty and the sexist implications of the Moynihan report—which blamed black women for poverty in the black community—forced them to realize that they would have to fight for themselves.[98] Friedan, beginning work on another book, contacted Martha Griffiths, Pauli Murray (whom she read about in the newspaper), members of the President's Commission, and staff members of the EEOC and discovered "a seething underground of women in the Government, the press and the labor unions."[99] The new consciousness and renewed activity even brought together opponents. In 1966, Esther Peterson arranged a luncheon meeting to discuss the implications of *White* v. *Crook*, a case that accepted the Fourteenth Amendment as grounds for ending discrimination against women in jury service. The guests included Margaret Hickey, Marguerite Rawalt, Mary Eastwood, Catherine East, Dorothy Kenyon, and the heads of the AAUW, BPW, and the Woman's Party. Woman's Party president Emma Guffey Miller, who was kept from attending by a broken hip, wrote to Peterson that she hoped the luncheon "was the means of bringing your organization and ours into closer relations and thus advance greatly the cause for which we are both working—complete equality of women."[100]

It was women from the "seething underground," gathered at the Third National Conference of Commissions on the Status of Women in June 1966, who decided to form a new organization, modeled on the NAACP, to fight for women's rights. Friedan suggested the name "National Organization for Women." The early leadership of NOW included women from the United Auto Workers, women like Pauli Murray from the President's Commission, women from the state commissions, government feminists such as Mary Eastwood and Catherine East, and women like Marguerite Rawalt who had long been involved in women's rights activity. The importance of labor women to this new coalition is especially striking. Caroline Davis and Dorothy Haener, who had worked for the UAW Women's Bureau since 1960, played a crucial role in the establishment of NOW. Their experiences with the enforcement of Title VII convinced them of the need for a new organization. Haener credits Dollie Lowther Robinson, a black attorney for the Department of Labor's Women's Bureau who had spoken to a conference sponsored by the UAW Women's Bureau, with originally suggesting a NAACP for women.[101]

The feminists in government positions also played a major role in the 1960s, but they had to be very careful; Mary Eastwood, for example, told Alice Paul that her superiors would consider her suggesting an amendment to the Constitution highly improper, so that anything she wrote for the Woman's Party or NOW would have to be "*very anonymous.*"[102] Eastwood, like Catherine East and Caruthers Berger, did all of her feminist work secretly. The government unwittingly contributed to NOW's early work, since the women who wrote the early sex discrimination briefs made use of government offices, typewriters, and copying machines at nights, unbeknownst to their superiors.[103] As Pauli Murray explained it, "women did not come to the women's movement without bringing an awful lot of resources: resources in organizations we hadn't even thought about, borrowed techniques of research; and I think a lot of this had to do with why the thing exploded so, because in other words unconsciously we had been trained in everybody else's business and when we finally turned to ourselves we had all of this conditioning and training and so forth."[104] This access to existing resources proved crucial in the development of the movement in the 1960s.[105]

For some of the women involved in the founding of NOW, the new organization filled an important need. Helen Schleman, who had been active on behalf of women in the field of education, commented that the growth of feminist activism meant that "You don't feel you're out in left field by yourself as you did in the early period."[106] Pauli Murray, who would not have used the word "feminist" to describe herself until about the time of the founding of NOW, said that she was "standing on the corner waiting for the women's movement to come along." From her perspective, "most of us who were sensitive to women's issues [in the 1940s and 1950s] were involved in isolated, very often ineffectual protests. It wasn't that we weren't aware, but it was that we had no instrument, no consciousness. And among ourselves we would gripe and so this is why I say that I was standing on the corner waiting."[107]

Although most of the women involved in the founding of NOW had not identified with the women's rights movement in the 1940s and 1950s, there were connections. Woman's Party member Caruthers Berger became a charter member of NOW. Although she welcomed NOW since the Woman's Party had been struggling with very little support, she later expressed some irritation with the new group: "I don't really have very much patience with them but I try to control myself because I feel that Johnnys-come-lately to the feminist movement are better than other people that never, never came at all."[108] Ruth Gage-Colby, a progressive Woman's Party member, described NOW as picking up the cause, giving it a different character, expanding its purpose, and making it a living thing.[109] Margaret Hickey, who chaired the meeting of state commissions on the status of women where NOW was born, declined an invitation to join the new organization on the grounds that she belonged to the BPW, which had been working the whole century for women.[110]

Alice Paul received a special invitation to join NOW, since the new officers hoped that the two groups could cooperate. Paul did join, but relations between NOW and the Woman's Party were uneasy. A friend of Paul's remembers that the Woman's Party leader "scorned the NOW people. She said they act as if they discovered the whole idea, it's been going for 40 years but they act as if they just discovered it."[111] Betty Friedan appointed a committee to draw up a new constitutional amendment, believing that a new organization needed a new amendment not saddled with the opposition of the old, but Alice Paul convinced Mary Eastwood, a member of the committee who soon joined the Woman's Party, of the need to stick with the existing amendment.[112] The issue of the ERA came up at NOW's national convention in 1967 and NOW voted to support it, despite the fact that such a position meant the loss of clerical services and facilities donated by the

United Automobile Workers, since the UAW women had not yet managed to wrest from the union support for the amendment.

NOW's endorsement of the ERA did result largely from Woman's Party prompting.[113] The Woman's Party took the same approach toward NOW that it took toward other older women's organizations: it sought endorsement of the ERA, it used infiltration, and it sought to recruit members from NOW's ranks. Alice Paul reported that a great many Woman's Party members joined NOW and voted for endorsement of the ERA at the convention; in her mind, "we captured the NOW people."[114] One California Woman's Party member wrote to NOW specifying that she would join if the group supported the ERA.[115] Another member reported to Alice Paul that NOW, at her constant urging, had formed a committee to decide to endorse "our Amendment." As national membership chairman for NOW, she declined to serve as local membership chairman for the Woman's Party, believing that she could not split herself that way and could help the ERA most by working toward organizational cooperation.[116] Other NWP members reported participating in consciousness-raising on the ERA among NOW members. Elizabeth Boyer, who later left NOW to form the Women's Equity Action League, told Alice Paul that she "picked up another NWP potential member last night, and will continue to attempt to line up women who I encounter in my NOW work."[117] Some NOW members did join the National Woman's Party. Mary Eastwood joined in 1967, then left NOW in 1968 and continued her activity in the Woman's Party. In 1970, the Woman's Party secretary wrote to one NOW member that "[t]he larger our 'family' grows, the happier it makes us."[118]

As NOW grew and established itself on the national scene, the two organizations tried to cooperate. Officers of NOW visited Woman's Party chairman Emma Guffey Miller to discuss the groups' common purposes and to establish lines of communication on ERA policy. NOW sent representatives to the Woman's Party annual celebration on Susan B. Anthony's birthday in 1968 and made use of the Woman's Party list of ERA sponsors in Congress. One NOW member who sought information and support from the Woman's Party for her project of placing an ERA advertisement in the New York Times expressed her admiration for the Woman's Party, remarking that she was made aware of its existence only after the establishment of NOW.[119] But even when the Woman's Party tried to cooperate, it maintained a proprietary attitude toward the ERA. Alice Paul described the new women's groups as a welcome addition in a 1970 issue of Newsweek, but she made clear her understanding of the relationship between the Woman's Party and groups like NOW: "They are behind our efforts for equal rights. I hope they are going to help us actively."[120]

As NOW grew into a vigorous organization that took over the leadership of the ERA struggle, the Woman's Party wariness turned into overt hostility. For one thing, the Woman's Party leadership disapproved of NOW's multi-issue approach to women's equality. From Alice Paul's perspective, NOW "made us really quite a lot of trouble" by bringing up issues such as abortion.[121] The abortion issue for one member pointed up the wisdom of Paul's insistence that the Woman's Party "remain pure in its purpose."[122] When Hazel Hunkins Hallinan, an eighty-seven-year-old suffragist who had moved to London and remained active there in feminist efforts, paid tribute to Alice Paul in 1977, she agreed that the ERA should not be "muddled up with anything else—no side issues."[123] In 1968, Betty Friedan, as president of NOW, appeared at the Democratic National Convention to support both the ERA and abortion rights, outraging the Woman's Party leadership, who considered it sabotage of the ERA.[124] One longtime activist disapproved of NOW's attention to "problems of sexual or psychological supremacy," which she felt debased the struggle for equal rights to the "battle of the sexes."[125] Mabel Vernon, Alice Paul's comrade from her Swarthmore days, told a reporter that she had no opinion on abortion and lesbianism, that she was in favor of equality and did not go into those other issues.[126] Elizabeth Chittick, the president of the Woman's Party since 1972, blamed some of the post-1972 opposition to the ERA on women's liberationists because of their positions on abortion, child care, and lesbianism.

But even stronger than the conviction that a multi-issue approach would harm the cause of the ERA was the resentment that NOW received so much credit for work on the amendment. Ruth Gage-Colby was irked that NOW's literature never mentioned the long hard struggle of the Woman's Party, although she was glad for the added number of women, especially young women, who joined "the Cause" through NOW.[127] Some younger women did recognize the contributions of the Woman's Party. One woman active in the Pittsburgh chapter of NOW assured Alice Paul that the Woman's Party had been the most important group in ERA work.[128] When the amendment finally passed Congress in 1972, the Woman's Party sponsored a victory celebration at Belmont House. According to the Washington *Post*, Alice Paul, president Elizabeth Chittick, a few other "old faithfuls," and a small group of young women from NOW gathered for tea.[129] One Woman's Party member present for the occasion reported that Paul was depressed because she feared, presciently as it turned out, that the seven-year deadline would make ratification impossible.[130] Despite the celebration at Belmont House, *Newsweek* gave Women United, another new organization, credit for the passage of the amendment, provoking one Woman's Party member to comment simply, "Can you imagine?"[131]

By the time Congress approved the ERA in 1972, the character of the debate— although not the arguments—had changed in significant ways. Title VII, by undermining protective legislation, had eliminated the major problem for those opponents of the ERA who were also seriously committed to women's equality. By the end of the 1960s, a number of federal court decisions clearly held that protective legislation that singled out women violated Title VII. As a result, former opponents began to endorse the ERA.

The Women's Bureau, for example, recommended passage of the ERA at its fiftieth anniversary conference in 1970. A representative from an organization known as the Council for Women's Rights wrote to Alice Paul after attending the Women's Bureau conference to report that many women there had shared the conviction that women needed to organize politically in their own cause and to ask about the Woman's Party, particularly "what possiblities [sic] there are of our building from the NWP base."[132] Esther Peterson reported in 1978 that her opposition to the ERA had stemmed chiefly from mistrust of the ERA's supporters.[133] As liberal and labor groups jumped on the ERA bandwagon, a new anti-ERA coalition of right-wing groups emerged. The ERA became increasingly linked to issues such as abortion and gay rights. One longtime ERA activist described the amendment as a "political football" and commented: "if you want to know the truth, I think that they [opponents] put on whatever front they want to, and that the truth is that there are both conservative and liberal men that very much want to keep women in their places."[134]

In any case, in the context of the burgeoning women's movement, the Woman's Party was an anachronism. A 1962 UPI story described the group as "a band of militant, white-haired indomitable women."[135] In 1963, Esther Peterson evoked an image of decay by describing Belmont House as "paint-peeled, mossy green brick, and ivy over-grown with weeds in front."[136] Anna Lord Strauss, ex-president of the League of Women Voters and, like Peterson, no friend of the Woman's Party, wrote to an official of the International Alliance of Women in 1968 that the Woman's Party "has been fading quietly away."[137] An article written in the 1970s, entitled "The Last Hurrah of the Woman's Party," described the group as "little known—even to many feminists—and . . . not one of the organizations that young feminists join."[138] The Woman's Party's lack of interest in any goal other than the ERA—not to mention the reactionary attitudes of many members on other issues—had isolated the group.

Women already involved in women's rights activity made connections with the new organizations, but they did not always approve of the new brand of feminism. Doris Fleischman Bernays, for example, a Lucy Stone League member, told a reporter in 1971 that she had been concerned with the status of women long before it was popular and had upstaged Betty Friedan by almost a decade. But she defined a liberated woman as one who had servants to do the housework: "Women's lib wastes energy coming out for nonessential, picky things. Whether one wears a bra or not is not important. What is important is fighting discrimination laws and not having to do housework."[139] Recommending servants—a blatantly class-biased proposal—could not have endeared Bernays to young feminists in the 1970s. Margaret Hickey disapproved of the "abrasive feminist elements which now show signs of revival," and she later accused the "liberationists" of creating an antiwoman, antihomemaker feeling.[140] Susan B. Anthony II told the *New York Times* in 1971 that the original Susan B. Anthony would not have approved of the women's liberation leaders.[141] Hazel Hunkins Hallinan compared old and new feminists and accused the new ones of being antimale.[142] A BPW leader believed that the ERA would have passed if NOW had stayed out of it: "if that aggressive attitude had not been developed . . . we would have had an Equal Rights Amendment. That turned people off. That aggressive bra-waving and all that sort of stuff."[143] Even Ruth Gage-Colby, with her progressive critique of society, longed for television coverage of the Woman's Party—"not all mixed up with Liberation women and the witches but just our own National Woman's Party story"—because of the "continuity, dignity, moral and legal rightness of it!"[144]

But older women's rights supporters were not always completely hostile to younger women. Gage-Colby attended the National Conference for New Politics in Chicago in 1967 and worked on the women's resolution, a quite radical document that the Woman's Party endorsed. Lucy Somerville Howorth of the BPW expressed her satisfaction that a library fund named in her honor would be used to purchase women's studies books.[145] BPW leader and federal judge Sarah Hughes told the newspaper that she favored women's liberation.[146] Second generation and longtime Woman's Party member Mary Kennedy went to hear Kate Millett speak at Purdue University and took ERA literature to hand out to the crowd. Although, because of a hearing impairment, she missed most of Millett's talk, she did report to Alice Paul that Millett favored the ERA and that the audience responded enthusiastically to her speech. Vividly contrasting the old suffragists of the women's rights movement with the new feminists of the 1970s, Kennedy also noted that she was the only one there in a hat and that everyone else, including Millett, had long hair.[147] Yet even Alice Paul, who criticized the new women's movement for taking up too many controversial issues, acknowledged that the women's liberationists "are a little like we were."[148]

Some of the organizations that had been part of the women's rights movement in the 1940s and 1950s faded away in the 1970s. Others became part of the liberal branch of the women's movement. The Woman's Party remained important in women's movement activity largely because of its ties to the suffrage movement and its ownership of Belmont House—in essence, its historical value. This significance of the Woman's Party came to mean more and more as younger women discovered the legacy of militant suffragism. Even in 1981, as we have seen, the Woman's Party's symbolic importance remained great enough to have inspired an attempted takeover by the Congressional Union. Throughout the 1970s, women's groups held functions at Belmont House because of its historical significance and prime location. Women of America for the Women of America sponsored a silent vigil during the 1970 Senate vote on the ERA and used Belmont House for headquarters. NOW sponsored a birthday benefit at Belmont House in 1977 to honor Alice Paul. By the 1970s, Alice Paul was no longer an enemy to the women's organizations that had long opposed the ERA, in part because she outlived most of her foes and in part because the traditional anti-ERA groups switched to support of the amendment by the early 1970s. Established groups such as the AAUW, League of Women Voters, National Council of Jewish Women, and National Council of Negro Women joined new groups such as the Coalition of Labor Union Women, the National Women's Political Caucus, the Women's Equity Action League, and NOW in celebrating

Paul's birthday. As the women's movement grew stronger and more militant, it claimed the suffragists, particularly the militants of the Woman's Party, as foremothers. In the process, the actual heirs of the early militants became more and more isolated and less and less important in the new movement.

Conclusion

The women's movement in the years between 1945 and the mid-1960s did make connections with the resurgent women's movement in the 1960s and 1970s. But as the history of the President's Commission, the Equal Pay Bill, the inclusion of sex discrimination in Title VII of the Civil Rights Act, and the early years of NOW make clear, the movement that grew in the mid-1960s consisted of new organizations and individuals who brought fresh perspectives to the question of women's status. At the same time, the women's rights movement as it had survived the 1950s played a role in the key events that marked the transition from one stage of the movement to another.

Some of the older organizations, and especially the National Woman's Party, functioned as what Aldon Morris, in his study of the civil rights movement, calls "movement halfway houses." Morris defines movement halfway houses as established organizations that lack a mass base and are relatively isolated from the larger society because of their activity on behalf of social change. Such organizations lack broad support but are able to mobilize a variety of resources that a developing mass movement can use.[149] The National Woman's Party, for example, brought the goal of the ERA, experience in lobbying, government contacts, and a historical tradition to the resurgent movement of the late 1960s.

Although the new feminists quickly outstripped the older women's rights supporters in their challenge to the basic structure of American society, and began to confront class bias, racism, and heterosexism in a way that would have been unimaginable in the 1950s, it is no coincidence that groups and individuals throughout the movement—from radical to liberal, from moderate to militant—continued to fight for passage of the ERA. Above all, the women's rights movement handed its major goal over to younger hands, and this legacy is symbolic of the continuity in the history of the American women's movement.

9

Theoretical Overview and Conclusions

"We were, in a way, a little bit like the story about Sergeant York, the famous hero of World War I. He hid behind a log and he was such a crack shot that he annihilated a whole regiment. As they came by the log he got one, and pow!, another. In other words, he was such a dead aim shot they thought he was a regiment because he was hitting everything that he shot at. I think for the short period of time that it took for us to get protection for women in Title VII, we were able to get the Congress to take women seriously enough to do it."

CARUTHERS BERGER, 1982

Caruthers Berger's comparison of the women's rights movement to the famous Sergeant York suggests the importance that women attributed to their work on behalf of equality. Although women's rights advocates throughout the years after 1945 could not always claim such success, they did function very much like the sharpshooter pretending to be a regiment. We expect some scholars will accuse us of mistaking the efforts of a few individuals for the activity of a social movement, but we are not alone in believing that the size of a movement's following is not the only criterion by which we can judge its significance. Addressing this same question, social movement scholar Paul Wilkinson states: "The assumption that only numbers count is in harmony with recent fashions in quantification in the social sciences and humanities. Some find it reassuring to have something concrete to measure and upon which to found explanations. However, the belief that 'you cannot argue with numbers' has never convinced the leaders and followers of movements. Some of the most interesting movements started as small minorities and remained small minorities. Their 'significance' does not lie in the successful mustering of large battalions, but as carriers of ideas, as harbingers of cultural and intellectual developments."[1]

We began this book by asking what it might mean to take seriously the perceptions of women's rights advocates that their efforts to improve the status of women in the 1940s and 1950s could be viewed as a social movement. Through an exploration of the activities of the National Woman's

187

Party and the other groups and networks of women that joined it to continue the struggle begun by women in the suffrage days, we have been led to the conclusion that the feminist activity that survived in the post-1945 years can, indeed, be viewed as a cycle in the history of the women's movement. The ultimate question that must be asked of any social movement, however, is what, if any, consequences it had for social change.

In this chapter, we broaden our focus to consider the women's rights movement of the post-1945 period in terms of its role in the larger history of the women's movement. Our discussion begins by considering the impact that the movement had in its own time as well as for the revitalized movement that took up the fight for gender equality by the end of the 1960s. We then present the theoretical insights from social movement theory that have influenced our analysis and apply them to understand the stage of the women's movement that we have described as "elite-sustained." To conclude, we explore what it means to suggest that the American women's movement has a continuous existence.

Consequences of the Movement

Few social movement scholars have addressed the question of movement consequences, despite its importance.[2] Most often, social movement outcomes are analyzed in terms either of success or failure in the achievement of stated goals, or in terms of patterns of decline.[3] Although most researchers see social movement decline as inevitable, this does not necessarily spell failure. Movements sometimes die because they have achieved success. Our findings suggest that a fruitful way to analyze social movements based on fundamental social cleavages—like the women's movement—is as entities that persist throughout the years and have consequences in their own time as well as for later stages of mobilization. Our discussion will focus on two kinds of consequences: the impact of this elite-sustained stage on both the larger society and movement participants at the time; and the consequences of feminist activism in the 1940s and 1950s for the women's movement of the late 1960s and 1970s.

As we have shown, the women's rights movement in the post-1945 years had three basic goals: passage of the Equal Rights Amendment, placement of women in policy-making positions, and recognition of women's history. Women's rights supporters never fully achieved their objectives in these three areas, but their activities did have some impact. Although the ERA did not pass Congress until 1972, it did make headway in the fifties. In 1945 the House Judiciary Committee reported the amendment favorably for the first time since its introduction in 1923. In 1946, the Senate Judiciary Committee followed suit and a majority—although not the necessary two-thirds—of the Senators voted for the amendment. In 1950 and 1953, the ERA passed the Senate, but in a version that included the nullifying Hayden rider. Although action on the ERA does not mean that the amendment had a real chance of passage in the fifties or that there was wide public support for the measure, it is testimony to the activity and effectiveness of the ERA coalition. As we have seen, opponents of the ERA, especially the labor unions and the Women's Bureau, were forced to expend effort to counteract the activity of the amendment's supporters, even to the extent of establishing a presidential commission to stall the ERA and find an alternative means of improving women's status.

With regard to women in policy-making, the women's rights movement could claim some victories, such as the appointment of Perle Mesta as U.S. minister to Luxembourg and Burnita Shelton Matthews as U.S. district court judge, even if feminists were not alone responsible for such appointments. Furthermore, the rosters of qualified women existed as a potential resource for government officials, and the activities of women's organizations and individual women in the political parties had some impact in winning recognition, if only grudging, of the importance of women in politics.

Likewise, advocates of women's history made some limited progress. We have seen that women's rights supporters could point to the existence of a few courses in women's history at universities, the flawed but nevertheless educational Seneca Falls postage stamp issued in 1948, occasional publicity about and celebrations of the history of the women's movement, and a growing body of literature about women's history as the products of its efforts.

Although social movements exist primarily to achieve their expressed aims, in addition they have latent goals and unintended consequences.[4] Individuals in movements must experience some sense of gratification and reward in order to maintain their commitment to the cause. Participation in the movement provided a range of benefits: for a few women, careers, experience in lobbying, and even a home and surrogate family. More typically, the movement provided organizational and leadership skills and expertise, the satisfaction of working for a worthwhile cause, personal relationships formed in the course of feminist work, and a sense of belonging. For many women the rewards of participation originated in the primary groups that developed among participants. In some ways, the primary groups that women formed were the very essence of the women's rights movement in this period.[5] If the movement had minimal impact on the larger society, it played a significant role in the lives of participants.

Apart from providing a supportive community for women who were already committed to the feminist cause, the women's rights movement also influenced some other individuals and groups. Because feminists put little or no effort into direct mobilization and conversion but instead directed their activities toward influencing elites and mobilizing ideological commitments held by existing groups, their effect on collective consciousness in the period was probably minor. But legislators who sponsored the ERA or whom the women's rights supporters lobbied, candidates for office who received material on the ERA, government officials responsible for appointing individuals to office who received the rosters of potential women appointees, and publishers and other key media figures whose support for a range of goals feminists solicited, were exposed to the feminist program. One representative from Connecticut, for example, was lobbied so persistently by a National Woman's Party member that he wondered "whether or not the Congress meets for any other purpose but to consider the Equal Rights Amendment."[6] In the same vein, the chairman of the National Committee to Defeat the Un-Equal Rights Amendment worried that "the Woman's Party is as active as ever, and is trying to persuade labor organizations to change their stand, as well as trying to line up the nominees for the new Congress."[7]

The women's rights movement in the post-1945 period did not always reach the wider public with its program, but probably the best measure of its influence was the establishment of the President's Commission on the Status of Women in 1961. That the activity of pro-ERA feminists instigated the creation of such a body and entered into its deliberations is clear. Women's rights supporters had little to do with the final Commission report, but Kennedy might never have agreed to lend the prestige of the presidency to such an undertaking without the persistent efforts of ERA supporters. The Commission, sparked by women's rights activity, ultimately served as a stimulus for the growth of a new and more broad-based cycle of the women's movement.

The effects of feminist activism in this period for the next phase of the women's movement were at least as important as the consequences it had in its own time. To understand the ways in which the women's rights movement was connected to the later feminism, it is useful to think of this cycle as providing crucial resources that molded the strategies and possibilities for action in subsequent years. Following Jo Freeman, movement resources can be categorized as tangible or intangible. Tangible resources include such things as the money, space, and publicity a movement needs to maintain an existence and make its program known, while intangible resources consist of such things as specialized expertise, access to networks, access to decision-makers, status in the larger society, status within the movement, and the time and commitment of supporters.[8]

The women's rights movement of the 1940s and 1950s provided the resurgent movement one particular tangible resource—space—as a result of the fact that most women active in the post-1945 years were relatively affluent, capable of paying membership dues and leaving bequests to support their organizational headquarters. For example, once the National Organization for Women endorsed the ERA in 1967, Belmont House was used as a centralized location for lobbying by groups other than the Woman's Party, for celebrations of women's history, and as a temporary residence for scholars and students engaged in feminist research in Washington. The attempt by the Congressional Union to take over the Woman's Party in 1981 is testimony to the historical and practical value of Belmont House as well as the name and tradition of the Woman's Party itself. Other organizations, too, such as the BPW and AAUW, had headquarters in Washington that became focal points for activity, information, and support as the movement coalesced around the ERA and other activities in the 1970s.

The women's rights movement also had access to a different kind of publicity than was available from other sources. As Freeman points out, the radical branch of the women's movement in the late 1960s had relied on the network of underground newspapers of the New Left to publicize its ideas and actions. In contrast, the emerging liberal branch for the most part did not have available to it any established or new means of communication. But groups and organizations that had been involved in women's rights activities in the 1940s and 1950s did have channels that they could activate on behalf of the ERA and other issues. The newsletters of the BPW, Woman's Party, professional women's organizations, and women's service organizations continued into the 1980s to reach a sector of women that otherwise could never have been influenced by the publications of NOW, much less by the newsletters and writings of the more radical groups of younger women. In addition, groups and networks of women that were part of the women's rights movement had access to established media that helped to publicize feminist activity when new groups such as NOW had yet to form such ties. We have seen, for example, that the *Christian Science Monitor* and the New York *Herald Tribune* both supported the ERA from the 1940s on, and that journalists looked to groups like the National Woman's Party and the BPW for information about the ERA before NOW gained visibility and wider support.

Far more important than the tangible resources, however, were the intangible ones that preexisting feminist groups provided the new organizations that emerged in the late 1960s. As we have seen, women's rights activists gave generously of their time and commitment in the 1940s and 1950s. Some of these women, such as Alice Paul, Marguerite Rawalt, Pauli Murray, Caruthers Berger, Margaret Hickey, Catherine East, and Mary Eastwood, continued to devote themselves to the women's movement into the 1980s. Others committed themselves to trying to get newly emerging organizations, such as NOW, and even coalitions of radical young women such as the women's caucus of the Conference on New Politics, to support the ERA.

In addition, women's rights supporters, active for many years on behalf of the ERA, brought specialized expertise in the form of regular procedures for lobbying senators and representatives and a system of record-keeping on supporters and opponents of the amendment. Feminists also had some access to key decision-makers in government and politics who proved useful in the struggle for the ERA and in the inclusion of sex discrimination in Title VII of the Civil Rights Act. Not always did this access prove useful to the newer organizations. For example, Democratic committeewoman Emma Guffey Miller used her influence within the Democratic Party to attempt to keep Betty Friedan from speaking in favor of the ERA and abortion rights at the 1968 convention. She feared that Friedan's support, since it would be viewed as coming from a militant and radical feminist, would harm the

cause of the ERA. But the list of ERA sponsors that the Woman's Party
had already built, as well as other contacts in government, was an important
resource in the ongoing effort to win passage of the ERA.

Women's rights activists also had access to existing networks of women
within a variety of established organizations, such as the AAUW, the BPW,
and other professional women's groups that might otherwise have remained
isolated from feminism and the form it took in the 1970s. When Helen
Schleman invited Betty Friedan to address the convention of the National
Association of Women Deans in the early 1960s, she made it possible for
Friedan to reach an audience that might not otherwise have been exposed
to new feminist ideas. By 1970, most women's organizations supported the
ERA and frequently took up support of some of the newer goals of the
movement.

Another resource that feminist activists in the fifties brought to the
movement in subsequent years was status within the feminist world. Alice
Paul, in particular, became a legendary figure whom feminists in later years
elevated to the level of heroine, chiefly to reinforce their own dedication
and to arouse commitment among others. Although the existence of a movement in the 1940s and 1950s has been almost totally overlooked, when
women active in those years surfaced in the 1970s and 1980s they were often
greeted as heroines. For example, a *Ms*. magazine article and a newspaper
column by Ellen Goodman in the 1970s lauded the activities of Alma Lutz.
Ruth Gage-Colby, who traveled to attend the 1983 National Women's Studies Association conference in Columbus, Ohio, received a standing ovation
when introduced at the opening session. Individual heroines from the past
were important because they served as testimony to the persistence and
validity of the feminist cause.

Most important, however, was the impact of Title VII of the Civil Rights
Act and the ERA on the revitalized movement. Title VII and the issue of
its enforcement by the Equal Employment Opportunity Commission helped
to spark the founding of NOW, shaped its earliest actions, and rendered
the political system vulnerable to the feminist challenge in the decades that
followed by increasing the opportunities for collective action around the
issue of sex discrimination.[9] Whatever else can be said about the significance
of the 1940s and 1950s women's rights movement for the later movement,
one thing is clear: the Equal Rights Amendment, authored in 1923 by the
Woman's Party, was handed to a new generation of women and became for
almost a decade one of the major issues around which the women's movement mobilized. To be sure, this legacy had negative as well as positive
consequences for the resurgent movement. In the early years of NOW, the
emphasis on the ERA and the strategy of attempting to mobilize what Paula
Giddings calls "the unreconstructed elements of the *Feminine Mystique*"
alienated black and union women.[10] This suggests one way in which feminists
in a later stage of the women's movement were unable to shake the class
and race limitations of their feminist predecessors.

Thus, the feminist activism that survived in the doldrums, even if it did not leave an indelible mark in history, had some impact in its own time as well as for the resurgent women's movement. While new organizations emerged in the late 1960s and 1970s and took up a broader range of issues, mobilized a larger and more diverse constituency, and developed new and varied strategies, the previous cycle of the movement served as a source of mobilizable resources for this later round of feminist activism. The women's movement of the 1960s might have exploded upon the American scene even if there had been no organized feminist activity in the post-1945 period. But in any case the nature and direction of the movement of the 1960s was influenced to some extent by what preceded it.

Social Movement Theory and the Research Perspective

The findings presented here challenge the widespread assumption that the women's movement died in the 1920s and was not resurrected until the mid-1960s. Underlying our argument that feminist activity from 1945 to the mid-1960s is best understood as a stage in the longer history of the American women's movement are specific ideas about the nature and development of social movements that can be traced to two sociological theories of social movements: classical collective behavior theory and resource mobilization theory.

The classical perspective on social movements arises out of a research tradition that views social movements as one form of collective behavior that shares certain common features with other more ephemeral forms of social action, including panics, crowds, protests, mobs, cults, fads, mass hysteria, crusades, and revivals.[11] Although the writings of classical collective behavior theorists are not in full agreement about the characteristics and explanations of the phenomena conceptualized as belonging to the family of collective behavior, several common assumptions have emerged from this theoretical tradition.

The first is that collective behavior is activity that arises out of a condition of stress, strain, or severe disruption of the social order, a view that has led scholars to look to shifting levels of deprivation and discontent among aggrieved groups in explaining the occurrence of collective behavior. Second, because collective behavior results from an impaired social order, it has generally been depicted as discontinuous with ordinary forms of social life. Hence, it is spontaneous rather than planned, unstructured rather than organized, guided by emotion rather than reason, and governed by non-rational rather than rational beliefs.[12] The third, and perhaps most important, assumption found in most writings on collective behavior is that the ultimate significance of these forms of behavior lies in their role in bringing about innovation and social change.

Our analysis generally agrees with classical formulations in viewing social movements as deliberate conscious efforts to bring about change that are rooted in a dissatisfaction with the status quo. In other important ways, our analytical perspective departs from the classical tradition and draws heavily on the newer resource mobilization perspective.

To understand the formation and development of a movement, collective behavior theory relies, ultimately, on social psychological factors such as shared discontent, ideology, emotionality, and irrational beliefs. Resource mobilization theory, in contrast, tends to regard the processes that govern social movements as similar to those of other types of social organization. As a result, the analytical focus shifts from individual and psychological factors to the role that organizational and structural factors play in determining the nature, course, and outcome of a movement.

Resource mobilization theory is so named because of the emphasis it places on a group's ability to organize, mobilize, and manage valuable resources—money, expertise, leadership, communication networks, access to publicity, and the support of influential outside groups and individuals—in explaining movement formation and development. Certainly, strain, social discontent, and grievances play a part in the rise of collective action. But, for resource mobilization theorists, such factors are not sufficient to explain social movement activity.[13] In most instances, grievances among groups whose discontent derives from basic conflicts of interest rooted in the institutional framework of society—as in the case of gender, race, ethnicity, and class—are relatively constant over time. What changes, giving rise to collective action, is not the degree of dissatisfaction among discontented groups, but the amount of social resources available to them that makes it possible to launch an organized demand for change.[14] An important factor influencing a group's ability to mobilize resources is the climate of the larger political environment, specifically the extent to which it inhibits or promotes the development of collective action around the group's objectives. Put simply, resource mobilization theorists suggest that there are particular periods during which a group's actions will be more or less likely to be supported and, therefore, meet with success or failure.

The resource mobilization model is particularly suited to the kinds of analytical questions that we raised about the post-1945 women's rights movement. Given the growing role strain and inconsistency women faced in the 1940s and 1950s, the argument that the small size and limited scope of the women's rights movement in these years was a manifestation of the absence of any deeply felt injustice on the part of American women is not only simplistic but inaccurate. It is true that the three social trends that scholars have identified as the major preconditions for a massive women's movement—an increase in women's educational level, a decline in fertility, and both a rise in women's labor force participation rate and a change in the patterns of that participation—did not come together until the 1960s, largely because of the temporary fertility increase in the immediate postwar decade. Demographer Abbott Ferriss, however, after examining long-term social trends affecting women's lives from 1950 to 1970, concluded that there were not enough indicators of social strain to explain on the basis of these factors alone why the women's movement erupted when it did in 1970 rather than at any other time in the previous two decades.[15] If stress and strain alone could fuel feminist protest, then there is reason to believe that the 1950s should have witnessed a mass uprising of women.

It was never the intent of this research, however, to analyze in detail the structural sources of women's discontent or the preconditions for a women's rights movement in the period. Rather, our primary aim was to describe the organizational patterns and strategies of the women's rights movement that managed to survive in the doldrums and to explain why the movement developed as it did. Adopting a resource mobilization perspective directs our attention to specific factors in answering this question: what was the interplay between the internal dynamics of the movement itself and the ways in which societal support and constraint, in particular the extreme antifeminism of the postwar decades, affected the women's rights movement's strategies, growth, and success?

The resource mobilization approach makes a major contribution by providing a framework that emphasizes the dynamic nature of social movements.[16] Earlier work had assumed that social movements inevitably pass through a standard "life cycle" from their emergence to their decline either as a result of institutionalization and bureaucratization or failure and collapse.[17] Recent approaches, instead, see social movements as taking diverse paths that lead to different outcomes.[18] Put simply, the history of any social movement is shaped by an interactive process in which the movement pursues a course of action, and the societal response to its actions, in turn, modifies the movement's structure, goals, and strategies. By studying movements in different stages of development as well as in different historical contexts, it can be shown that the same social movement adopts different structural forms and mobilization strategies at various periods in its history depending upon whether it is in a stage of formation, success, continuance and survival, or decline.[19]

Scholars who have applied the concepts of resource mobilization theory to the American civil rights movement have found this to be true. For example, Aldon Morris characterizes the civil rights movement in its formative stages, from 1953 to 1963, as an indigenous movement.[20] An indigenous movement owes its existence to preexisting institutions—in the case of the civil rights movement, the black church, black colleges, and other black organizations—that provide the leadership, ideas, communication network, money, membership, and specialized expertise necessary to mobilize. Having strong ties to its mass base through indigenous institutions, this kind of movement primarily draws upon resources *within* the previously organized dominated group. By the early 1970s, a period that marked the decline of widespread black protest, the civil rights movement could no longer be characterized as an indigenous movement, but functioned more like what McCarthy and Zald term a professional social movement that "speaks for" rather than attempts to mobilize those who would benefit from its actions.[21] Lacking strong ties to its mass base but having convinced the larger society of the seriousness of its challenge, civil rights activism increasingly depended for its existence upon a core of trained movement entrepreneurs capable of securing financial resources and other kinds of support for civil rights from government agencies, foundations, political elites, the mass media, and other influential groups. Not surprisingly, the nature and direction of the movement shifted as it was increasingly shaped by resources obtained *outside* the dominated group.

Our analysis follows in the tradition begun by scholars of the civil rights movement.[22] It emphasizes the historical continuity of the American women's movement and introduces the concept of an elite-sustained movement to characterize the stage that existed from 1945 to the mid-1960s. These years marked a period of decline in the history of the American women's movement. The movement that survived was largely, although not exclusively, a remnant of the twentieth-century suffrage movement and provided an environment in which already committed feminists could pursue their goals. In a political climate that offered feminists few opportunities to mobilize, organizations and networks of women that continued feminist activity in this period directed little effort toward expanding the base of support for women's rights, as the women's movement of the late 1960s and early 1970s did.[23] Instead, the movement functioned primarily through channeling preexisting group and individual commitment and the other resources necessary to survive when, for the most part, odds were against its success. To the extent that it relied primarily on resources from within its own ranks, the women's rights movement of the post-1945 period was an indigenous movement. But its base of support was restricted mainly to women of elite social status with limited ties to a mass base.

An Elite-Sustained Movement: Organizational Dilemmas and Tactical Choices

British sociologist Olive Banks argues that the American women's movement is best understood as "one single historical process."[24] Our findings here suggest that viewing the women's movement as a continuous phenomenon from the nineteenth century into the present requires that we recognize significant variations in its organizational form and strategies at different periods in its history. To summarize what is meant by an elite-sustained stage of the women's movement, we apply the concepts and perspective of the resource mobilization theory of social movements to analyze the characteristics of the women's rights movement from 1945 to the mid-1960s. Our discussion will be organized around four analytical themes that parallel major organizational dilemmas and tactical choices any movement confronts in attempting to mobilize. First, the analysis will examine the ways in which the movement sought, in a social climate inhospitable to its aims, to attain legitimacy. Second, the consequences of adopting a decentralized structure rather than a centralized one will be discussed in light of the movement's objectives. Third, the nature of the infrastructural supports available to the movement and the ways they shaped its mobilization strategies will be examined. Finally, the discussion focuses on major constraints on the tactical choices that could be made by the movement.

Movement Legitimacy

There are two principal kinds of legitimacy that a movement strives to achieve: legitimacy of numbers and legitimacy of means.[25] Legitimacy of numbers is achieved by mobilizing a large number of people committed to challenging the existing distribution of power. The rationale behind this strategy is that the polity only accepts among its ranks challenging groups that are able to muster significant numbers of supporters.[26] Legitimacy of means is achieved by convincing the public, through particular aims and strategies, that the movement is the most appropriate vehicle for achieving the goals of its constituency. This kind of legitimacy is necessary for recruiting new members and for obtaining access to the mass media and other elites. To achieve legitimacy of means, a movement must make certain that its grievances are seen as credible and serious enough to warrant its existence as a social movement.[27]

By the post-1945 period, the women's rights movement had dwindled in size so that it no longer could claim a large and active base of supporters. From the perspective of the mass media, the public, and others that the movement hoped to influence, the small size and the selective composition of the movement made it easy to question whether those active on behalf of women's rights legitimately represented the interests of American women. Because the majority of women active in the period were white, middle or upper class, well-educated, and employed in professional or semi-professional occupations, they could be seen as having access to other more institutionalized means of influence. As a result, the movement for women's rights had difficulty convincing the public that it truly represented a disadvantaged group for whom a social movement was the appropriate vehicle for expressing its grievances.

Recognizing that it was important that the movement appear to represent a larger sector of women than it actually was able to mobilize, feminists used strategies available to them because of their elite social status in order to give the appearance of having a mass constituency. Among the relatively small core of intensely committed members of the movement were a number of well-known and admired women who headed national women's organizations. These women typically were able to command a following when support for a particular goal became essential. Thus, when a national organization supported an issue such as the ERA, movement leaders claimed the entire membership of that organization in listing numbers of women who supported the movement's position. The BPWs "Buttonhole Campaign," in which local and state members were urged to "buttonhole" their senators and representatives to ask for passage of the ERA, is a good example of mobilization of members by the national leadership. Another

tactic used to inflate the movement's numbers was seeking paper membership by listing as members those who paid one-time nominal dues, as the
National Woman's Party did, or establishing state branches with a single
active member.

To try to counter the image that the movement was a small, exclusive,
and eccentric group whose claims of injustice and discrimination might not
be warranted, movement leaders sought to establish front organizations that
would give the impression of a large and broad-based membership. Such
groups included the Women's Joint Legislative Committee for the ERA,
the Women's Industrial League for Equality, and the St. Joan Society. The
most important strategies devised to establish legitimacy of means were those
that attempted to give the appearance that the movement had wide grass-
roots appeal among less privileged women, such as working-class, union,
and black women.

The support of women from labor unions was extremely important
because a primary source of opposition to the ERA in this period was
organized labor. In an effort to try to persuade the public and elite groups
that women in labor favored equal rights for women, the few working-class
women and union women who were active in the women's rights movement
were used as spokespersons whenever possible. The National Woman's Party
also established and financed the Women's Industrial League for Equality
to give the appearance of a coalition between labor and the women's rights
movement.

The same kind of tokenism was evident in the way in which black women
were dealt with in order to advance the cause of the ERA. Although women's
rights movement organizations made no large-scale efforts to open their
membership to black women, they did encourage the participation of particular women like Mary Church Terrell, founder of the National Association
of Colored Women, in order to obtain her group's endorsement of the ERA.
The kind of actions taken by women's rights movement organizations with
regard to union women and black women suggest that feminists were not
committed to creating truly meaningful political alliances with other constituencies. Instead, they tended to use individuals and groups for their own
ends, to give the impression that the movement represented a truly aggrieved
group and was legitimate in making its demands for change.

Movement Decentralization

Considerable debate exists in the social movement literature over the relative
effectiveness of centralized bureaucratic movement structures, as compared
with decentralized structures, for achieving a movement's goals.[28] Scholars
generally agree that each type of movement structure has its relative advantages. Centralized bureaucratic structures tend to produce the organizational stability, technical expertise, and coordination necessary for
achieving their goals. Decentralized structures, on the other hand, maximize

participation benefits, such as commitment and community, and encourage the development of multidirected strategies for change. But they can also contribute to factionalism and intense conflict among movement organizations, which can be detrimental to a social movement's overall strategic effectiveness.[29]

The structure of the women's rights movement was decentralized. Although the National Woman's Party formed the core of the movement and attempted, whenever possible, to direct all efforts on behalf of women's rights, other movement organizations and groups took leadership on various issues and actions, and leadership by individual women able to command a following also changed depending on the goal and strategy. Alice Paul, even when she was not the formal head of the Woman's Party, directed ERA lobbying activities, but when the BPW made the ERA a priority, it worked independently of the Woman's Party. Support of the appointment of women to policy-making positions was coordinated by the heads of organizations such as the BPW or AAUW and often relied on the efforts of women in politics, such as India Edwards. Individuals, for example Rose Arnold Powell, sought to coordinate campaigns to win recognition for women's history.

The decentralized structure benefitted the movement in two ways. First, it encouraged the development of varied goals and strategies in the pursuit of women's rights. State and local groups engaged in independent activities that could not have been undertaken by the national leadership. For example, Florence Kitchelt of the Connecticut Committee, because of close personal ties to members of the American Civil Liberties Union, worked to influence that organization to change its position on the ERA. The Little Below the Knee Club in Dallas protested the exploitation of women by the fashion industry. In addition, strategies were used by movement organizations that would not have been encouraged under the tightly directed leadership of Alice Paul and the National Woman's Party because of the Party's view that the ERA was the only means of winning equality for women. For instance, Rose Arnold Powell, to Alice Paul's dismay, drafted as an alternative to the ERA a Declaratory Act that would guarantee that the generic "man" in the constitution included women. Neither did the Woman's Party trust or expend much energy on tactics like those employed by the Women's Division of the Republican National Committee, which fought for a series of all-women breakfasts with Eisenhower.

Second, decentralization functioned to sustain individual commitment by creating a strong sense of community among those working toward common goals, and thus helped to ensure the survival of feminist activity in the face of strong opposition from the larger society.[30] Many women's rights supporters literally devoted their lives to feminist activity. Yet because the separate organizations and networks of women that comprised the movement were held together by rather loose ties and often were hierarchical organizations in which the leaders enjoyed tremendous respect and influence, the movement as a whole was ridden with conflict and competition.[31] To a large extent, this conflict centered around the fact that the National

Woman's Party, throughout its history, believed that no other organization could be trusted to devote itself wholeheartedly to the ERA. In effect, the continuing competition and conflict between different movement organizations, together with the emphasis placed on moral commitment to the cause and on the maintenance of close personal ties among core members, contributed to a lack of emphasis on the movement's larger goals for institutional change.

Even though the movement's decentralized structure promoted conflict and continuing factionalism among movement organizations, this type of configuration had one important advantage in the struggle for the ERA. Each of the separate organizations, groups, and networks of women who comprised the movement had a different base of support and sphere of influence and was able to make use of its ties to and status in established institutions to promote the amendment in ways that might not have been possible had the Woman's Party remained the sole representative of the movement.

Movement Infrastructure

Social movement scholars generally agree that the mobilization potential of a movement is affected by the degree of preexisting organization among dominated groups.[32] If the social base of a movement is already linked together in some sort of primary network so that people have common values and moral commitments, interpersonal ties, and communication on a regular basis, then this prior group solidarity can serve as a basis for mobilization.

The infrastructure or organizational base of the post-1945 women's rights movement can be traced to three main sources: a shared bond of participation in the suffrage struggle, business and professional ties, and friendship and other close personal relationships. It is not surprising that participation in the suffrage struggle helped to hold together a community of women across the years, since a commitment to militancy had required great sacrifices and forged strong bonds. Professional ties also played a part in recruiting women to feminist activity and sustaining their commitment. Since most of the feminists about whom information was available held professional positions, it was not uncommon for them to have become involved in women's rights work through professional associations such as the BPW, AAUW, or the National Association of Women Lawyers. Margaret Hickey, for example, attributed her interest in the women's rights movement to her contact with a network of powerful women lawyers involved in feminist activities. Other women came to feminist work out of personal relationships: family ties, friendship, or committed couple relationships between women. Primary group networks, then, were at the root of most women's involvement in the movement. Whether these grew out of a common suffrage heritage, professional ties, or other strong emotional and social bonds, the community of women that formed the core of the movement was held together by close and loving friendships.

Women involved in the movement in this period created a homosocial world through which they carried out their political work and maintained a female public sphere that facilitated the survival of feminism.[33] Only women without family obligations, with supportive husbands, or in relationships with other women who shared their commitment to women's rights were likely to find that their personal lives meshed well with their work in the movement. Thus, those who carried forward the feminist challenge in this period were not, on the whole, the women who suffered the greatest tension between work and family roles.

Myra Marx Ferree and Frederick Miller discuss three general strategies a movement can use to mobilize participants: conversion that creates new ideological commitments among individuals; coalition formation that mobilizes ideological commitments already held by individuals and groups; and direct mobilization that relies on network ties and situational factors to bring about participation without ideological commitment.[34] Because the women's rights movement in this period was rooted in strong primary relationships and intense personal ties, the mobilization strategies the movement relied on were coalition formation among previously connected networks and groups and, to a lesser extent, direct mobilization of the membership of large organizations by the leadership. Movement organizations used newsletters to keep their own members informed about their programs and actions. When the Lucy Stone League was revived in New York, one of its most important activities was publication of a newsletter, and when the National Woman's Party journal, *Equal Rights*, ceased publication in 1954, many viewed it as a sign of decline. In addition, multiple organizational memberships were common. Many National Woman's Party members belonged to the BPW, the AAUW, and later to NOW; some even joined the anti-ERA League of Women Voters. They used these multiple memberships to try to win or maintain endorsements of the ERA.

Preexisting organizations and networks of women, although they were unable to influence wide-scale change or disseminate their views to large audiences, had access to important resources that made it possible to maintain a commitment to women's rights in a climate of antifeminism. Belmont House was used by members and nonmembers of the Woman's Party as a feminist hotel, meeting place, and center of organizational activities. Leaders of women's organizations had access to politicians and other decision-makers and to existing networks of women who could be mobilized for activities such as letter-writing. And, because feminist activists were largely middle- and upper-class women with the financial means and time to devote themselves to the cause, many were able to make women's rights their life work. Yet the very factors that served to integrate feminists into this close-knit community of women and made it possible to sustain commitment also isolated movement organizations and made it difficult to recruit younger and less privileged women.

Preexisting structures, then, played a larger part in developing and maintaining the movement in this period than did spontaneous individual and group actions aroused by a sense of injustice. At a time when large numbers of American women were not drawn to feminism, the women's rights movement relied on available resources from preexisting organizations and networks. It is important, however, in examining the historical development of a movement, to distinguish between different types of resources necessary for movements mobilizing for survival and continuance compared with movements undergoing mass or "white-hot mobilization."[35] The women's rights movement, from the end of World War II until the resurgence of more widespread feminist activism in the 1960s, was in a continuance stage, which required a highly dedicated core of activists held together by a willingness to devote themselves almost single-mindedly to the cause, close interpersonal ties, a shared moral commitment, and strong leadership capable of coordinating and devising multiple strategies and actions.

Constraints and Strategic Choices

The actions and strategies of social movements are, by definition, noninstitutionalized.[36] There is a tendency, however, to think of only those actions that are dramatic and discontinuous from the everyday acts of social institutions as constituting social movement activity. But social movements can and do employ a range of tactics—respectable and nonrespectable, disruptive and nondisruptive, threatening and nonthreatening, violent and nonviolent, covert and overt—depending on whether their aims are bargaining, persuasion, or coercion.[37] Jo Freeman identifies four factors that affect the strategies adopted by social movements: available resources, constraints on the use of those resources, a movement's own organization, and its expectations about the groups it hopes to influence.[38]

Given the composition of the women's rights movement in the post-1945 period and the kinds of support and other resources feminists could realistically expect to mobilize in an inhospitable political climate, only certain kinds of strategies were available. On the whole, groups adopted persuasive strategies aimed at affecting national issues and actions and pursued tactics that could be accomplished with the support of elites in government and politics, while avoiding actions such as public demonstrations and protests that required the involvement of large numbers of participants. For example, the National Woman's Party instigated the inclusion of the prohibition on sex discrimination in Title VII of the Civil Rights Act through its contacts with conservative Representative Howard Smith. Women's organizations worked for the ERA by lobbying Congress, seeking endorsements from candidates for political office and other organizations, establishing coalitions to support the amendment, and educating the public through newspaper and magazine articles, letters to the editor, and radio and television appearances.

Women organized letter-writing campaigns on every issue the movement supported—not only the ERA, but the election of pioneer feminists like Susan B. Anthony to the Hall of Fame and the support of women candidates for elected offices and government appointments. Sometimes these campaigns were undertaken by a single individual, but attempts were almost always made to make the efforts of individuals appear to represent a wider constituency, for example, by printing up organizational letterhead even if only one individual wrote all the letters and constituted the entire membership. We have seen that Florence Kitchelt at times did almost all of the work of the Connecticut Committee for the ERA, as did Fannie Ackley for the Spokane branch of the National Woman's Party.

The core activists in the movement, although relatively homogeneous in terms of social characteristics, were diverse politically. They were Democrats and Republicans, liberals and conservatives, integrationists and segregationists, McCarthyites and anti-McCarthyites, who agreed upon little else politically except their feminist goals. Despite the fact that these women's efforts on behalf of feminism went against the broad social trends of the times, the movement as a whole sought respectability in its relationship to almost all other social and political issues. Because the ERA was in this period opposed by labor, liberals, left-wing organizations, and most women's groups, feminists did not generally align their efforts with other progressive interests. When they did, it was out of self-interest, as was the case with the attempt to include sex discrimination in civil rights legislation, a campaign which was viewed with suspicion by many civil rights groups. Similarly, the National Woman's Party tried to give the appearance that women in labor were involved in women's movement activities through spotlighting token union women who supported the ERA, but it never really addressed the concerns of women in labor over the issue of protective legislation and how it would be affected by the ERA.

At the same time, feminists were willing to use almost any strategy, no matter how contradictory to the basic principles of social equality, to further their goals as long as it did not involve engaging in mass protest and other disruptive tactics. For example, the National Woman's Party attempted to win support for the ERA by using the fact that the Communist Party opposed the amendment to threaten vulnerable groups and individuals. In sum, the relationships that feminists established with other major social movements of the period were limited and undertaken only to advance the women's rights movement's own narrowly defined feminist goals.

Consistent with what other scholars have found, the strategies and tactics of the women's rights movement in this period were influenced not only by its limited opportunities for collective action but by its base of mobilization and its position in the power structure.[39] The movement operated from a small social base of well-educated, economically privileged, and already committed women with ties to established institutions. Recognizing the unlikelihood of generating wide support for its aims, the movement had little choice but to adopt legal change and institutionalized reform as its ultimate goals, and to use strategies that were respectable and non-

disruptive to persuade political elites and other established groups of the
righteousness of its cause.

Conclusion

For many scholars, the most interesting and debated research question about
the contemporary American women's movement has been why it emerged
when it did. This question, however, presupposes only two waves of the
movement, the suffrage movement of the late nineteenth and early twentieth
centuries and the resurgent movement of the late 1960s and 1970s. As we
have suggested, such a view of the women's movement ignores the existence
of the feminist challenge in periods when its proponents have been unable
to mobilize a large base of support. This research began with an entirely
different question. We asked what had happened to the groups and individuals who had earlier worked for women's rights when they faced a period
known for its virulent opposition to feminism, a period that nevertheless
stood on the eve of the rebirth of a massive women's movement. The answer
to this question emerges as the story of the women's rights movement in a
stage of survival and continuance.

By tracing the history of the women's rights movement in a period in
which it has been assumed by all but a handful of scholars that no feminist
activity existed, this research provides evidence for the historical continuity
of the American women's movement. Arguing for an approach to social
movements that recognizes movement continuity, Joseph Gusfield calls for
the study of movements that "remain dormant and reemerge with their
rhetoric shining and seemingly unused."[40] The findings presented here suggest that in periods that are unfavorable to their aims, social movements
develop certain kinds of structures and strategies that ensure their continuation until the equation of forces changes, making it possible to resume
more overt mass-based protest.

After the suffrage victory had been won, it was only a minority of
women—variously defined as "hard-line feminists" or "radicals"—who,
seeing the vote as only the first step toward achieving full equality for women,
attempted to enlist support to continue the struggle for equal rights.[41] The
more these feminists insisted on the need to organize to promote women's
own special interests, the more they were ridiculed as "old-fashioned."
Subsequently, by the post-1945 period, the women's rights movement had
dwindled to a handful of organizations and individuals spearheaded by the
National Woman's Party, heir to the militant branch of the suffrage
movement.

We have chosen the term "elite-sustained" to describe the women's
rights movement throughout the 1940s and 1950s because the movement
was comprised of a small, relatively affluent, well-educated, and highly
dedicated core of women who, despite little support and strong opposition
from the wider public, carried on the struggle for women's rights first begun

by the suffragists. Yet, as encapsulated as the movement might have been in its own time and as little progress as it made toward its major goal, the ERA, the activities of feminists in the post-1945 period had an impact on the goals and strategies, ideology, and composition of the resurgent women's movement that grew in the late 1960s.

The question of when and why a movement takes this shape is critical to understanding different cycles of movement activity. Proponents of the resource mobilization perspective suggest three factors that affect the nature and development of a movement: (1) the structure of political opportunities, especially events external to the movement, that inhibit or contribute to the availability of resources and the success of tactics; (2) the organizational strength of movement forces that influence its base and breadth of mobilization and tactics; and (3) the nature of the group's own resources, including its ideology, which influence its tactical and structural choices.

Applying these factors to this case, it is evident, first, that groups and individuals that attempted to sustain the struggle for women's rights from 1945 to the 1960s recognized that they were working in a social climate that was generally hostile to feminism. They expected and found little external support—publicity, financial assistance, or favorable sentiment—for the cause of women's rights. Second, the movement that managed to survive the antifeminist backlash suffered organizational problems that prevented it from broadening its base of support. Its core group, the National Woman's Party, was, in essence, a remnant of a previous stage of more successful mobilization that had no mass base. Born out of the factionalism that split the suffrage movement, the Woman's Party and the women's rights movement as a whole continued to be plagued with conflict and factionalism over goals, strategies, and leadership throughout the post-1945 years. In addition to internal problems, women's rights groups lacked sufficient ties to other organized constituencies, such as women in labor, socialist women, and black women, which prevented them from launching a unified and broad-based challenge on behalf of women's rights.

The third factor that contributed to the movement's structural and tactical choices was the nature of the resources that its constituent groups had available to them. Most of the organizations and networks of women who comprised the movement were privileged groups that, because of the sponsorship of a loyal supporter or number of supporters, pursued limited goals consistent with the kinds of resources they were able to obtain on the basis of the high status of movement participants. Thus time, money, informal connections to decision-makers in politics and government, legal expertise, and communication networks with mainly middle-class professional women were the major resources the movement was able to mobilize.

These findings suggest a different model for understanding recurrent collective action around a common set of interests than the one proposed by traditional scholars of social movements. Too often, a preference for studying movement emergence rather than continuity has led to an analysis of each cycle of growth and decline of a social movement as though it were independent of prior stages of collective action around the same grievances. Our research questions the traditional life-cycle model that conceives of social movements as having discernible beginnings and endings. Perhaps social movements based on long-standing grievances are better understood in terms of their ebb and flow. To the extent that the women's rights movement of the 1940s and 1950s was shaped by the suffrage cause, it was a product of another stage of feminist mobilization. By exploring this link as well as the connections between the resurgent women's movement of the 1960s and that of the earlier decades, we have demonstrated that in the case of the American women's movement, " 'emergence' does not imply discontinuity with the past."[42]

We end by asking how this study of the past feminist struggle helps us to understand the contemporary women's movement. Much has been written in the 1980s about the decline of the women's movement and the "post-feminist generation." While few would claim that the feminist movement of today has suffered the same devastating loss of participants and influence that decimated the movement in the fifties, most observers find the 1980s less supportive of feminist activism than the previous two decades.[43] By illustrating one way in which an unfavorable political climate can affect the structural and strategic options of the movement, this case study provides some insight into potential negative and positive consequences of feminist survival in the doldrums. For the community that feminists built in the fifties isolated them from younger and less privileged women, and the elite-sustained movement's focus on the ERA to some extent bequeathed a legacy of limitation to the resurgent movement. At the same time, to return to Caruthers Berger's image of the women's rights movement as the legendary Sergeant York, we should not overlook the potential contributions of the isolated sharpshooter, even if it requires a feminist army to win the war.

Notes

Abbreviations

AAUW	American Association of University Women
BPW	National Federation of Business and Professional Women's Clubs
JFKL	John F. Kennedy Library
LC	Library of Congress
LWV	League of Women Voters
NFBPWC	National Federation of Business and Professional Women's Clubs
NWP	National Woman's Party
NWTUL	National Women's Trade Union League
SL	Schlesinger Library
UMSL	University of Missouri, St. Louis
WJLC	Women's Joint Legislative Committee
YWCA	Young Women's Christian Association

Chapter 1. Introduction

1. Anita Pollitzer to Miriam Holden, Sept. 27, 1947, National Woman's Party papers, reel 92.

2. Caroline Babcock, Sept. 20, 1948, Babcock-Hurlburt papers, box 9 (111), Schlesinger Library, Radcliffe College.

3. Marie Moore Forrest and Cornelia Carter to Tiera Farrow, Jan. 1, 1949, NWP papers, reel 94.

4. Florence Kitchelt to Ethel Ernest Murrell, Sept. 7, 1953, Kitchelt papers, box 4 (87), SL.

5. Rose Arnold Powell, diary entry, July 19, 1955, Powell papers, box 1, v. 7, SL.

6. Anita Pollitzer to Warren R. Austin, Feb. 5, 1948, NWP papers, reel 92.

7. Rose Arnold Powell to Cecil B. DeMille, Aug. 27, 1957, Powell papers, box 2 (21), SL.

8. Charles Tilly, *From Mobilization to Revolution* (Reading, Mass.: Addison-Wesley, 1978), p. 61.

9. Florence Kitchelt to Katharine Ludington, Aug. 14, 1950, Kitchelt papers, box 6 (175), SL.

208   NOTES

10. Katharine Hepburn, speech to the Connecticut Committee, n.d. [1946?], Kitchelt papers, box 6 (153), SL.

11. Charlotte Perkins Gilman, "Women are Free at Last in All the Land," *Suffragist*, Sept. 1920.

12. "Colby Proclaims Woman Suffrage," *New York Times*, Aug. 27, 1920.

13. Scholarship on the women's movement in the interwar years includes: Susan D. Becker, *The Origins of the Equal Rights Amendment: American Feminism Between the Wars* (Westport, Conn.: Greenwood Press, 1981); William H. Chafe, *The American Woman: Her Changing Social, Economic, and Political Roles, 1920–1970* (New York: Oxford University Press, 1972); Nancy F. Cott, "Feminist Politics in the 1920s: The National Woman's Party," *Journal of American History* 71 (June 1984), 43–68; Peter Geidel, "The National Woman's Party and the Origins of the Equal Rights Amendment, 1920–1923," *Historian* 42 (1980), 557–582; Susan M. Hartmann, "Women's Organizations During World War II: The Interaction of Class, Race and Feminism," in *Woman's Being, Woman's Place*, ed. Mary Kelley (Boston: G. K. Hall, 1979), pp. 313–328; J. Stanley Lemons, *The Woman Citizen: Social Feminism in the 1920's* (Urbana: University of Illinois Press, 1973); Marjory Nelson, "Ladies in the Streets: A Sociological Analysis of the National Woman's Party, 1910–1930," Diss., SUNY-Buffalo, 1976; Cynthia M. Patterson, "New Directions in the Political History of Women: A Case Study of the National Woman's Party's Campaign for the Equal Rights Amendment, 1920–1927," *Women's Studies International Forum* 5 (1982), 585–597; Lois Scharf, *To Work and to Wed: Female Employment, Feminism and the Great Depression* (Westport, Conn.: Greenwood Press, 1980); Lois Scharf and Joan M. Jensen, *Decades of Discontent: The Women's Movement, 1920–1940* (Westport, Conn.: Greenwood Press, 1983).

14. The most relevant work on the post–1945 period is Cynthia E. Harrison, "Prelude to Feminism: Women's Organizations, the Federal Government and the Rise of the Women's Movement, 1942 to 1968," Diss., Columbia University, 1982, which examines the work of three groups struggling to improve women's position in society and in the work force during this period. Harrison does not, however, consider this activity to have been part of a social movement. Although our interpretations differ, we see her work as complementary to ours. Two volumes in a new series on twentieth-century American women's history recognize "seeds of change" in the 1940s and 1950s. While they do not focus on the women's movement, both Susan M. Hartmann, *The Home Front and Beyond: American Women in the 1940's* (Boston: Twayne, 1982); and Eugenia Kaledin, *Mothers and More: American Women in the 1950's* (Boston: Twayne, 1984), point to the conditions which, in Hartmann's words, "worked a deeper transformation in women's consciousness, aspirations, and opportunities a generation or so later" (p. 215).

In addition, scholars interested in the origins of the contemporary women's movement have given some consideration to the 1950s and early 1960s. Jo Freeman, *The Politics of Women's Liberation* (New York: David McKay, 1975); Sara Evans, *Personal Politics: The Roots of Women's Liberation in the Civil Rights Movement and the New Left* (New York: Knopf, 1979); and Ethel Klein, *Gender Politics: From Consciousness to Mass Politics* (Cambridge: Harvard University Press, 1984), explore the roots of the contemporary movement.

Finally, Olive Banks, *Faces of Feminism: A Study of Feminism as a Social Movement* (Oxford: Martin Robertson, 1981), argues for the historical continuity of an organized women's movement in both the United States and Britain, although she does not discuss the post–1945 period in any detail.

15. June Sochen, *Movers and Shakers: American Women Thinkers and Activists, 1900–1970* (New York: Quadrangle/The New York Times Book Co., 1973).

16. On the civil rights movement, see Aldon Morris, *The Origins of the Civil Rights Movement: Black Communities Organizing for Change* (New York: The Free Press, 1984); Doug McAdam, *Political Process and the Development of Black Insurgency, 1930–1970* (Chicago: University of Chicago Press, 1982); Harvard Sitkoff, "American Blacks in World War II: Rethinking the Militancy-Watershed Hypothesis," in *The Home Front and War in the Twentieth Century: The American Experience in Perspective*, ed. James Titus (Washington, D.C.: U.S. Government Printing Office, 1984), pp. 147–155; and Peter J. Kellogg, "Civil Rights Consciousness in the 1940's," *Historian* 42 (1979), 18–41.

On the gay rights movement, see John D'Emilio, *Sexual Politics, Sexual Communities: The Making of a Homosexual Minority in the U.S., 1940–1970* (Chicago: University of Chicago Press, 1983); and Phyllis Gorman, "An Analysis of the Social Movement Organization the Daughters of Bilitis, 1955–1963," M.A. thesis, Ohio State University, 1985.

17. We are grateful to Laurel Richardson for suggesting the term "elite-sustained."

18. Mayer N. Zald and Roberta Ash, "Social Movement Organizations: Growth, Decay and Change," *Social Forces* 44 (1966), 327–341; Luther P. Gerlach and Virginia H. Hine, *People, Power, Change: Movements of Social Transformation* (Indianapolis: Bobbs-Merrill, 1970); John D. McCarthy and Mayer N. Zald, *The Trend of Social Movements in America: Professionalization and Resource Mobilization* (Morristown, N.J.: General Learning Press, 1973); Anthony Oberschall, *Social Conflict and Social Movements* (Englewood Cliffs, N.J.: Prentice-Hall, 1973); William A. Gamson, *The Strategy of Social Protest* (Homewood, Ill.: Dorsey, 1975); Charles Tilly et al., *The Rebellious Century, 1830–1930* (Cambridge: Harvard University Press, 1975); Charles Tilly, *From Mobilization to Revolution* (Reading, Mass.: Addison-Wesley, 1978); Charles Perrow, "The Sixties Observed," in *The Dynamics of Social Movements*, eds. Mayer N. Zald and John D. McCarthy (Cambridge, Mass.: Winthrop Publishers, 1979), pp. 192–211; McAdam, *Political Process*; J. Craig Jenkins, "Resource Mobilization Theory and the Study of Social Movements," *Annual Review of Sociology* 9 (1983), 527–553.

19. David G. Bromley and Anson D. Shupe, Jr., *"Moonies" in America: Cult, Church, and Crusade* (Beverly Hills, Calif.: Sage Publications, (1979); David G. Bromley and Anson D. Shupe, Jr., "Repression and the Decline of Social Movements: The Case of the New Religions," in *Social Movements of the Sixties and Seventies*, ed. Jo Freeman (New York: Longman, 1983), pp. 335–347.

20. McAdam, *Political Process*; Morris, *Origins of the Civil Rights Movement*.

Chapter 2. The Social Context of Women's Rights Activism

1. Edna Capewell to Alice Paul, Nov. 24, 1945, NWP papers, reel 88.

2. Doris Stevens to Betty Gram Swing, Jan. 8, 1946, Stevens papers, carton 4, SL.

3. Mary Sinclair Crawford to Florence Armstrong, Aug. 30, 1946, Armstrong papers, box 3 (42), SL.

4. Jo Freeman, *The Politics of Women's Liberation* (New York: David McKay, 1975); Joan Huber and Glenna Spitze, *Sex Stratification: Children, House-*

work, and Jobs (New York: Academic Press, 1983); Ethel Klein, *Gender Politics: From Mass Consciousness to Mass Politics* (Cambridge: Harvard University Press, 1984).

5. Valerie Kincade Oppenheimer, "Demographic Influence on Female Employment and the Status of Women," in *Changing Women in a Changing Society*, ed. Joan Huber (Chicago: University of Chicago Press, 1973), pp. 184–199.

6. Huber and Spitze, *Sex Stratification*; Oppenheimer, "Demographic Influence."

7. Figures from the U.S. Bureau of the Census, 1983b, cited in Myra Marx Ferree and Beth B. Hess, *Controversy and Coalition: The New Feminist Movement* (Boston: Twayne, 1985), p. 4.

8. Freeman, *Politics of Women's Liberation*; Klein, *Gender Politics*.

9. Eugenia Kaledin, *Mothers and More: American Women in the 1950's* (Boston: Twayne, 1984), p. 41.

10. See Lyn Goldfarb, *Separated and Unequal: Discrimination Against Women Workers After World War II* (Washington, D.C.: Union for Radical Political Economics Political Education Project, n.d.); Karen Beck Skold, "The Job He Left Behind: American Women in the Shipyards During World War II," in *Women, War, and Revolution*, eds. Carol R. Berkin and Clara M. Lovett (New York: Holmes and Meier, 1980), pp. 55–75; Sheila Tobias and Lisa Anderson, *What Really Happened to Rosie the Riveter*, MSS Modular Publication, Module 9 (1974); and Susan M. Hartmann, *The Home Front and Beyond: American Women in the 1940's* (Boston: Twayne Publishers, 1982).

11. The use of the term "snow-blindness" to describe the invisibility of women of color to white society comes from Adrienne Rich, " 'Disloyal to Civilization:' Feminism, Racism, and Gynephobia," *Chrysalis* 7 (1979), 9–27.

12. The classic description and analysis of the nineteenth-century image is Barbara Welter, "The Cult of True Womanhood, 1820–1860," *American Quarterly* 18 (1966), 151–174. See Betty Friedan, *The Feminine Mystique* (New York: Dell, 1963).

13. This analysis is based on a survey of women's and other popular magazines as well as the findings of other researchers. Douglas T. Miller and Marion Nowak, *The Fifties: The Way We Really Were* (Garden City, N.Y.: Doubleday, 1977), is a popular survey that includes a chapter on women. Susan Hartmann's book, *Home Front and Beyond*, includes a chapter on women in popular culture. Kathryn Weibel, *Mirror, Mirror: Images of Women Reflected in Popular Culture* (Garden City, N.Y.: Doubleday, 1977), surveys a variety of media from the nineteenth century to the present. L. Ann Geise, "The Female Role in Middle Class Women's Magazines from 1955 to 1976: A Content Analysis of Nonfiction Selections," *Sex Roles* 5 (1979), 51–62, is a systematic study of *Redbook* and *Ladies' Home Journal*. Studies of the image of women in other media include: Stuart Ewen, *Captains of Consciousness: Advertising and the Social Roots of the Consumer Culture* (New York: McGraw-Hill, 1976); Janice Welsch, "Actress Archetypes in the 1950's: Doris Day, Marilyn Monroe, Elizabeth Taylor, Audrey Hepburn," in *Women and the Cinema*, eds. Karyn Kay and Gerald Peary (New York: Dutton, 1977), pp. 99–111; and Kay F. Reinartz, "The Paper Doll: Images of American Women in Popular Song," in *Women: A Feminist Perspective*, ed. Jo Freeman (Palo Alto, Calif.: Mayfield Publishing Co., 1975), pp. 293–308.

14. Judith Tarcher, "Most Likely to Succeed," *Good Housekeeping* 126 (June 1948), 33.

15. Bea Carroll, "How to Get Along With Women," *Ladies Home Journal* 67 (January 1950), 73.

16. Elaine Whitehall, "I'd Hate to be a Man," *Coronet* 37 (January 1955), 27–30; "I'm Lucky! Lucky!" *Ladies' Home Journal* 77 (February 1960) 63. See also Poppy Cannon, "It's a Man's World . . . Maybe," *Woman's Home Companion* 74 (January 1947), 40; Fredda Dudley Balling, "The Meeker Sex," *Good Housekeeping* 132 (March 1951), 64; "The New (?) Woman," *Ladies' Home Journal* 69 (April 1952), 54–55; Benjamin Spock, "What's She Got That I Haven't?" *Ladies' Home Journal* 69 (October 1952), 56–57; Michael Drury, "The Women in My Life," *Good Housekeeping* 144 (January 1957), 55.

17. See Friedan, *Feminine Mystique*, pp. 117–141; and Miller and Nowak, *The Fifties*, pp. 150–151.

18. Quoted in Klein, *Gender Politics*, p. 38.

19. William H. Chafe, *The American Woman: Her Changing Social, Economic, and Political Role, 1920–1970* (New York: Oxford University Press, 1972), pp. 183, 192.

20. See Hazel Erskine, "The Polls: Women's Role," *Public Opinion Quarterly* 35 (Summer 1971), 282–287; and Barbara McGowan, "Postwar Attitudes Toward Women and Work," in *New Research on Women and Sex Roles*, ed. Dorothy McGuigan (Ann Arbor: University of Michigan Center for Continuing Education of Women, 1976). Ethel Klein, "A Social Learning Perspective on Political Mobilization: Why the Women's Movement Happened When It Did," Diss., University of Michigan, 1979, cites a *Fortune* poll from 1946 that shows a significantly higher proportion of women and men approving of work for married women with no children under sixteen. Klein's dissertation is the basis of her book, *Gender Politics*.

21. See Klein, "Social Learning Perspective," pp. 72–77; and D'Ann Campbell, "Wives, Workers and Womanhood: America During World War II," Diss., University of North Carolina, 1979. Campbell's dissertation has been published as *Women At War With America: Private Lives in a Patriotic Era* (Cambridge: Harvard University Press, 1984).

22. Klein, "Social Learning Perspective," p. 37.

23. Barbara Benson, "Do Men or Women Lead the Harder Life," *Ladies' Home Journal* 64 (May 1947), 44–45.

24. See Herbert J. Gans, *The Levittowners* (New York: Random House, 1967).

25. Elizabeth Douvan, "Differing Views on Marriage—1957 to 1976," *Newsletter*, Center for Continuing Education on Women, 12 (Spring 1979), 1–2; quoted in Klein, "Social Learning Perspective," p. 120.

26. Michael Harrington, *The Other America* (New York: Macmillan, 1962), p. 9.

27. See Miller and Nowak, *The Fifties*, p. 155.

28. Klein, "Social Learning Perspective," pp. 83–96. See also Robert H. Bremner, "Families, Children, and the State," in *Reshaping America*, ed. Robert H. Bremner and Gary Reichard (Columbus: Ohio State University Press, 1982), pp. 3–32.

29. Alfred C. Kinsey et al., *Sexual Behavior in the Human Female* (Philadelphia: W. B. Saunders Co., 1953). See Miller and Nowak, *The Fifties*, p. 157.

30. Marilyn French, *The Women's Room* (New York: Summit Books, 1977).

31. "American Woman's Dilemma," *Life* 22 (June 16, 1947), 101–111.

32. "The Girls," *Life* 27 (Aug. 15, 1949), 39–40.

33. "An Introduction by Mrs. Peter Marshall," *Life* 41 (Dec. 24, 1956), 2–3.

See also Max Lerner, "Let's Play Square With Women Who Choose to Defy Tradition," New York *Star*, Sept. 23, 1948; Margaret Mead, "What Women Want," *Fortune*, n.d., clipping in Florence Kitchelt papers, box 2 (30), SL; Edward L. Bernays, "The Two Lives of Women," *McCall's*, n.d., clipping in Florence Kitchelt papers, box 3 (40), SL. For a similar analysis of public expressions of discontent, see Marty Jezer, *The Dark Ages: Life in the U.S., 1945–1960* (Boston: South End Press, 1982), pp. 231–234.

34. Lynn White, Jr., *Educating Our Daughters* (New York: Harper, 1950).

35. See Miller and Nowak, *The Fifties*, chapt. 13.

36. For a discussion of the relationship between images of women and business, see Friedan, *Feminine Mystique*, pp. 197–223; Weibel, *Mirror, Mirror*, pp. 135–173; and especially Ewen, *Captains of Consciousness*, pp. 159–184.

37. Kaledin, *Mothers and More*, p. 214.

38. See, for example, David Riesman, *The Lonely Crowd* (New Haven: Yale University Press, 1961); William F. Whyte, *The Organization Man* (New York: Simon and Schuster, 1956); David Potter, *People of Plenty* (Chicago: University of Chicago Press, 1954); Louis Hartz, *The Liberal Tradition in America* (New York: Harcourt, Brace, 1955); and Daniel J. Boorstin, *The Americans* (New York: Random House, 1958).

39. "American Women," *Life* 18 (Jan. 29, 1945), 28.

40. See, for example, John P. Dolch, "American Girls Are Swell, But . . . ," *American Magazine* 141 (January 1946), 45; Victor Dallaire, "The American Woman? Not for this GI," *The New York Times Magazine*, March 10, 1946; Agnes E. Meyer, "A Challenge to American Women," *Collier's* 117 (May 11, 1946), 15. On antifeminism in the 1920s and 1930s, see Susan D. Becker, *The Origins of the Equal Rights Amendment* (Westport, Conn.: Greenwood Press, 1981), pp. 57–64.

41. Philip Wylie, *Generation of Vipers* (New York: Rinehart & Co., 1942); Edward A. Strecker, *Their Mothers' Sons: The Psychiatrist Examines an American Problem* (Philadelphia: Lippincott, 1946). See Chafe, *The American Woman*, p. 201.

42. Agnes E. Meyer, *Out of These Roots: The Autobiography of an American Woman* (Boston: Little, Brown, 1953).

43. Ferdinand Lundberg and Marynia F. Farnham, *Modern Woman: The Lost Sex* (New York: Harper and Brothers, 1947). See Chafe, *The American Woman*, pp. 202–207; and Miller and Nowak *The Fifties*, pp. 153–155.

44. Arnold W. Green and Eleanor Melnick, "What Has Happened to the Feminist Movement?" in *Studies in Leadership*, ed. Alvin W. Gouldner (New York: Russell and Russell, 1950), p. 183.

45. Struthers, Burt, "Women, Dog Dab 'Em!" *Ladies' Home Journal* 64 (November 1947), 11.

46. Interview no. 1 and interview no. 2.

47. Mary Beard to W. K. Jordan, April 4, 1947, Mary Beard papers, box 2 (29), SL.

48. W. K. Jordan to Mary Beard, April 7, 1947, Mary Beard papers, box 2 (29), SL.

49. Ralph S. Banay, "The Trouble With Women," *Collier's* 118 (Dec. 7, 1946), 21.

50. Ann MacLeod, "Should Women Vote?" *Ladies Home Journal* 75 (February 1958), 6.

51. Willard Waller, "The Coming War on Women," *This Week*, n.d., clipping in Florence Kitchelt papers, box 3 (36), SL.

52. Letter to the editor of the *Star*, Nov. 3, 1946; carbon in League of Women Voters papers, box 700, Library of Congress.

53. Robert C. Ruark, quoted in Marguerite Rawalt, *History of the National Federation of Business and Professional Women's Clubs* (Washington, D.C.: NFBPWC, 1969), pp. 46–47.

54. Dorothy Thompson, "Are Women Different?" *Woman's Home Companion* 68 (November 1951), 11.

55. Cornelia Otis Skinner, "Women are Misguided," *Life* 41 (Dec. 24, 1956), 73.

56. Doris E. Fleischman (Bernays), "Notes of a Retiring Feminist," *American Mercury* 68 (February 1949), 161–168.

57. Doris Fleischman Bernays, *A Wife Is Many Women* (New York: Crown Publishers, 1955).

58. Jane Grant to Florence Kitchelt, Sept. 16, 1944, Kitchelt papers, box 6 (147), SL.

59. Rose Arnold Powell to Alice Paul and Anita Pollitzer, Aug. 3, 1948, NWP papers, reel 94.

60. Ella Sherwin to Caroline Babcock, July 13, 1945, NWP papers, reel 87; Caroline Babcock to Lucy Rice Winkler, Sept. 1, 1945, NWP papers, reel 87.

61. Margaret Hickey, remarks as chairman of panel on "Women's Responsibility in Public Affairs," n.d., Hickey papers, University of Missouri, St. Louis.

62. Susan B. Anthony II, "Is It True What They Say About Women?," *Woman's Home Companion* 72 (June 1945), 22; "Do Women Have Equal Rights?" *Good Housekeeping* 127 (December 1948), 38–39; Ashley Montagu, "The Natural Superiority of Women," *Ladies' Home Journal* 69 (July 1952), 36–37.

63. Jane Grant to Caroline Babcock, n.d. (January 1945), NWP papers, reel 84; also telegram from Jane Grant to Alice Paul, Jan. 27, 1945, NWP papers, reel 84; Jane Grant to Florence Kitchelt, Jan. 27, 1945, Kitchelt papers, box 3 (38), SL.

64. Florence Kitchelt to Ashley Montagu, March 4, 1952, Kitchelt papers, box 7 (191), SL; Florence Kitchelt to Mrs. Harrison Smith, March 5, 1952, Kitchelt papers, box 7 (191), SL; Florence Kitchelt to Alma Lutz, March 9, 1952, Kitchelt papers, box 7 (178), SL; Alma Lutz to Florence Kitchelt, March 12, 1952, Kitchelt papers, box 7 (191), SL; Jane Grant to Florence Kitchelt, n.d. (1952), Kitchelt papers, box 7 (191), SL; Alma Lutz to Rose Arnold Powell, May 29, 1953, Powell papers, box 3 (43), SL.

65. Alma Lutz to Florence Kitchelt, Feb. 26, 1953, Kitchelt papers, box 7 (178), SL; Alma Lutz to Florence Kitchelt, March 9, 1953, Kitchelt papers, box 7 (178), SL; Florence Kitchelt to Alma Lutz, March 2, 1953, Kitchelt papers, box 7 (178), SL; Florence Kitchelt to Esther Caukin Brunauer, April 1, 1953, Kitchelt papers, box 5 (122), SL.

66. Ruth Herschberger, *Adam's Rib* (New York: Pellegrini & Cudahy, 1948).

67. AAUW General Director's Letter, vol. 27 (November 1959).

68. Marjorie Longwell to Emma Guffey Miller, April 12, 1963, NWP papers, reel 108.

69. Alma Lutz to Emma Guffey Miller, Feb. 23, 1963, NWP papers, reel 108; also Mary F. Anderson to Alice Paul, Sept. 23, 1963, NWP papers, reel 108.

70. Margaret Hickey, "What Next for Women?" speech to BPW, July 7, 1946, Hickey papers, box 28, UMSL; Margaret Hickey, "What's Ahead for the Woman Who Earns," speech to the Women's Bureau Conference, March 14, 1946, Hickey papers, box 30 (112), UMSL. See "Women Report Job Chances on Tobaggan,"

Washington *Post*, n.d. (1945), clipping in Somerville-Howorth papers, box 6 (120), SL.

71. Margaret Culkin Banning, quoted in Rawalt, *History of NFBPWC*, p. 48.

72. Alma Lutz to Emma Guffey Miller, Sept. 3, 1960, NWP papers, reel 106; Alma Lutz to the editor of the *Nation*, Sept. 18, 1948, clipping in Florence Kitchelt papers, box 6 (177), SL.

73. Caroline Babcock to Anna Kelton Wiley, Dec. 6, 1947, Babcock-Hurlburt papers, box 8 (106), SL.

74. Florence Kitchelt to Annie Goodrich, May 4, 1945, Kitchelt papers, box 4 (70), SL; Florence Kitchelt to Anita Pollitzer, May 4, 1945, Kitchelt papers, box 8 (209), SL.

75. Fannie Ackley to Caroline Babcock, Nov. 29, 1945, NWP papers, reel 88.

76. Margaret Barnard Pickel, "There's Still a Lot for Women to Learn," *New York Times*, Nov. 11, 1945, VI, p. 14; Margaret Hickey, quoted in *New York Times*, Feb. 26, 1945, p. 9.

77. Green and Melnick, "What Has Happened," p. 278.

78. Interview no. 13.

79. Mirra Komarovsky, "Women Then and Now: A Journey of Detachment and Engagement," *Women's Studies Quarterly* 10 (Summer 1982), 5–9; Mirra Komarovsky, *Women in the Modern World* (Boston: Little, Brown, 1953).

80. Joan Huber, "Ambiguities in Identity Transformation: From Sugar and Spice to Professor," in *Feminist Frontiers: Rethinking Sex, Gender, and Society*, eds. Laurel Richardson and Verta Taylor (Reading, Mass.: Addison-Wesley, 1983), p. 330.

Chapter 3. The National Woman's Party

1. On the early years of the National Woman's Party, see Inez Haynes Irwin, *The Story of Alice Paul and the National Woman's Party* (1921; rep. Fairfax, Va.: Denlinger's Publishers, 1977); and Doris Stevens, *Jailed for Freedom* (1920; rep. New York: Schocken Books, 1976), both written by participants. See also Eleanor Flexner, *Century of Struggle*, rev. ed. (Cambridge, Mass.: Belknap Press, 1975); Amelia R. Fry, "The Divine Discontent: Alice Paul and Militancy in the Suffrage Campaign," unpublished paper presented at the Berkshire Conference of Women Historians, 1981; and Aileen Kraditor, *The Ideas of the Woman Suffrage Movement, 1890–1920* (New York: Columbia University Press, 1965).

2. On the Woman's Party in the 1920s, see Susan D. Becker, *The Origins of the Equal Rights Amendment* (Westport, Conn.: Greenwood Press, 1981); Nancy F. Cott, "Feminist Politics in the 1920s: The National Woman's Party," *Journal of American History* 71 (June 1984), 43–68; Peter Geidel, "The National Woman's Party and the Origins of the ERA, 1920–1923," *Historian* 42 (1980), 557–582; Marjory Nelson, "Ladies in the Streets: A Sociological Analysis of the National Woman's Party, 1910–1930," Diss., SUNY-Buffalo, 1976; Cynthia M. Patterson, "New Directions in the Political History of Women: A Case Study of the NWP's Campaign for the ERA, 1920–1927," *Women's Studies International Forum* 5 (1982), 585–597.

3. See Susan M. Hartmann, *The Home Front and Beyond* (Boston: Twayne Publishers, 1982).

4. Olive Beale to Marye Seabrook, May 28, 1945, NWP papers, reel 86.

5. See, for example, Flexner, *Century of Struggle*; J. Stanley Lemons, *Woman*

Citizen (Urbana: University of Illinois Press, 1973); and William O'Neill, *Everyone Was Brave* (Chicago: Quadrangle Press, 1971).

6. See, for example, Arnold W. Green and Eleanor Melnick, "What Has Happened to the Feminist Movement?" in *Studies in Leadership*, ed. Alvin W. Gouldner (New York: Russell and Russell, 1950), pp. 277–302; Peggy Capewell to Jeannette Marks, Oct. 1, 1945, Babcock-Hurlburt papers, box 3 (29), SL; Alma Lutz to the editor of the New England *Teamster*, Oct. 17, 1951; Helen Thomas, UPI story, Feb. 13, 1962, NWP papers, reel 107; Maud Jamison to NWP, May 28, 1949, NWP papers, reel 95.

7. For an analysis of membership in the early years, see Nelson, "Ladies in the Streets."

8. Alice Paul, "Conversations with Alice Paul: Woman Suffrage and the Equal Rights Amendment," an oral history conducted in 1972 and 1973 by Amelia R. Fry, Regional Oral History Office, University of California, 1976, pp. 179–183, 188–189. This view was also espoused by Inez Haynes Irwin in the official history, *The Story of Alice Paul.*

9. Interview no. 12.

10. Paul, "Conversations," pp. 441–442; interview no. 1 and interview no. 5.

11. Interview no. 2.

12. In 1955, 1956, and 1957, a number of bills introduced in the Senate provided for the appropriation of the entire block in which Belmont House was located in order to build a parking lot. The National Woman's Party leadership believed that the opponents of the ERA instigated these bills in order to hamper the work of the organization.

13. See Becker, *Origins of the ERA*, pp. 89–92.

14. Marie T. Lockwood to Emma Guffey Miller, Jan. 24, 1945, NWP papers, reel 84.

15. Doris Stevens to Katherine Callery, June 23, 1946, Stevens papers, SL.

16. Doris Stevens to Betty Gram Swing, Jan. 8, 1946, Stevens papers, SL; Doris Stevens, diary entry, Sept. 28, 1945, Stevens papers, SL; Doris Stevens to Florence [no last name], Nov. 13, 1945, NWP papers, reel 88.

17. Doris Stevens to Betty Gram Swing, Jan. 18, 1946, Stevens papers, SL.

18. Olive Beale to Clara Snell Wolfe, May 6, 1947, NWP papers, reel 91; Elizabeth S. Rogers to Alice Paul, May 6, 1947, NWP papers, reel 91; Anna Kelton Wiley, Laura Berrien, Doris Stevens to Alice Paul, April 29, 1947, NWP papers, reel 91.

19. Laura Berrien and Doris Stevens, "An Open Letter to Miss Alice Paul," Committee on Information, Bulletin No. 4, July 30, 1947, Katharine A. Norris papers, box 2 (7), SL.

20. Ella M. Sherwin to Caroline Babcock, Oct. 9, 1945, Babcock-Hurlburt papers, box 7 (96), SL.

21. Paul, "Conversations," pp. 551–591.

22. Mabel [no last name] to Anita Pollitzer, April 13, 1947, NWP papers, reel 91; "To All Members and Friends of the National Woman's Party," Aug. 18, 1947, NWP papers, reel 91.

23. Lavinia Dock to Alice Paul, Feb. 26, 1946, NWP papers, reel 90; Lavinia Dock to Party, May 26, 1947, NWP papers, reel 91; Lavinia Dock to Party, May 30, 1947, NWP papers, reel 91; Lavinia Dock to Olive Lacy, July 22, 1947, NWP papers, reel 91; Lavinia Dock, "The Case of the Dissidents," Sept. 10, 1947, NWP papers, reel 92.

24. Jane Norman Smith to Anita Pollitzer, Nov. 27, 1946, NWP papers, reel 90; telegram from Nina Horton Avery to Anita Pollitzer, Jan. 8, 1947, NWP papers, reel 90; Frances Green to Agnes Wells, Aug. 20, 1947, NWP papers, reel 91; Alice Paul to Mrs. George Howard, Aug. 4, 1948, NWP papers, reel 94; Jane Grant to Florence Kitchelt, n.d., Kitchelt papers, box 6 (147), SL; James Colligan, "Rebels 'Boring From Within,' Says Woman's Party Founder," newspaper clipping, Babcock-Hurlburt papers, box 8 (102), SL.

25. Paul, "Conversations," pp. 551–591.

26. Mildred Taylor to Caroline Babcock, Sept. 1, 1948, Babcock-Hurlburt papers, box 9 (111), SL.

27. Interview no. 10.

28. Article by Cookie Wascha in the Flint *Journal*, Feb. 14, 1980.

29. Alma Lutz to Florence Kitchelt, Feb. 11, 1949, Kitchelt papers, box 6 (177), SL.

30. Gertrude Crocker to Caroline Babcock, Dec. 20, 1947, Babcock-Hurlburt papers, box 9 (108), SL.

31. Paul, "Conversations," pp. 591–594; Burnita Shelton Matthews, "Pathfinder in the Legal Aspects of Women," an oral history conducted by Amelia R. Fry, Regional Oral History Office, University of California, 1975.

32. See, for example, "Secession Splits the Woman's Party?" *New York Times*, Jan. 14, 1947; James Colligan, "Reinforcements Arrive to Aid Woman's Party Combatants," Washington *Times-Herald*, Jan. 14, 1947.

33. [Mildred Palmer] to Neenah Hastings Lessemann, Oct. 30, 1951, NWP papers, reel 98.

34. Elizabeth Forbes to Mary Sinclair Crawford, July 27, 1951, NWP papers, reel 98.

35. Revised prospectus, Nov. 18, 1952, NWP papers, reel 99; Christine Coombes to Helen Vanderbur, Sept. 29, 1952, Florence Armstrong papers, v. 3, SL.

36. Alma Lutz to Florence Kitchelt, Oct. 6, 1952, Kitchelt papers, box 7 (178), SL.

37. Florence Kitchelt to Alma Lutz, Oct. 7, 1952, Kitchelt papers, box 7 (178), SL.

38. Interview no. 2.

39. This incident involved an accusation by Helena Hill Weed that Murrell took improper credit for a pamphlet that Weed wrote and Murrell revised.

40. Ethel R. Coombes to Alice Paul, Dec. 24, 1952, NWP papers, reel 99.

41. Open letter from Ernestine Bellamy to Ethel Ernest Murrell, May 24, 1953, NWP papers, reel 99.

42. Ethel Ernest Murrell to Cecil Norton Broy, April 2, 1953, NWP papers, reel 99.

43. Open letter from Ernestine Bellamy to Ethel Ernest Murrell, May 24, 1953, NWP papers, reel 99.

44. Ethel Ernest Murrell, Biennial Report, n.d. (June 12, 1953), NWP papers, reel 99.

45. Irwin, *Story of Alice Paul*, p. 12.

46. Ethel Ernest Murrell to Alice Paul, Dec. 8, 1952, NWP papers, reel 99.

47. Alma Lutz to Florence Kitchelt, Oct. 10, 1953, Kitchelt papers, box 4 (87), SL; Alma Lutz to Caroline Babcock, Oct. 23, 1953, Babcock-Hurlburt papers,

box 9 (117), SL; Alma Lutz to Florence Kitchelt, Nov. 2, 1953, Kitchelt papers, box 7 (178), SL.

48. Ethel Ernest Murrell to Florence Kitchelt, Sept. 5, 1953, Kitchelt papers, box 4 (87), SL.

49. Paul, "Conversations," pp. 304–305.

50. Alma Lutz to Florence Kitchelt, Sept. 8, 1953, Kitchelt papers, box 7 (178), SL; also Alma Lutz to Caroline Babcock, Oct. 23, 1953, Babcock-Hurlburt papers, box 9 (117), SL.

51. Jeannette Marks to Helen Paul, April 16, 1947, NWP papers, reel 91.

52. See Irwin, *Story of Alice Paul*; Doris Stevens, *Jailed for Freedom*; Becker, *Origins of the ERA*. Amelia R. Fry, who interviewed Alice Paul for the Regional Oral History Office at the University of California, is currently working on a biography of Paul. See also Jean L. Willis, "Alice Paul: The Quintessential Feminist," in *Feminist Theorists*, ed. Dale Spender (New York: Pantheon Books, 1983), pp. 285–295.

53. Maria Karagianis, "Alice Paul Still Waits for Equal Rights," Boston *Globe*, Feb. 1, 1976.

54. [Florence Armstrong], "What Alice Paul Has Done for ME," Feb. 11, 1954, Armstrong papers, box 1 (1), SL; [Florence Armstrong], notes on a telephone call from Ethel Johnson, May 20, 1957, Armstrong papers, box 3 (42), SL.

55. Florence Armstrong to Alice Paul, Aug. 28, 1959, NWP papers, reel 105.

56. Alma Lutz to Florence Kitchelt, Nov. 30, 1949, Kitchelt papers, box 6 (177), SL; Alma Lutz to Florence Kitchelt, April 15, 1955, Kitchelt papers, box 7 (178), SL.

57. Stevens, *Jailed for Freedom*; Doris Stevens to Westbrook Pegler, May 3, 1946, Stevens papers, SL.

58. Doris Stevens, diary entry, Feb. 11, 1946, Stevens papers, SL.

59. Ethel R. Coombes to Alice Paul, Nov. 24, 1952, NWP papers, reel 99; Rosalind G. Bates to Alice Paul, June 27, 1957, NWP papers, reel 103; Amelia Himes Walker to Alice Paul, Jan. 31, 1958, NWP papers, reel 104; Essie Newton to Alice Paul, Feb. 22, 1965, NWP papers, reel 109.

60. Irwin, *Story of Alice Paul*, p. 25.

61. Anna Kelton Wiley to Lena Madesin Phillips, Dec. 22, 1945, Phillips papers, SL.

62. Clara Snell Wolfe to Emma Guffey Miller, July 25, 1962, NWP papers, reel 107; Emma Guffey Miller to Clara Snell Wolfe, July 28, 1962, NWP papers, reel 107; Mary Glenn Newell to Emma Guffey Miller, Aug. 6, 1962, NWP papers, reel 107.

63. Lena Madesin Phillips to Anna Kelton Wiley, Jan. 7, 1946, Phillips papers, SL.

64. Interview no. 12.

65. Paul, "Conversations," p. 277.

66. Interview no. 2. Kathryn Paulsen, "The Last Hurrah of the Woman's Party," NWP papers, reel 111.

67. Elizabeth Sudler, "Reclaiming Our Heritage: A Trip to the NWP Convention," *Congressional Union* 1 (December 1981), 4; Betty Cuniberti, "Young, Old Feminists Split," *Los Angeles Times*, Oct. 6, 1981. We are grateful to Pam Elam for sending us this material.

218   NOTES

68. Nelson, "Ladies in the Streets," analyzes what she calls the feminist sub-culture of the Woman's Party in its early years, and her analysis is consistent with what we have found in the post–1945 period.

69. Caroline Babcock to Grace Cook Kurz, July 8, 1946, NWP papers, reel 89.

70. Mabel Griswold to Agnes Wells, Aug. 4, 1950, NWP papers, reel 97.

71. Mabel Van Dyke Baer to Anna Kelton Wiley, Aug. 12, 1947, NWP papers, reel 91; also Mabel Griswold to L. J. D. Daniels, May 24, 1948, NWP papers, reel 93; and Mabel Griswold to Mrs. G. M. Fuller, May 24, 1948, NWP papers, reel 93.

72. Quoted in Lisa Sergio, *A Measure Filled* (New York: Robert B. Luce, 1972), p. 135.

73. Communication from Frances Kolb, Winter 1980. Kolb is currently writing a history of the first ten years of the National Organization for Women.

74. Anita Pollitzer to Mrs. Betts, Nov. 20, 1957, NWP papers, reel 104.

75. Sarah M. Chorm to Mr. R. R. Bowker, Sept. 25, 1961, NWP papers, reel 107.

76. Paul, "Conversations," pp. 189, 198–199. Zilfa Estcourt, "Women in War," San Francisco *Chronicle*, Feb. 14, 1945.

77. Interview no. 1, interview no. 5, and interview no. 6.

78. Claire Mish to Mary Elizabeth Nye, Feb. 11, 1948, NWP papers, reel 92.

79. Interview no. 2.

80. Interview no. 5.

81. Elizabeth Osgood to Caroline Babcock, Sept. 4, 1945, NWP papers, reel 87.

82. Martha Souder to Ethel Adamson, April 18, 1948, NWP papers, reel 93.

83. Mary Alice Matthews to Alice Paul, March 24, 1945, NWP papers, reel 85; also Martha Souder to Agnes Wells, n.d. [rec. May 21, 1950], NWP papers, reel 96; Mary Kennedy to Alice Paul, Dec. 12, 1957, NWP papers, reel 104; Ruth Gage-Colby to Alice Paul, June 11, 1959, NWP papers, reel 105.

84. Alice Paul to Nina Allender, Nov. 20, 1954, NWP papers, reel 100; Kay Boyle to Alice Paul, Dec. 5, 1954, NWP papers, reel 100; Alice Paul to Nina Allender, Dec. 6, 1954, NWP papers, reel 100.

85. Dorothy Ogle Graham to Marion Sayward, July 14, 1960, NWP papers, reel 106.

86. Edith Goode to Alice Paul, Feb. 17, 1959, NWP papers, reel 105; also Alice Paul to Alice Morgan Wright, Oct. 23, 1945, NWP papers, reel 88.

87. Elsie Hill to Alice Paul, Feb. 10, 1945, NWP papers, reel 84; Anita Pollitzer to Alice Paul, Feb. 14, 1945, NWP papers, reel 84; Elsie Hill to Florence Armstrong, Feb. 28, 1945, NWP papers, reel 84; Marion Mary to Alice Paul, n.d. [May 1947], NWP papers, reel 91; Dora Ogle to Florence Bayard Hilles, Dec. 29, 1947, NWP papers, reel 92; Mary Burt Messer to Alice Paul, Jan. 3, 1948, NWP papers, reel 92.

88. Interview no. 7.

89. [Mildred Palmer] to Lucy Rice Winkler, Sept. 5, 1951, NWP papers, reel 98. See also Peggy Hayes to Alice Paul, May 27, 1967, NWP papers, reel 110; Mattie Russel to Alice Paul, April 9, 1957, NWP papers, reel 103.

90. Ruth Gage-Colby to Alice Paul, Jan. 16, 1959, NWP papers, reel 105.

91. Interview no. 7, interview no. 11, and interview no. 12.

92. Nelson, "Ladies in the Streets," suggests that the original move into Belmont House marked a shift in political priorities as energy and money went into furnishing the house rather than passing the ERA.

93. Appeal for contributions from Ethel Ernest Murrell, n.d. (December 1951), NWP papers, reel 98.

94. Laura Berrien to Anita Pollitzer, Oct. 18, 1946, NWP papers, reel 90; also unsigned memo, n.d. [November 1946], NWP papers, reel 90; Winifred Mallon to Doris Stevens, Dec. 2, 1946, NWP papers, reel 90; Helena Hill Weed to Anita Pollitzer, Jan. 15, 1947, NWP papers, reel 90.

95. Alice Paul to Laura Berrien and Dorris Stevens, Dec. 30, 1946, NWP papers, reel 90.

96. Alma Lutz to Florence Kitchelt, May 24, 1945, Kitchelt papers, box 6 (177), SL.

97. William A. Gamson, *The Strategy of Social Protest* (Homewood, Ill.: Dorsey, 1975); Robert J. Ross, "Generational Change and Primary Groups in a Social Movement," in *Social Movements of the Sixties and Seventies*, ed. Jo Freeman (New York: Longman, 1983), pp. 177–189; Alan E. Gross et al., "The Men's Movement: Personal Versus Political," in Freeman, *Social Movements*, pp. 71–81.

98. Gamson, *Strategy of Social Protest*: see also Joan Cassell, *A Group Called Women* (New York: David McKay, 1977).

Chapter 4. The Women's Rights Movement

1. John D. McCarthy and Mayer N. Zald, "Resource Mobilization and Social Movements: A Partial Theory," *American Journal of Sociology* 82 (1977), 1212–1421.

2. Luther P. Gerlach and Virginia H. Hine, *People, Power, and Change: Movements of Social Transformation* (Indianapolis: Bobbs-Merrill, 1970); McCarthy and Zald, "Resource Mobilization;" Jo Freeman, *Social Movements of the Sixties and Seventies* (New York: Longman, 1983).

3. Remarks of Margaret Hickey, Chairman, Women's Responsibility in Public Affairs, n.d., Hickey papers, UMSL.

4. Arnold W. Green and Eleanor Melnick, "What Has Happened to the Feminist Movement?" in *Studies in Leadership*, ed. Alvin W. Gouldner (New York: Russell & Russell, 1950), pp. 277–302.

5. [Norma Lindsay] to Gladys [no last name], March 5, 1956, NWP papers, reel 102.

6. Florence Kitchelt to Evelyn Samras, Nov. 23, 1945, Kitchelt papers, box 8 (224), SL.

7. Babette Kass and Rose C. Feld, *The Economic Strength of Business and Professional Women* (New York: NFBPWC, 1954), p. iii.

8. Local interview no. 2.

9. Local interview no. 6.

10. Anna Kelton Wiley to Anita Pollitzer, Nov. 7, 1945, Babcock-Hurlburt papers, box 7 (96), SL.

11. Anna Lord Strauss to office, n.d. [rec. Aug. 19, 1946], League of Women Voters papers, box 751, LC.

12. Anna Lord Strauss, "Arguments for and against changeing [*sic*] name of

League of Women Voters to League of Active Voters (or some such)," April 21, 1953, LWV papers, box 1198, LC.

13. Dorothy Kenyon to Anna Lord Strauss, Oct. 5, 1946, LWV papers, box 696, LC.

14. Memo to Anna Lord Strauss from Mrs. Stough, Jan. 29, 1947, LWV papers, box 700, LC.

15. Anna Lord Strauss to Jane Barus, Jan. 14, 1949, LWV papers, box 738, LC. Also Anna Lord Strauss to Mrs. M. I. Corbett Ashby, March 8, 1949, LWV papers, box 744, LC; report from Thelma M. Burgess to Anna Lord Strauss, Sept. 29, 1949, LWV papers, box 743, LC; Katharine Blunt to Mrs. John G. Lee, Aug. 4, 1950, LWV papers, box 744, LC.

16. Memo to Anna Lord Strauss from Mrs. Stough, Jan. 29, 1947, LWV papers, box 700, LC.

17. Statement of Miss Frieda S. Miller to a Special Subcommittee of the Senate Judiciary Committee on Proposed Amendments to the Judicial Code for the Improvement of the Jury System in the Federal Courts, April 24, 1947, NWP papers, reel 91.

18. Winifred G. Helmes to Helen Loy, Oct. 15, 1954, Alice Leopold papers, box 2 (8), SL.

19. This profile of the membership is based on an analysis of the social characteristics of 108 individuals whose national participation was substantial enough to give them visibility. These are individuals whose names appeared regularly in correspondence and for whom information about social characteristics could be located in biographical or archival materials. We recorded any information related to race, class, education, occupation, age, place of residence, political affiliation, political views, marital status, presence and number of children, living situation, and time of first involvement in the women's rights movement. In addition, we took note of any comments made by participants about the characteristics of the membership.

20. Interview no. 5.

21. Kass and Feld, *Economic Strength*.

22. Interview no. 8.

23. Robert Coughlan, "Changing Roles in Modern Marriage," *Life* 41 (Dec. 24, 1956).

24. Brochure for Elizabeth Hatfield Schnabel in Babcock-Hurlburt papers, box 9 (114), SL.

25. Lucy Somerville Howorth, "From a One-Horse Surrey into the Atomic Age," *Alumnae Bulletin of Randolph-Macon Woman's College*, June 1948.

26. Doris Stevens, diary entry, April 5, 1956, Stevens papers, SL.

27. Florence Kitchelt to Jane Grant, July 20, 1951, Kitchelt papers, box 3 (50), SL; Florence Kitchelt to Alma Lutz, Dec. 8, 1947, Kitchelt papers, box 6 (177), SL; Florence Kitchelt to Jane Grant, n.d. [1946], Kitchelt papers, box 6 (147), SL; Florence Kitchelt to editor of the New Haven *Journal Courier*, Jan. 12, 1944, Kitchelt papers, box 3 (42), SL.

28. Florence Kitchelt to Nora Stanton Barney, Feb. 5, 1945, NWP papers, reel 84; Florence Kitchelt to Emma Guffey Miller, Sept. 17, 1945, Kitchelt papers, box 7 (189), SL.

29. Speech to NWP by Perle Mesta, May 26, 1949, NWP papers, reel 95.

30. Howorth, "From a One-Horse Surrey."

31. Interview no. 1, interview no. 5, and interview no. 6.

32. Interview no. 3, interview no. 4, and interview no. 8.

33. Local interview no. 5.

34. Quoted in an interview by Suzanne Colezal, "The Contemporary Susan B. Anthony," Boston *Globe*, May 22, 1979.

35. Alma Lutz to Florence Kitchelt, July 11, 1951, Kitchelt papers, box 7 (178), SL.

36. *Webster's Collegiate Dictionary*, 5th ed. (Springfield, Mass.: G. & C. Merriam Co., 1942). We are indebted to Kathryn Kish Sklar for bringing this definition to our attention.

37. Nora Stanton Barney to Bella Dodd, Jan. 17, 1945, NWP papers, reel 84; Nora Stanton Barney to Alice Paul, Feb. 9, 1945, NWP papers, reel 84.

38. Florence Kitchelt to [Rachel Conrad Nason], Jan. 5, 1944 [*sic*–1945], Kitchelt papers, box 6 (168), SL.

39. Mary Beard to Florence Kitchelt, March 20, 1950, Kitchelt papers, box 5 (113), SL; Mary Beard to Rose Arnold Powell, July 3, 1948, Powell papers, box 2 (27), SL.

40. Katie Louchheim, speech at 1960 Democratic National Convention, NWP papers, reel 106.

41. Quoted in Marguerite Rawalt, *History of the NFBPWC*, vol. 2 (Washington, D.C.: NFBPWC, 1969), p. 24.

42. Rose Arnold Powell, diary entry, July 4, 1949, Powell papers, box 1, v. 5, SL.

43. Lena Madesin Phillips to Carola Redlich, Jan. 31, 1948, Phillips papers, SL; [no signature] to Lena Madesin Phillips and Marjory Lacey-Baker, July 25, 1945, Phillips papers, SL; Olivia Rossetti Agresti to Lena Madesin Phillips, Sept. 15, 1952, Phillips papers, SL.

44. Alice Paul, "Conversations With Alice Paul: Woman Suffrage and the Equal Rights Amendment," an oral history conducted in 1972 and 1973 by Amelia R. Fry, Regional Oral History Office, University of California, 1976, pp. 401–403.

45. India Edwards, *Pulling No Punches: Memoir of a Woman in Politics* (New York: Putnam's, 1977), p. 109.

46. Rose Arnold Powell to Alice Stone Blackwell, June 3, 1945, Powell papers, box 2 (25), SL.

47. Florence Kitchelt to Alice Hamilton, Aug. 6, 1945, Kitchelt papers, box 6 (151), SL; Florence Kitchelt to Ethel Ernest Murrell, Sept. 17, 1953, Kitchelt papers, box 4 (87), SL.

48. Andy Logan, "New Yorker Profile," *New Yorker*, Feb. 3, 1951, 36–50.

49. Jane Grant, "Confessions of a Feminist," *American Mercury*, December 1943.

50. Myrtle Cain to Alice Paul, Aug. 7, 1958, NWP papers, reel 104.

51. Florence Kitchelt to Jane Grant, n.d., Kitchelt papers, box 6 (147), SL; Florence Kitchelt to Jane Grant, July 20, 1951, Kitchelt papers, box 3 (50), SL.

52. Florence Kitchelt to Mrs. Warren J. Woodward, Sept. 1, 1947, Kitchelt papers, box 6 (169), SL.

53. Florence Kitchelt to the editor of *Life*, May 17, 1954, Kitchelt papers, box 3 (38), SL; Florence Kitchelt to Alma Lutz, June 17, 1954, Kitchelt papers, box 7 (178), SL; Florence Kitchelt to Caroline Babcock, June 5, 1954, Babcock-Hurlburt papers, box 9 (109), SL; "Picklers' Pick," the *Union Times*, Dec. 7, 1946; Florence Kitchelt to Eric Sevareid, July 9, 1959, Kitchelt papers, box 3 (59), SL.

54. Rose Arnold Powell to Francis W. Reichelderfer, Oct. 1, 1949, Powell papers, box 2 (19), SL.

55. Fannie Ackley to Mildred Palmer, Nov. 20, 1953, NWP papers, reel 99.

56. Mabel Griswold to Helena Ratterman, April 23, 1947, NWP papers, reel 91.

57. Florence Kitchelt to Charlotte Kimball Kruesi, Feb. 17, 1950, Kitchelt papers, box 1 (1), SL; Jane Grant to Doris Stevens, March 14, 1951, Stevens papers, SL.

58. Rose Arnold Powell, diary entry, Nov. 2, 1960, Powell papers, box 1, v. 8, SL; Rose Arnold Powell to Mary Beard, June 23, 1948, Powell papers, box 2 (27), SL; Rose Arnold Powell to Anita Pollitzer, Jan. 13, 1949, Powell papers, box 3 (50), SL; Rose Arnold Powell, diary entry, April 15, 1947, Powell papers, box 1, v. 5, SL.

59. Fannie Ackley to Mildred Palmer, Sept. 19, 1950, NWP papers, reel 97.

60. Olga K. Robinson to Caroline Babcock, Jan. 10, 1945, NWP papers, reel 84.

61. Luella Huggins to Alice Paul, Aug. 21, 1954, NWP papers, reel 100.

62. Olive Hurlburt to Anna Kelton Wiley, Caroline Babcock, and Laura Berrien, April 20, 1950, Babcock-Hurlburt papers, box 9 (113), SL.

63. Leslie Black to Burnita Shelton Matthews, May 4, 1945, NWP papers, reel 86.

64. Rose Arnold Powell, diary entry, June 6, 1955, Powell papers, box 1, v. 7, SL.

65. Nancy F. Cott, "The Problem of Feminism in the 1920's," 1981, makes this point with regard to the National Woman's Party in the 1920s. See also Cott's article, "Feminist Politics in the 1920s: The National Woman's Party," *Journal of American History* 71 (June 1984), 43–68.

66. Paul, "Conversations," p. 197.

67. Ella M. Sherwin to Katharine Norris, Feb. 3, 1946, Norris papers, box 1 (4), SL.

68. Caroline Katzenstein to Alice Paul and Anita Pollitzer, June 4, 1948, NWP papers, reel 93.

69. Florence Kitchelt to Geneva McQuatters, July 29, 1948, Kitchelt papers, box 3 (50), SL.

70. Martha Taylor Howard to Florence Kitchelt, Aug. 25, 1950, Kitchelt papers, box 3 (60), SL.

71. Jane Grant to Florence Kitchelt, n.d. [July 1955], Kitchelt papers, box 5 (133), SL.

72. Florence Kitchelt to Alma Lutz, May 29, 1947, Kitchelt papers, box 6 (177), SL; Florence Kitchelt to Nina Horton Avery, July 22, 1946, Kitchelt papers, box 5 (105), SL; Florence Kitchelt to Sally Butler, July 23, 1946, Kitchelt papers, box 5 (127), SL.

73. Jeannette Marks to Doris Stevens, April 10, 1946, Stevens papers, SL.

74. Marion May to Alice Paul, May 19, 1949, NWP papers, reel 95.

75. Elsie Hill to Alice Paul, Aug. 8, 1951, NWP papers, reel 98; Lucy Rice Winkler to Ethel Ernest Murrell, Aug. 30, 1951, NWP papers, reel 98; Nellie Parr to Mildred Palmer, Aug. 30, 1951, NWP papers, reel 98; Ethel Ernest Murrell to Lucy Rice Winkler, Sept. 11, 1951, NWP papers, reel 98; By-Laws of Council for

Legal Equality, Southern California Branch of the NWP, n.d., NWP papers, reel 98.

76. Nancy Jewell Cross to Marion Sayward, Sept. 13, 1959, NWP papers, reel 105; Marion Sayward to Alice Paul, Sept. 22, 1959.

77. Letter from organizations opposed to ERA to Maurice J. Tobin, Jan. 27, 1949, LWV papers, box 1131, LC.

78. Mary Anderson to Lena Madesin Phillips, Aug. 5, 1937, Phillips papers, SL.

79. Mary Beard to Florence Kitchelt, March 17, 1950, Kitchelt papers, box 5 (113), SL.

80. Mary Van Kleeck to Alice Paul, Aug. 5, 1948, NWP papers, reel 94.

81. Dorothy Kenyon to Joseph Guffey, April 25, 1946, NWP papers, reel 89; Dorothy Kenyon to Florence Kitchelt, Sept. 3, 1948, Kitchelt papers, box 6 (162), SL.

82. Florence Kitchelt to Alma Lutz, July 4, 1948, Kitchelt papers, box 6 (177), SL.

83. Edith Goode to Katharine Arnett, Feb. 18, 1958, NWP papers, reel 104.

84. Mary Markajani to Alice Paul, March 1, 1958, NWP papers, reel 104; Peggy Capewell to Caroline Babcock, May 27, 1946, NWP papers, reel 89; Fannie Ackley to Caroline Babcock, Nov. 29, 1945, NWP papers, reel 88.

85. Alice M. Birdsall to W. J. Eden, Jan. 22, 1945, NWP papers, reel 84.

86. See Lois Scharf, *To Work or To Wed* (Westport, Conn.: Greenwood Press, 1980).

87. Emma Guffey Miller to Theodore G. Green, April 16, 1945, LWV papers, box 1131, LC.

88. Caroline Babcock to Florence Kitchelt, Jan. 10, 1945, NWP papers, reel 84.

89. Florence Kitchelt, "Red Herring Tactics Held Used Against the Equal Rights Bill," New York *Herald Tribune*, July 1, 1945.

90. Ella Sherwin to Alice Paul, July 4, 1945, NWP papers, reel 87.

91. Anna Kelton Wiley to Gertrude Baer, Feb. 10, 1945, NWP papers, reel 84.

92. Esther Peterson to Dorothy McDiarmid, Jan. 9, 1973, Peterson papers, SL.

93. Emma Guffey Miller to Stuart Symington, n.d. [Feb. 1960], Miller papers, box 4 (63), SL.

94. R. St. George to National Women's Trade Union League, Feb. 3, 1950, NWTUL papers, reel 14, LC.

95. Memo to member organizations, National Committee to Defeat the Un-Equal Rights Amendment, April 3, 1947, LWV papers, box 698, LC.

96. Bruce Bliven to Florence Kitchelt, Sept. 7, 1945, Kitchelt papers, box 3 (44), SL.

97. Harrison Smith to Florence Kitchelt, Jan. 9, 1950, Kitchelt papers, box 8 (231), SL.

98. Helena Ratterman to Amelia Walker, Aug. 15, 1959, NWP papers, reel 105; Emma Guffey Miller to Katharine St. George, Sept. 8, 1961, NWP papers, reel 107; Notes, n.d. [1961], Esther Peterson papers, SL.

99. Mabel Griswold to Mrs. Will Sheffer, Oct. 10, 1948, NWP papers, reel 94.

100. Anna Kelton Wiley to Gertrude Baer, Feb. 10, 1945, NWP papers, reel 84.

101. Mary Sinclair Crawford to Katharine A. Norris, April 7, 1945, Norris papers, box 1 (3), SL.

102. Molly Dewson to Lucy Somerville Howorth, Aug. 21, 1949, Sommerville-Howorth papers, box 7 (135), SL.

103. Quoted in Rawalt, *History of the NFBPWC*, vol. 2, p. 70.

104. Anna M. Kross to Lucy Somerville Howorth, Sept. 28, 1949, Somerville-Howorth papers, box 5 (106), SL.

105. Genevieve Blatt to Emma Guffey Miller, Nov. 8, 1958, Miller papers, box 4 (59), SL.

106. Alice Paul to Burnita Shelton Matthews, Dec. 26, 1949, Matthews papers, SL.

107. See Eleanor F. Straub, "United States Government Policy Toward Civilian Women During World War II," *Prologue* 5 (Winter 1973), 240–254.

108. Lucy Somerville Howorth, "Memo to Accompany Three Clippings," n.d. [1958], Somerville-Howorth papers, box 6 (120), SL.

109. India Edwards, *Pulling No Punches*. On the role of Edwards, see Cynthia E. Harrison, "Prelude to Feminism: Women's Organizations, the Federal Government and the Rise of the Women's Movement, 1942 to 1968," Diss., Columbia University, 1982.

110. Doris Fleeson, "Influence of Women Seen Nil in Washington Events Today," Washington *Star*, Oct. 4, 1945, Somerville-Howorth papers, box 6 (120), SL.

111. Doris Fleeson, "Equal Rights for Women?", n.d. [1957], Somerville-Howorth papers, box 6 (120), SL.

112. Emma Guffey Miller to John F. Kennedy, Feb. 21, 1961, Miller papers, box 4 (68), SL; Genevieve Blatt to Emma Guffey Miller, Dec. 5, 1961, Miller papers, box 4 (68), SL.

113. Burnita Shelton Matthews to Sylvia Bacon, Jan. 18, 1967, Matthews papers, SL.

114. Letter from Minnie Maffett, Kathryn McHale, Charl Ormond Williams to Elizabeth [*sic*] Christman, April 28, 1944, NWTUL papers, reel 19, LC; Mrs. James Austin Stone to Rose Schneiderman, May 11, 1944, NWTUL papers, reel 19, LC; Charl Ormond Williams to Continuation Committee, Dec. 11, 1944, NWTUL papers, reel 19, LC.

115. Rawalt, *History of the NFBPWC*, vol. 2, p. 11.

116. See, for example, the January and March 1953 issues of *Independent Woman*, the BPW journal, for coverage of the women appointees of the Eisenhower administration.

117. Caroline Babcock to Mary Woolley, April 22, 1946, NWP papers, reel 89; Edith Goode to Nina Horton Avery, Aug. 27, 1946, Katharine Norris papers, box 1 (4), SL; AAUW General Director's Letter, 26 (November 1958).

118. Sophia Yarnall Jacobs to Doris Stevens, Aug. 7, 1961, Stevens papers, SL.

119. Rawalt, *History of the NFBPWC*, vol. 2, p. 12.

120. Edwards, *Pulling No Punches*, p. 158.

121. Ibid., p. 109.

122. " 'Miss Republican' Goes 'Politikin' Eastern Shore Style," Salisbury *Advertiser* and the Wicomico *Countian*, March 20, 1952; " 'Breakfast With Ike'—Here's

What It's Like," Washington *Sunday Star*, May 29, 1955. On Bertha Adkins, see Harrison, "Prelude to Feminism."

123. Emma Guffey Miller to Stephen Mitchell, Jan. 31, 1953, Miller papers, box 3 (40), SL; Emma Guffey Miller to John M. Bailey, Sept. 3, 1961, Miller papers, box 4 (68), SL.

124. Molly Dewson to Lucy Somerville Howorth, Nov. 13, 1954, Somerville-Howorth papers, box 5 (117), SL.

125. India Edwards' Christmas card, 1959, Emma Guffey Miller papers, box 4 (62), SL.

126. Edwards, *Pulling No Punches*, p. 203.

127. Belle Everett to Emma Guffey Miller, n.d. [July 1962], Miller papers, box 4 (69), SL.

128. Margaret Hickey to Lucy Somerville Howorth, Jan. 15, 1948, Somerville-Howorth papers, box 8 (151), SL.

129. Quoted in Rawalt, *History of the NFBPWC*, vol. 2, p. 46.

130. Sarah T. Hughes to Lucy Somerville Howorth, Jan. 18, 1949, box 5 (104), SL.

131. Agnes Wells and Mamie Sydney Mizen to Tom Connally, May 18, 1949, NWP papers, reel 95.

132. Mary Kennedy to Lena Madesin Phillips, June 30, 1948, Phillips papers, SL.

133. Paul, "Conversations," pp. 474–475. Woman's Party members Mary Kennedy and Alma Lutz agreed; Mary C. Kennedy to Alice Paul, March 25, 1945, NWP papers, reel 85; Alma Lutz to Florence Kitchelt, Nov. 24, 1953, Kitchelt papers, box 7 (178), SL.

134. Alma Lutz to Rose Arnold Powell, Aug. 18, 1948, Powell papers, box 3 (43), SL; Alma Lutz to Florence Kitchelt, Aug. 4, 1948, Kitchelt papers, box 6 (177), SL.

135. Ernestine Powell to Clara Wolfe, July 26, 1947, NWP papers, reel 91.

136. Emma Guffey Miller to Jeannette Marks, April 13, 1948, NWP papers, reel 93.

137. Ethel Adamson to Alice Paul and Anita Pollitzer, June 13, 1948, NWP papers, reel 93.

138. [Martha Souder] to Ethel Adamson, June 14, 1948, NWP papers, reel 93; Alice Paul to Martha Taylor Howard, Aug. 4, 1948, NWP papers, reel 94.

139. Interview no. 2.

140. Florence Kitchelt to Martha Taylor Howard, Nov. 3, 1950, Kitchelt papers, box 3 (60), SL.

141. Rose Arnold Powell to Elizabeth B. Borden, April 13, 1958, Powell papers, box 1, SL.

142. See *Mary Ritter Beard: A Sourcebook*, ed. Ann J. Lane (New York: Schocken Books, 1977), pp. 53–59.

143. Florence Kitchelt to Katharine St. George, Nov. 4, 1948, Kitchelt papers, box 8 (222), SL; Miriam Y. Holden, "Women in History," Feb. 25, 1949, NWP papers, reel 95.

144. Miriam Y. Holden, "Women in History," Feb. 25, 1949, NWP papers, reel 95.

145. Alma Lutz to Rose Arnold Powell, May 29, 1953, Powell papers, box 3 (43), SL.

146. Marjorie Barstow Greenbie, May 15, 1950, NWP papers, reel 96.

147. Mary Beard to Lena Madesin Phillips, June 3, 1951, Phillips papers, SL.

148. Mary R. Beard, *Woman as Force in History* (New York: Macmillan, 1946).

149. Mary Beard to Margaret Storrs Grierson, Feb. 27, 1948, Smith College Library, quoted in Lane, *Mary Ritter Beard*, p. 58.

150. Mary Beard to Rose Arnold Powell, July 3, 1948, Powell papers, box 2 (27), SL.

151. Sarah T. Hughes to Lucy Somerville Howorth, Jan. 18, 1949, Somerville-Howorth papers, box 5 (104), SL.

152. Susan B. Riley to Lucy Somerville Howorth, July 2, 1951, Somerville-Howorth papers, box 7 (147), SL.

153. Peggy Capewell to Caroline Babcock, n.d., Babcock-Hurlburt papers, box 3 (929), SL.

154. Florence Kitchelt to Louise S. Earle, Nov. 27, 1944, Kitchelt papers, box 6 (141), SL.

155. Minutes of meeting, July 19, 1943, Florence Kitchelt papers, box 4 (70), SL; Josepha Whitney to Florence Kitchelt, July 16, 1943, Kitchelt papers, box 4 (70), SL; Florence Kitchelt to Louise S. Earle, Nov. 27, 1944, Kitchelt papers, box 6 (141), SL.

156. "Sally Butler Calls Gathering of Club Leaders," Indianapolis *Star*, Feb. 15, 1948, NWP papers, reel 92.

157. Libby Sachar to Florence Kitchelt, April 18, 1960, Kitchelt papers, box 8 (221), SL.

158. Florence Kitchelt to Anita Pollitzer, Nov. 19, 1946, Kitchelt papers, box 4 (74), SL.

159. Alma Lutz to Florence Kitchelt, June 17, 1952, Kitchelt papers, box 7 (178), SL; Alma Lutz to Florence Kitchelt, March 10, 1954, Kitchelt papers, box 7 (178), SL; Alma Lutz to Florence Kitchelt, July 17, 1954, Kitchelt papers, box 7 (178), SL; Alma Lutz to Florence Kitchelt, Jan. 29, 1955, Kitchelt papers, box 7 (178), SL.

160. Florence Kitchelt to Margaret Hickey, June 19, 1946, Kitchelt papers, box 6 (154), SL.

161. Quoted in Rawalt, *History of the NFBPWC*, vol. 2, p. 43; Florence Kitchelt to Nina B. Horton Avery, April 25, 1947, Kitchelt papers, box 5 (105), SL.

162. Quoted in Rawalt, *History of the NFBPWC*, vol. 2, p. 44; Helen Paul to Ernestine Powell, June 10, 1954, NWP papers, reel 100.

163. Florence Kitchelt to Barbara Woollcott, March 5, 1945, Kitchelt papers, box 3 (36), SL.

164. Alma Lutz to Florence Kitchelt, Feb. 11, 1949, Kitchelt papers, box 6 (177), SL; Alice Paul to Florence Kitchelt, March 4, 1954, Kitchelt papers, box 7 (198), SL; Alma Lutz to Florence Kitchelt, March 10, 1954, Kitchelt papers, box 7 (178), SL.

165. Alice K. Leopold to Florence Kitchelt, Sept. 5, 1952, Kitchelt papers, box 4 (87), SL; Florence Kitchelt to Alice K. Leopold, Dec. 19, 1953, Kitchelt papers, box 6 (169), SL; Florence Kitchelt to Alma Lutz, Jan. 2, 1954, Kitchelt papers, box 7 (178); SL; Florence Kitchelt to Alma Lutz and Jane Grant, Jan. 10, 1954, Kitchelt papers, box 7 (178), SL.

166. Helen G. Irwin to Nina Horton Avery, June 21, 1954, NWP papers, reel 100.

167. Priscilla Wagoner to NWP, Aug. 25, 1954, NWP papers, reel 100.

168. Mary Kennedy to Alice Paul, Sept. 7, 1954, NWP papers, reel 100.

169. Alice Paul to Florence Kitchelt, Oct. 12, 1954, Kitchelt papers, box 7 (198), SL.

170. Zelma L. Huxtable to Alice Paul, May 27, 1945, NWP papers, reel 86.

171. Lewis A. Coser, *The Functions of Social Conflict* (Glencoe, Ill.: Free Press, 1956).

172. Katharine Norris to Nina Horton Avery, July 20, 1946, Norris papers, box 1 (4), SL; Katharine Norris to Nina Horton Avery, Jan. 14, 1947, Norris papers, box 1 (5), SL; Alice Paul to Florence Kitchelt, Oct. 12, 1954, Kitchelt papers, box 7 (198), SL; Alice Paul to Mary Kennedy, March 26, 1958, NWP papers.

173. Florence Kitchelt to Nina B. Horton Avery, Sept. 18, 1946, Kitchelt papers, box 5 (105), SL; Nina B. Horton Avery to Florence Kitchelt, Sept. 24, 1946, Kitchelt papers, box 5 (105), SL; Alma Lutz to Florence Kitchelt, July 29, 1954, Kitchelt papers, box 7 (178), SL; Alma Lutz to Florence Kitchelt, July 29, 1954, Kitchelt papers, box 7 (178), SL.

174. Alice Paul to Florence Kitchelt, Oct. 12, 1954, Kitchelt papers, box 7 (198), SL.

175. Jeannette Marks to Lena Madesin Phillips, Feb. 22, 1945, Phillips papers, SL.

176. Jeannette Marks to Dorothy Shipley Granger, Nov. 24, 1945, Babcock-Hurlburt papers, box 7 (96), SL.

177. Caroline Babcock to Alma Harrison Ambrose, Jan. 6, 1948, Babcock-Hurlburt papers, box 9 (111), SL.

178. Alma Lutz to Florence Kitchelt, Dec. 13, 1947, Kitchelt papers, box 6 (177), SL; Alma Lutz to Florence Kitchelt, July 1, 1948, Kitchelt papers, box 6 (177), SL; Alma Lutz to Florence Kitchelt, Sept. 8, 1948, Kitchelt papers, box 6 (177), SL; Alma Lutz to Florence Kitchelt, Sept. 27, 1954, Kitchelt papers, box 7 (178), SL.

179. Alma Lutz to Caroline Babcock, June 2, 1956, Babcock-Hurlburt papers, box 9 (117), SL.

180. [Alma Lutz] to Mrs. James Carmak, Jan. 14, 1949, Babcock-Hurlburt papers, box 9 (112), SL; Alma Lutz to Florence Kitchelt, April 21, 1949, Kitchelt papers, box 6 (177), SL.

181. Anna Kelton Wiley to Florence Kitchelt, Feb. 27, 1949, Kitchelt papers, box 9 (245), SL.

182. Alma Lutz to Florence Kitchelt, Nov. 12, 1954, Kitchelt papers, box 7 (178), SL; Florence Kitchelt to Alice Paul, Feb. 19, 1955, Kitchelt papers, box 7 (198), SL; Florence Kitchelt to Alma Lutz, July 7, 1955, Kitchelt papers, box 7 (178), SL.

183. Louise Earle to Florence Kitchelt, Sept. 3, 1946, Kitchelt papers, box 6 (141), SL.

184. Emily Hickman to Anna Lord Strauss, Dec. 2, 1946, LWV papers, box 701, LC; Charlotte Mahon to Percy Lee, March 1, 1951, LWV papers, box 741, LC.

185. Anna M. Kross to Lucy Somerville Howorth, March 29, 1946, Somerville-Howorth papers, box 13 (253), SL; Anna M. Kross to Lucy Somerville Howorth, May 13, 1946, Somerville-Howorth papers, box 13 (253), SL; Anna M. Kross to Lucy Somerville Howorth, May 17, 1946, Somerville-Howorth papers, box 13 (253), SL; Letha P. Scott to Lucy Somerville Howorth, Feb. 10, 1949, Somerville-Howorth papers, box 8 (151), SL.

186. Dorothy Kenyon to board members of Women in World Affairs, Dec. 15, 1953, Somerville-Howorth papers, box 7 (145), SL.

187. Lucy Somerville Howorth to Dorothy Kenyon, April 13, 1954, Somerville-Howorth papers, box 7 (145), SL.

188. Dorothy Kenyon to Lucy Somerville Howorth, April 22, 1954, Somerville-Howorth papers, box 7 (145), SL.

189. Information on the Assembly of Women's Organizations for National Security can be found in Rawalt, *History of the NFBPWC*, pp. 87, 100–101; and in the Somerville-Howorth papers, SL.

190. "Meeting of Women's Organizations on Mobilization," Oct. 6, 1950, LWV papers, box 775, LC; minutes of meeting of subcommittee of Washington members of the steering committee of the Clearinghouse of Women's Organizations for National Defense, Jan. 15, 1951, Florence Armstrong papers, box 3 (41), SL; Eleanor F. Dolan to Dorothy Houghton, March 8, 1951, Somerville-Howorth papers, box 7 (147), SL.

191. Pauline E. Mandigo to Lucy Somerville Howorth and Marguerite Rawalt, Feb. 2, 1953, Somerville-Howorth papers, box 7 (147), SL.

192. Lucy Somerville Howorth, notes for archives, July 1957, Somerville-Howorth papers, box 7 (147), SL.

193. Thomas C. Pardo, *The National Woman's Party Papers, 1913–1974: A Guide to the Microfilm Edition* (Sanford, N.C.: Microfilming Corporation of America, 1979); interview no. 2; Alma Lutz to Caroline Babcock, Aug. 24, 1946, Babcock-Hurlburt papers, box 9 (116), SL; Dorothy Dunbar Bromley to Florence Kitchelt, Aug. 7, 1945, Kitchelt papers, box 5 (120), SL; Florence Kitchelt to Dorothy Dunbar Bromley, Aug. 27, 1945, Kitchelt papers, box 5 (120), SL.

194. Katharine Norris to Laura Miller Derry, Aug. 1, 1946, Norris papers, box 1 (4), SL.

195. Alma Lutz to Florence Kitchelt, Feb. 6, 1954, Kitchelt papers, box 7 (178), SL.

196. Emma Guffey Miller to Perle Mesta, May 7, 1962, NWP papers, reel 107.

197. Caroline Babcock, "The NWP and the ERA," August 1962, Babcock-Hurlburt papers, box 7 (93), SL.

198. Amelia Himes Walker, vice president of the World Woman's Party, and Anna Lord Strauss, president of the League of Women Voters, debated the ERA on the Esther Van Wagoner Tufty show in 1946; Lucy Somerville Howorth appeared on the Mary Margaret McBride show in 1952, as did Alice Paul and Woman's Party chairman Ernestine Powell in 1954. Dorothy Fuldheim, a radio commentator, merited praise from Florence Kitchelt in 1948 and promised to speak on the ERA if the occasion arose. Sarah Hughes, president of the BPW, and Frances Maule, editor of *Independent Woman*, appeared on the "We the People" radio program in 1951. The papers of women active in the movement contain many other references to radio appearances by feminists.

199. Leslie Black to Lucy Rice Winkler, Sept. 16, 1952, NWP papers, reel 99; Amelia Himes Walker to NWP, Aug. 21, 1959, NWP papers, reel 105.

200. Emma Newton to Marjorie Longwell, May 30, 1958, NWP papers, reel 104.

201. Barbara Sapinsley to Alice Paul, June 30, 1961, NWP papers, reel 107; Anita Pollitzer to Alice Paul, Jan. 8, 1962, NWP papers, reel 107; Barbara Sapinsley to Alice Paul, Feb. 12, 1962, NWP papers, reel 107; Morag M. Simchak to Esther

Peterson, Feb. 14, 1962, Peterson papers, SL; CBS press release, Dec. 6, 1962, NWP papers, reel 108.

202. Anna Kelton Wiley to Anita Pollitzer, Nov. 14, 1945, Babcock-Hurlburt papers, box 7 (96), SL.

203. Anita Pollitzer to Jane Norman Smith, June 4, 1948, NWP papers, reel 93; Ella C. Werner to Inka O'Hanrahan, Sept. 12, 1967, NWP papers, reel 110; also Aurelle Burnside to Anita Pollitzer, June 5, 1948, NWP papers, reel 93.

204. Marjorie Longwell to Elizabeth Forbes, Feb. 16, 1960, NWP papers, reel 106; Marjorie Longwell to Alice Paul, July 16, 1960, NWP papers, reel 106.

205. Lillian H. Kerr to Alice Paul, April 12, 1956, NWP papers, reel 102.

206. Alma Lutz to Rose Arnold Powell, June 21, 1955, Powell papers, box 3 (43), SL.

207. May G. Schaeffer to Caroline Babcock, March 14, 1945, NWP papers, reel 85.

208. Freda M. Klauden to Dora Ogle, Sept. 10, 1946, NWP papers, reel 89.

209. Rose Arnold Powell to Hubert H. Humphrey, Aug. 30, 1957, Powell papers, box 2 (21), SL; also Rose Arnold Powell to Una Winter, Feb. 26, 1949, Powell papers, box 4 (57), SL; Rose Arnold Powell to Sara Whitehurst, June 21, 1953, Powell papers, box 2 (20), SL; Rose Arnold Powell to Alma Lutz, July 1, 1957, Powell papers, box 3 (43), SL; Rose Arnold Powell to Mildred Sandison Fenner, Jan. 15, 1958, Powell papers, box 3 (47), SL.

210. Neenah Hastings Lessemann to Alice Paul, May 16, 1945, NWP papers, reel 86; [Nora Stanton Barney] to Alice Paul, May 7, 1945, NWP papers, reel 86.

211. Barbara Westebbe to Miss Hiatt, Aug. 25, 1949, NWP papers, reel 95.

212. Ernestine Bellamy to Elizabeth Forbes, April 14, 1950, NWP papers, reel 96; Elizabeth Kent to Amelia Himes Walker and Ernestine Bellamy, June 28, 1950, NWP papers, reel 96; Agnes Wells to Amelia Himes Walker, Dec. 20, 1950, NWP papers, reel 97; Ernestine Bellamy to Ethel Ernest Murrell, Aug. 8, 1951, NWP papers, reel 98.

213. Marjorie Barstow Greenbie to Emma Guffey Miller, July 10, 1961, NWP papers, reel 107.

214. Patricia A. McDonald to Helen Elizabeth Brown and Rose Zetzer, Sept. 2, 1969, NWP papers, reel 111; Patricia A. McDonald to Frances Kamm, Aug. 30, 1969, NWP papers, reel 111.

215. Jane Grant to Florence Kitchelt, n.d. [September 1948], Kitchelt papers, box 6 (147), SL; Jane Grant to Florence Kitchelt, n.d. [1948], Kitchelt papers, box 6 (147), SL; Alma Lutz to Florence Kitchelt, Sept. 8, 1948, Kitchelt papers, box 6 (177), SL.

216. Florence Kitchelt to Charlotte Kimball Kruesi, Feb. 17, 1950, Kitchelt papers, box 1 (1), SL; Jane Grant to Doris Stevens, Feb. 19, 1951, Stevens papers, SL; Jane Grant to Doris Stevens, March 14, 1951, Stevens papers, SL; Jane Grant to Doris Stevens, Nov. 30, 1951, Stevens papers, SL; Jane Grant to Doris Stevens, Jan. 3, 1961, Stevens papers, SL.

217. Lena Madesin Phillips to Marion Llewellyn, Dec. 3, 1948, Phillips papers, SL.

218. Florence Kitchelt to Misses Estin, Fisher, Gilbert, and Schoenfield, May 22, 1948, Kitchelt papers, box 3 (38), SL.

219. Florence Armstrong to "Stedfast Class," September 23, 1953, Armstrong papers, box 1 (17), SL.

220. Florence Kitchelt to Warren Seavey, March 8, 1945, Kitchelt papers, box 4 (93), SL.

221. Anna M. Kross to Lucy Somerville Howorth, March 29, 1946, Somerville-Howorth papers, box 13 (253), SL.

222. Mrs. Harold Jons to NWP, Dec. 9, 1945, NWP papers, reel 88.

223. Betty Gram Swing to Alice Paul, April 1, 1948, NWP papers, reel 93.

224. Marion Sayward to Alice Paul, March 10, 1958, NWP papers, reel 104.

225. Marjorie Longwell to Mary Glen Newell, June 4, 1966, NWP papers, reel 110.

226. Alma Lutz to Florence Kitchelt, Aug. 8, 1951, Kitchelt papers, box 7 (178), SL.

227. Ruth Nelson Edelman to Caroline Babcock, June 28, 1945, NWP papers, reel 86; M. E. Owens to Alice Paul, Aug. 12, 1957, NWP papers, reel 104.

228. Aline Mackenzie Taylor to Alice Paul, April 19, 1965, NWP papers, reel 109.

229. Alma Lutz to Rose Arnold Powell, May 29, 1953, Powell papers, box 3 (43), SL.

230. Rose Arnold Powell to Adelaide Johnson, May 24, 1945, Powell papers, box 3 (40), SL; Rose Arnold Powell to Adelaide Johnson, Oct. 26, 1945, Powell papers, box 3 (40), SL; Rose Arnold Powell, diary entry, March 2, 1947, Powell papers, box 1, v. 5, SL.

231. Mary C. Kennedy to Florence Armstrong, March 23, 1945, NWP papers, reel 85.

232. Peggy Capewell to Caroline Babcock, Jan. 3, 1945, Babcock-Hurlburt papers, box 3 (29), SL.

233. Amy R. Juengling to Dora Ogle, Feb. 25, 1945, NWP papers, reel 84.

234. Elsie Hill to Florence Armstrong, Feb. 28, 1945, NWP papers, reel 84.

235. Mary Sinclair Crawford to Katharine Norris, April 7, 1945, Norris papers, box 1 (3), SL.

Chapter 5. Commitment and Community

1. Mary Daly, *Gyn/Ecology: The Metaethics of Radical Feminism* (Boston: Beacon Press, 1978), p. 379.

2. Mabel Vernon, "Speaker for Suffrage and Petitioner for Peace," an oral history conducted in 1972 and 1973 by Amelia R. Fry, Regional Oral History Office, University of California, 1976, p. 191.

3. Rosabeth Moss Kanter, *Commitment and Community: Communes and Utopias in Sociological Perspective* (Cambridge: Harvard University Press, 1972), p. 66.

4. James Q. Wilson, *Political Organizations* (New York: Basic Books, 1973); adopted from Jo Freeman, "Crises and Conflicts in Social Movement Organizations," *Chrysalis* 5 (1977), 43–51.

5. Mary Birckhead to Emma Guffey Miller, Aug. 11, 1965, Miller papers, box 5 (91), SL.

6. Alice Paul, "Conversations with Alice Paul: Woman Suffrage and the Equal Rights Amendment," an oral history conducted in 1972 and 1973 by Amelia R. Fry, Regional Oral History Office, University of California, 1976, pp. 277–278.

7. Mary Elizabeth Nye to Agnes Wells, March 1, 1950, NWP papers, reel 96; Margaret Davis to Alice Paul, Feb. 5, 1956, NWP papers, reel 102.

8. Mary Elizabeth Nye to Amelia Himes Walker, Oct. 20, 1955, NWP papers, reel 101.

9. Elsie M. Wood to Alice Paul, Feb. 1, 1955, NWP papers, reel 101.

10. Anna Kelton Wiley to Caroline Babcock, Oct. 21, 1945, Babcock-Hurlburt papers, box 7 (96), SL; Anna Kelton Wiley to Guy M. Gillette, Oct. 2, 1949, Florence Kitchelt papers, box 9 (245), SL; Anna Kelton Wiley to Florence Kitchelt, March 19, 1948, Kitchelt papers, box 9 (245), SL.

11. Florence Kitchelt to Margaret Bruton, July 11, 1947, Kitchelt papers, box 4 (93), SL.

12. Helen B. Bruyere to Alice Paul, Aug. 5, 1948, NWP papers, reel 94.

13. Julia Pembleton to Alice Paul, n.d. [November 1955], NWP papers, reel 101.

14. Lillian V. Hulse to Dora Ogle, April 14, 1952, NWP papers, reel 98.

15. Lorraine Bentley to Alice Paul, April 21, 1956, NWP papers, reel 102; Mamie Sydney Mizen to Florence Armstrong, Oct. 25, 1948, NWP papers, reel 94.

16. Hubert H. Humphrey to Emma Newton, March 19, 1957, Emma Guffey Miller papers, box 3 (53), SL.

17. Sally Luther, "Susan B. Anthony Disciple Still Battles for Leader," Minneapolis Sunday *Tribune*, Jan. 23, 1949; also Rose Arnold Powell to Lincoln Borglum, June 25, 1956, Powell papers, box 2 (29), SL.

18. Rose Arnold Powell, diary entry, April 28, 1947, Powell papers, box 1, v. 5, SL; Rose Arnold Powell, diary entry, April 6, 1947, Powell papers, box 1, v. 5, SL; Rose Arnold Powell, diary entry, Feb. 23, 1949, Powell papers, box 1, v. 5, SL; Rose Arnold Powell, diary entry, Feb. 11, 1946, Powell papers, box 1, v. 4, SL.

19. Martha Taylor Howard to Rose Arnold Powell, Dec. 3, 1950, Powell papers, box 4 (58), SL.

20. Lena Madesin Phillips to Florence Allen, Aug. 25, 1953, Phillips papers, SL.

21. Lena Madesin Phillips to Annabelle Rankin, Jan. 21, 1947, Phillips papers, SL.

22. Betty Gram Swing to Ethel Ernest Murrell, Oct. 3, 1952, NWP papers, reel 99.

23. Elsie Hill to Ethel Ernest Murrell, Oct. 2, 1951, NWP papers, reel 98.

24. Helen Hunt West to Agnes Wells, June 23, 1949, NWP papers, reel 95.

25. Dorothy Spinks to Anita Pollitzer, May 28, 1949, NWP papers, reel 95; Alice Morgan Wright to Anita Pollitzer, n.d. [July 1946], NWP papers, reel 89; Anita Pollitzer to Emma Guffey Miller, June 1, 1964, NWP papers, reel 109; Anita Pollitzer to Alice Paul and Nina Avery, Aug. 6, 1964, NWP papers, reel 109.

26. Mary Kennedy to Agnes Wells, July 12, 1950, NWP papers, reel 97.

27. Interview no. 8.

28. Dorothy Kenyon to Anna Lord Strauss, Feb. 27, 1950, LWV papers, box 744, LC.

29. Gillie A. Larew to Lucy Somerville Howorth, July 2, 1946, Somerville-Howorth papers, box 13 (253), SL.

30. Marguerite Rawalt, *History of the NFBPWC*, vol. 2 (Washington, D.C.: NFBPWC, 169), p. 237.

31. Dorothy McCullough Lee to Lucy Somerville Howorth, Feb. 21, 1951, Somerville-Howorth papers, box 5 (117), SL.

32. Margaret Hickey to Eva v. B. Hansl, April 15, 1966, Hickey papers, box 6 (6), UMSL.

232 NOTES

33. Dorothy Shipley Granger to Caroline Babcock, Dec. 1, 1946, Babcock-Hurlburt papers, box 7 (98), SL.

34. Charlotte Johnson Opheim to co-workers, Aug. 12, 1950, NWP papers, reel 97.

35. Interview no. 5.

36. Local interview no. 5.

37. Esther Cloward to NWP, Dec. 5, 1946, NWP papers, reel 90.

38. Barbara Westebbe to Miss Hiatt, Aug. 25, 1949, NWP papers, reel 95.

39. Draft of Jane Grant's will, Jan. 3, 1963, Doris Stevens papers, SL.

40. Draft of Doris Stevens' will, n.d., Stevens papers, SL.

41. Norma Lindsay to Alice Paul, Nov. 22, 1954, NWP papers, reel 100.

42. Lena Madesin Phillips to Carola Redlich, Jan. 31, 1948, Phillips papers, SL.

43. Katherine Callery to Doris Stevens, Oct. 11, 1944, Stevens papers, SL; Katherine Callery to Doris Stevens, Sept. 28, 1944, Stevens papers, SL.

44. Caroline Babcock to Nora Stanton Barney, July 25, 1946, NWP papers, reel 89.

45. Mabel Griswold to Clara Snell Wolfe, May 7, 1947, NWP papers, reel 91.

46. Amelia Himes Walker to Emma Guffey Miller, Sept. 5, 1956, Miller papers, box 3 (510), SL.

47. Charl Ormond Williams to Rose Arnold Powell, May 17, 1946, Powell papers, box 4 (60), SL.

48. Doris Stevens, diary entry, April 11, 1951, Stevens papers, SL.

49. Katharine Norris to Caroline Babcock, Sept. 25, 1946, Babcock-Hurlburt papers, box 7 (98), SL.

50. Dorothy Shipley Granger to Caroline Babcock, Dec. 1, 1946, Babcock-Hurlburt papers, box 7 (98), SL.

51. Lavinia Dock to Party, Dec. 17, 1948, NWP papers, reel 94.

52. Mary Kennedy to Alice Paul, Aug. 12, 1964, NWP papers, reel 109.

53. Mary Murray to Anita Pollitzer, n.d. [October 1947], NWP papers, reel 92.

54. Margaret Ball to Florence Kitchelt, Oct. 30, 1947, Kitchelt papers, box 5 (109), SL.

55. Lillian H. Kerr to Alice Paul, May 15, 1959, NWP papers, reel 105.

56. Mary Glenn Newell to Mrs. Grenville D. Braman, April 14, 1959, NWP papers, reel 105.

57. Nina [Allender] to Martha [Souder], Jan. 7, 1949, NWP papers, reel 94.

58. "Vicky" Vickers to "my very dear lady" [Alice Paul], n.d. [February 1950], NWP papers, reel 96.

59. Helen Vanderburg to Florence Armstrong, Feb. 4, 1956, Armstrong papers, box 2 (30), SL.

60. Dorothy Rogers to Elsie Hill, June 15, 1965, NWP papers, reel 109.

61. HQ Secy. to Ruth M. Knight, Dec. 13, 1954, NWP papers, reel 100; Helen Hunt West to Agnes Wells, May 16, 1949, NWP papers, reel 95; Ernestine Powell to Mildred Palmer, Oct. 13, 1954, NWP papers, reel 100; Mary F. Anderson to Helen Paul, May 23, 1956, NWP papers, reel 102.

62. Mary Markajani to Anita Pollitzer, Feb. 23, 1948, NWP papers, reel 92.

63. Myrtle Bell to Mary Church Terrell, March 29, 1951, Terrell papers, reel 11, LC.

64. Local interview no. 2.

65. Leslie Black to Caroline Babcock, June 25, 1945, NWP papers, reel 86; Rilla A. Nelson to Olive Hurlburt, May 1, 1947, Babcock-Hurlburt papers, box 8 (105), SL; Marie C. Armstrong to Elsie Hill, Aug. 13, 1947, NWP papers, reel 91; Marie Horton to Anita Pollitzer, Feb. 15, 1948, NWP papers, reel 92.

66. Ruth Roach to Caroline Babcock, July 22, 1946, NWP papers, reel 89.

67. Betty Gram Swing to Alice Paul, April 25, 1948, NWP papers, reel 93.

68. Florence Kitchelt to Misses Estin, Fisher, Gilbert, Shoenfield, May 22, 1948, Kitchelt papers, box 3 (38), SL.

69. Zelia Ruebhausen to Henry David, Feb. 21, 1956, LWV papers, box 1174, LC.

70. Anna Kelton Wiley to Lucy Rice Winkler, July 4, 1951, NWP papers, reel 98.

71. Margaret Hickey to Lucy Somerville Howorth, Aug. 15, 1946, Somerville-Howorth papers, box 13 (253), SL.

72. Charlotte N. Leyden to Rose Arnold Powell, April 12, 1960, Powell papers, box 2 (21), SL.

73. Evelyn Samras to Florence Kitchelt, Dec. 3, 1945, Kitchelt papers, box 8 (224), SL.

74. Peggy Capewell to Caroline Babcock, Jan. 3, 1945, Babcock-Hurlburt papers, box 7 (97), SL.

75. Dorothy Shipley Granger to Caroline Babcock, Jan. 7, 1946, Babcock-Hurlburt papers, box 7 (97), SL; Dorothy Shipley Granger to Caroline Babcock, Feb. 11, 1946, Babcock-Hurlburt papers, box 7 (97), SL; Dorothy Shipley Granger to Caroline Babcock, n.d. [1945], Babcock-Hurlburt papers, box 3 (29), SL.

76. Helen Hunt West to Dora Ogle, n.d. [April 1952], NWP papers, reel 98.

77. Rawalt, *History of the NFBPWC*, vol. 2, p. 158.

78. Interview with Margaret Hickey, Feb. 14, 1974, conducted by Ann Lever and Susan M. Hartmann, UMSL.

79. Local interview no. 8. See also local interview no. 11 and no. 35.

80. Interview no. 3.

81. Katharine Norris to Edith Goode, Sept. 3, 1946, Norris papers, box 1 (4), SL.

82. Interview no. 13.

83. Rose Arnold Powell to Charl Ormond Williams, June 12, 1946, Powell papers, box 4 (60), SL; Rose Arnold Powell to Charl Ormond Williams, June 30, 1946, Powell papers, box 4 (60), SL.

84. Margaret Hickey, "Never Underestimate the Power of Your Club," speech, 1950, Hickey papers, box 30 (113), UMSL.

85. Local interview no. 2. See also local interview no. 3.

86. Local interview no. 5.

87. Mary [no last name] to Lucy Somerville Howorth, Sept. 19, 1958, Somerville-Howorth papers, box 13 (257), SL.

88. Rose Arnold Powell to Dora Ogle, Feb. 16, 1950, NWP papers, reel 96; Rose Arnold Powell to Alma Lutz, July 7, 1956, Powell papers, box 3 (43), SL; Rose Arnold Powell to Grace Nicholas, Oct. 15, 1956, Powell papers, box 3 (37), SL; Rose Arnold Powell, diary entry, July 8, 1958, Powell papers, box 1, v. 8, SL.

89. Anne Carter to Betty [Gram Swing], Oct. 27, 1954, NWP papers, reel 100.

90. Lena Madesin Phillips to Marion Llewellyn, Dec. 3, 1948, Phillips papers, SL.

91. Rawalt, *History of the NFBPWC*, vol. 2, p. iv.

92. M. Eunice Hilton to Margaret Hickey, Feb. 20, 1962, Hickey papers, box 28 (99), UMSL.

93. Perle Mesta, *Perle: My Story* (New York: McGraw-Hill, 1960), pp. 77, 94.

94. [Mildred Palmer] to Amelia Himes Walker, Aug. 21, 1950, NWP papers, reel 97.

95. Marjorie Longwell to Alice Paul, May 5, 1963, NWP papers, reel 108.

96. Elizabeth Kent to Agnes Wells, Dec. 31, 1950, NWP papers, reel 97.

97. Janet Griswold to Alice Paul, Feb. 2, 1955, NWP papers, reel 101.

98. Inez Haynes Irwin to Alice Paul, July 5, 1956, NWP papers, reel 102.

99. Ruth S. Brumbaugh to Mary Church Terrell, July 13, 1950, Terrell papers, reel 11, LC.

100. Local interview no. 6.

101. Interview no. 10.

102. Thelma Davenport to Lucy Somerville Howorth, July 9, 1947, Somerville-Howorth papers, box 13 (254), SL.

103. Cookie Wascha, clipping from the Flint *Journal*, Feb. 14, 1980, in the Olive Hurlburt Biographical File, SL.

104. Gaeta Boyer to Caroline Babcock, n.d. [1960], Babcock-Hurlburt papers, box 9 (113), SL.

105. Lorraine L. Blair to Caroline Babcock, Dec. 19, 1946, Babcock-Hurlburt papers, box 8 (104), SL.

106. Florence Kitchelt to Jane Grant, Nov. 4, 1950, Kitchelt papers, box 6 (147), SL.

107. Florence Kitchelt to Alma Lutz, Feb. 27, 1955, Kitchelt papers, box 7 (178), SL.

108. Florence Kitchelt to Caroline Babcock, Jan. 2, 1949, Babcock-Hurlburt papers, box 9 (109), SL.

109. Perle Mesta to Emma Guffey Miller, Aug. 13, 1949, Miller papers, box 2 (37), SL.

110. Mary Norton to Emma Guffey Miller, April 6, 1950, Miller papers, box 3 (38), SL.

111. Mesta, *Perle: My Story*, pp. 127–128.

112. Corinne H. Klinzing to Emma Guffey Miller, Jan. 25, 1955, Miller papers, box 3 (42), SL.

113. Rose Arnold Powell, diary entry, Dec. 31, 1934, Powell papers, box 1, v. 1, SL.

114. Rose Arnold Powell to Martha Taylor Howard, Feb. 21, 1947, Powell papers, box 4 (58), SL.

115. Rose Arnold Powell to Una Winter, Jan. 11, 1954, Powell papers, box 4 (57), SL.

116. Rose Arnold Powell to Martha Taylor Howard, Nov. 18, 1953, Powell papers, box 4 (58), SL.

117. Rose Arnold Powell to Martha Taylor Howard, April 3, 1952, Powell papers, box 4 (58), SL.

118. Vicky Vickers to Alice Paul, n.d. [September 1954], NWP papers, reel 100.

119. Alice Paul to Elizabeth Rogers, June 4, 1950, NWP papers, reel 96.

120. Elizabeth Rogers to Alice Paul, July 4, 1945, NWP papers, reel 87.

121. See Blanche Wiesen Cook, "Female Support Networks and Political Activism," *Chrysalis* 3 (1977), 43–61.

122. [Alice Paul] to Lavinia Dock, May 5, 1945, NWP papers, reel 86.

123. Lavinia Dock to Alice Paul, May 9, 1945, NWP papers, reel 86.

124. Ruth S. Brumbaugh to Mary Church Terrell, n.d. [1950], Terrell papers, reel 11, LC.

125. Geneva K. Valentine to Mary Church Terrell, Oct. 16, 1950, Terrell papers, reel 11, LC.

126. Mary Waring to Mary Church Terrell, April 25, 1950, Terrell papers, reel 10, LC.

127. Ada S. Howell to Mary Church Terrell, May 9, 1952, Terrell papers, reel 11, LC; Ruth Davie to Mary Church Terrell, April 23, 1951, Terrell papers, reel 11, LC; Ruth Davie to Mary Church Terrell, March 19, 1951, Terrell papers, reel 1, LC.

128. Otelia Love Jackson to Mary Church Terrell, n.d. [February 1952], Terrell papers, reel 11, LC.

129. Geneva K. Valentine to Mary Church Terrell, Sept. 4, 1953, Terrell papers, reel 12, LC.

130. Marjory Lacey-Baker, "Chronological Record of Events and Activities for the Biography of Lena Madesin Phillips, 1881–1955," Phillips papers, SL.

131. Mary Kennedy to Lena Madesin Phillips, July 12, 1954, Phillips papers, SL.

132. Ruff [J. Margaret Warner] to Lena Madesin Phillips, April 5, 1945, Phillips papers, SL.

133. Lisa Sergio, *A Measure Filled: The Life of Lena Madesin Phillips Drawn From Her Autobiography* (New York: Robert B. Luce, 1972), p. 216.

134. Such relationships were often passionate and sensual, but historians disagree about labeling them sexual. See Carroll Smith-Rosenberg, "The Female World of Love and Ritual: Relations Between Women in Nineteenth-Century America," *Signs* 1 (1975), 1–29; Nancy F. Cott, *The Bonds of Womanhood* (New Haven, Conn.: Yale University Press, 1977), chapt. 5; Nancy Sahli, "Smashing: Women's Relationships Before the Fall," *Chrysalis* 8 (1979), 17–27; Cook, "Female Support Networks;" Lillian Faderman, *Surpassing the Love of Men* (New York: William Morrow, 1981).

135. See Vern Bullough and Bonnie Bullough, "Lesbianism in the 1920's and 1930's: A Newfound Study," *Signs* 2 (1977), 895–904; Cook, "Female Support Networks;" Blanche Wiesen Cook, "The Historical Denial of Lesbianism," *Radical History Review* 20 (1979), 60–65; Sahli, "Smashing;" Sasha Gregory Lewis, *Sunday's Women: A Report on Lesbian Life Today* (Boston: Beacon Press, 1979); Madeline Davis, Liz Kennedy, and Avra Michelson, "Aspects of the Buffalo Lesbian Community in the Fifties," Buffalo Women's Oral History Project, paper presented at the National Women's Studies Association conference, Bloomington, Indiana, May 1980; and Leila J. Rupp, " 'Imagine My Surprise:' Women's Relationships in Historical Perspective," *Frontiers* 5 (1980), 61–70.

136. Estelle Freedman, "Separatism as Strategy: Female Institution Building and American Feminism, 1870–1930," *Feminist Studies* 5 (1979), 512–529.

137. Anna Mary Wells, *Miss Marks and Miss Woolley* (Boston: Houghton Mifflin, 1978), p. 56.

138. Caroline Babcock to Jeannette Marks, Feb. 12, 1947, Babcock-Hurlburt papers, box 8 (105), SL.

139. Marjory Lacey-Baker, "Chronological Record of Events and Activities for the Biography of Lena Madesin Phillips, 1881–1955," Phillips papers, SL.

140. Marjory Lacey-Baker, "Chronological Record," Phillips papers, SL.

141. Lena Madesin Phillips to Audrey Turner, Jan. 21, 1948, Phillips papers, SL; Lena Madesin Phillips to Olivia Rossetti Agresti, April 26, 1948, Phillips papers, SL.

142. Gordon Holmes to Madesin [Phillips] and Maggie [Lacey-Baker], Dec. 15, 1948, Phillips papers, SL.

143. Lena Madesin Phillips to Gordon Holmes, March 28, 1949, Phillips papers, SL.

144. Sergio, *Measure Filled*, p. 60.

145. Alma Lutz to Florence Kitchelt, July 1, 1948, Kitchelt papers, box 6 (177), SL.

146. Alma Lutz to Florence Kitchelt, July 29, 1959, Kitchelt papers, box 7 (178), SL.

147. Alma Lutz to Florence Armstrong, Aug. 26, 1959, Armstrong papers, box 1 (17), SL.

148. Alma Lutz to Rose Arnold Powell, Dec. 14, 1959, Powell papers, box 3 (43), SL.

149. Press release from Mabel Vernon Memorial Committee, and obituary in Wilmington *Morning News*, Sept. 3, 1975, in Vernon, "Speaker for Suffrage."

150. Nora Stanton Barney to Alice Paul, n.d. [May 1945], NWP papers, reel 86.

151. Alice Morgan Wright to Anita Pollitzer, n.d. [July 1946], NWP papers, reel 89.

152. Doris Stevens, diary entry, Feb. 4, 1946, Stevens papers, SL.

153. Doris Stevens, diary entry, Dec. 1, 1945, Stevens papers, SL.

154. Katharine Callery to Doris Stevens, Aug. 17, 1944, Stevens papers, SL.

155. Doris Stevens to Westbrook Pegler, May 3, 1946, Stevens papers, SL.

156. Quoted in Richard Polenberg, *One Nation Divisible* (New York: Penguin Books, 1980), p. 124; and in Robert Griffith, "Harry S. Truman and the Burden of Modernity," *Reviews in American History* 9 (1981), 298. John D'Emilio's unpublished paper, "The Homosexual Menace: The Politics of Sexuality in Cold War America," presented at the Organization of American Historians conference, Philadelphia, April 1982, explores this connection. See also John D'Emilio, *Sexual Politics, Sexual Communities: The Making of a Homosexual Community in the U.S., 1940–1970* (Chicago: University of Chicago Press, 1983).

157. Agnes E. Meyer, *Out of These Roots: The Autobiography of an American Woman* (Boston: Little, Brown, 1953), p. 69.

158. Doris Stevens, diary entries, Aug. 30, 1953, Sept. 1, 1953, and Aug. 24, 1953, Stevens papers, SL.

159. India Edwards, *Pulling No Punches* (New York: Putnam's, 1977), pp. 189–190.

160. Doris Faber, *The Life of Lorena Hickok: E.R.'s Friend* (New York: William Morrow, 1980).

161. Interview no. 9 and local interview no. 1.

162. Edith Conway to Caroline Babcock, n.d. [March 1945], NWP papers, reel 85.

163. Mathe Eastlack Driscott to Mary Church Terrell, n.d. [September 1950], Terrell papers, reel 11, LC.

164. Minnie L. Maffett to Lena Madesin Phillips, Jan. 6, 1950, Phillips papers, SL.

165. Rawalt, *History of the NFBPWC*, vol. 2, p. iv.

166. Ethel Ernest Murrell to Anita Pollitzer, Nov. 19, 1946, NWP papers, reel 90.

167. Blanche LeBaron Gruver to Lucy Somerville Howorth, May 23, 1946, Somerville-Howorth papers, box 8 (151), SL.

168. See Alice S. Rossi, *The Feminist Papers* (New York: Bantam, 1973); and Sara Evans, *Personal Politics* (New York: Knopf, 1979), on the same phenomenon in the New Left.

169. Lucy H. Hedrick, "Four Generations of Stanton Spunk," *Ms.*, September 1984, 100–102.

170. Caroline Babcock to Gertrude Robbins, July 15, 1946, NWP papers, reel 89.

171. Telegram from Edna Capewell to Alice Paul, Sept. 14, 1945, NWP papers, reel 87.

172. Memo from Lucile W. Heming to Anna Lord Strauss, April 8, 1947, LWV papers, box 694, LC.

173. Polly Graham, "Nellie Nugent Somerville," July 31, 1957, Somerville-Howorth papers, box 2 (19), SL.

174. Pauline E. Mandigo to Lucy Somerville Howorth, June 26, 1951, Somerville-Howorth papers, box 7 (147), SL.

175. Kathryn McHale to Lucy Somerville Howorth, March 3, 1952, Somerville-Howorth papers, box 5 (108), SL.

176. NWP *Bulletin* 1 (January and February 1966).

177. Lucy Branham to Evelyn Mazumdar, April 3, 1959, NWP papers, reel 105.

178. Memo from Martha Souder to Ethel Ernest Murrell, Aug. 15, 1951, NWP papers, reel 98.

179. Charlotte Johnson Opheim to Anita Pollitzer, April 14, 1949, NWP papers, reel 95.

180. Grace Cook Kurz to Anne Whitehouse, Feb. 17, 1948, NWP papers, reel 92; Anne Whitehouse to Grace Cook Kurz, March 11, 1948, NWP papers, reel 92.

181. Alice Paul to Janet Griswold, Feb. 17, 1955, NWP papers, reel 101; Janet Griswold to Alice Paul, March 3, 1955, NWP papers, reel 101; Janet Griswold to Alice Paul, Feb. 15, 1963, NWP papers, reel 108.

182. Neenah Hastings Lessemann to Florence Armstrong, Aug. 8, 1945, NWP papers, reel 87.

183. Helen Paul to Mabel Griswold, June 21, 1954, NWP papers, reel 100; Helen Paul to Ernestine Powell, June 24, 1956, NWP papers, reel 102; Ernestine Powell to Alice Paul, Jan. 8, 1968, NWP papers, reel 110.

184. Local interviews no. 5, no. 6, and no. 1. See also local interview no. 10 and no. 16.

185. Mary Lee Mann to Alice Paul, May 6, 1965, NWP papers, reel 109.

186. Mrs. Walter Ferguson to Caroline Babcock, May 16, 1945, NWP papers, reel 86.

187. Kathryn Paulsen, "The Last Hurrah of the Woman's Party," n.d., NWP papers, reel 111.

188. Interview no. 7.

Chapter 6. Feminist Lives

1. Rose Arnold Powell, diary entry, May 15, 1946, Powell papers, box 1, v. 4, SL. This section is based on diaries and correspondence in the Powell papers.

2. Rose Arnold Powell to Elizabeth B. Borden, April 13, 1958, Powell papers, box 1, SL.

3. Rose Arnold Powell to Mary Beard, June 23, 1948, Powell papers, box 2 (27), SL.

4. Rose Arnold Powell to Frances T. Freeman, Aug. 13, 1957, Powell papers, box 2 (21), SL; Rose Arnold Powell, diary entry, Feb. 15, 1946, Powell papers, box 1, v. 4, SL.

5. Rose Arnold Powell, diary entry, March 2, 1947, Powell papers, box 1, v. 5, SL.

6. Alma Lutz to Rose Arnold Powell, Aug. 7, 1956, Powell papers, box 3 (43), SL.

7. Rose Arnold Powell to Charl Ormond Williams, June 30, 1946, Powell papers, box 4 (60), SL.

8. Rose Arnold Powell, diary entry, Dec. 13, 1934, Powell papers, box 1, v. 1, SL.

9. Rose Arnold Powell to Martha Taylor Howard, Nov. 18, 1953, Powell papers, box 4 (58), SL.

10. Rose Arnold Powell, diary entry, June 1, 1948, Powell papers, box 1, v. 5, SL.

11. "Some biographical material," May 11, 1951, Florence Kitchelt papers, box 1, SL. This section is based on correspondence and other papers in the Kitchelt collection.

12. Florence Kitchelt to Misses Estin, Fisher, Gilbert, Schoenfield, May 22, 1948, Kitchelt papers, box 3 (38), SL.

13. Holiday Greetings, December 1960, Kitchelt papers, box 9 (253), SL.

14. Florence Kitchelt to Katharine St. George, Feb. 20, 1954, Kitchelt papers, box 8 (222), SL.

15. "To Dear Friends of Florence Ledyard Cross Kitchelt from Her Sister, Dorothy Zeiger," n.d. [1961], Kitchelt papers, box 1, SL.

16. Florence Kitchelt to Secretary of NWP, April 1, 1943, Kitchelt papers, box 1 (1), SL.

17. Florence Kitchelt to Barbara Woollcott, March 5, 1945, Kitchelt papers, box 3 (36), SL.

18. Florence Kitchelt to Louise Earle, April 7, 1949, Kitchelt papers, box 6 (141), SL.

19. Florence Kitchelt to K. Frances Scott, June 15, 1949, Kitchelt papers, box 8 (226), SL.

20. Florence Kitchelt to Florence Ripley Mastin, Aug. 22, 1950, Kitchelt papers, box 7 (184), SL.

21. Florence Kitchelt to Warren Seavey, March 8, 1945, Kitchelt papers, box 4 (93); Florence Kitchelt to Ashley Montagu, March 19, 1952, Kitchelt papers, box 7 (191), SL.

22. Florence Kitchelt to Jane Grant, Sept. 20, 1955, Kitchelt papers, box 5 (133), SL.

23. This section is based on correspondence to, from, and about Ackley in the papers of the National Woman's Party.

24. Fannie Ackley to NWP, n.d. [April 1946], NWP papers, reel 89.

25. Fannie Ackley to Caroline Babcock, Nov. 29, 1945, NWP papers, reel 88; Fannie Ackley to Mildred Palmer, Sept. 19, 1950, NWP papers, reel 97.

26. Fannie Ackley to Caroline Babcock, Nov. 29, 1945, NWP papers, reel 88; Fannie Ackley to Caroline Babcock, Oct. 17, 1946, NWP papers, reel 90.

27. Fannie Ackley to Caroline Babcock, Nov. 6, 1946, NWP papers, reel 90; Fannie Ackley to Mabel Griswold, Jan. 5, 1949, NWP papers, reel 94.

28. Fannie Ackley to Olive Beale, May 18, 1947, NWP papers, reel 91.

29. Alma Lutz to Florence Kitchelt, Kitchelt papers, box 4 (87), SL.

30. Fannie Ackley to Caroline Babcock, Oct. 17, 1946, NWP papers, reel 90.

31. Fannie Ackley to Mabel Griswold, Jan. 5, 1949, NWP papers, reel 94.

32. Mary Elizabeth Nye to Amelia Himes Walker, Oct. 20, 1955, NWP papers, reel 101.

33. Note from Amelia Himes Walker to Mildred Palmer, on letter from Fannie Ackley to Amelia Himes Walker, Aug. 20, 1950, NWP papers, reel 97.

34. Alice Paul to Fannie Ackley, Nov. 14, 1955, NWP papers, reel 101.

35. This section is based on "The Unfinished Autobiography of Lena Madesin Phillips," notes on her life written by Marjory Lacey-Baker, and correspondence in the Phillips papers. See also the biographical sketch by J. Stanley Lemons in *Notable American Women: The Modern Period*, eds. Barbara Sicherman and Carol Hurd Green (Cambridge, Mass.: Belknap Press, 1980), pp. 544–545.

36. Lena Madesin Phillips to Annabelle Rankin, Jan. 21, 1947, Phillips papers, SL.

37. Lena Madesin Phillips to Mary C. Kennedy, Dec. 18, 1950, Phillips papers, SL.

38. "Feminism is Seen Losing Force Here," *New York Times*, Dec. 15, 1944.

39. Marianne Beth to Lena Madesin Phillips, n.d. [June 1951], Phillips papers, SL.

40. The description appears in James Herbert, "Democrat Gals Get a Fair Deal," clipping in the Emma Guffey Miller papers, box 4 (67), SL. This section is based on correspondence and other papers in the Miller collection and in the NWP papers. See also the article on Miller by Keith Melder in *Notable American Women*, pp. 476–478.

41. Emma Guffey Miller to Agnes Wells, Feb. 22, 1950, NWP papers, reel 96.

42. Emma Guffey Miller to Caroline Babcock, June 25, 1945, NWP papers, reel 86.

43. Emma Guffey Miller to Adlai Stevenson, Nov. 12, 1956, Miller papers, box 3 (191), SL.

44. Emma Guffey Miller to Elizabeth Carpenter, Jan. 16, 1964, Miller papers, box 5 (74), SL.

45. Emma Guffey Miller to Elizabeth Carpenter, June 9, 1964, Miller papers, box 5 (75), SL.

46. Lyndon B. Johnson to Emma Guffey Miller, June 13, 1964, Miller papers, box 5 (75), SL.

47. Marion J. Harron to Emma Guffey Miller, Feb. 28, 1961, NWP papers, reel 106; Anna Alpern to Emma Guffey Miller, April 8, 1963, Miller papers, box 4 (71), SL.

240 NOTES

48. Grace Sloan to Emma Guffey Miller, Aug. 30, 1962, Miller papers, box 4 (70), SL.

49. Anita Pollitzer to Emma Guffey Miller, Aug. 30, 1962, Miller papers, box 4 (70), SL.

50. Ruth [Gage-Colby] to Alice Paul, May 5, 1964, NWP papers, reel 109.

51. All of the information here comes from the interview conducted in December 1979 and from private correspondence in E. S. Pollock's possession.

52. Ellen Goodman's column and Judy Gurovitz, "Suffragists Still Going Strong," both in Alma Lutz Biography File, SL. This section is based on correspondence to and from Lutz in a number of collections at the Schlesinger Library and in the NWP papers.

53. Alma Lutz to Marion Sayward, May 18, 1963, NWP papers, reel 108.

54. Alma Lutz to Florence Kitchelt, July 27, 1945, Kitchelt papers, box 7 (178), SL.

55. Alma Lutz to Florence Kitchelt, June 2, 1956, Babcock-Hurlburt papers, box 9 (117), SL.

56. Alma Lutz, letter to the editor of the *Nation*, Sept. 18, 1948; Alma Lutz to Caroline Babcock, Jan. 27, 1954, Babcock-Hurlburt papers, box 9 (117), SL; Alma Lutz to Florence Kitchelt, Dec. 17, 1953, Kitchelt papers, box 7 (178), SL.

57. Alma Lutz to Florence Kitchelt, July 11, 1951, Kitchelt papers, box 7 (178), SL.

58. Alma Lutz to Florence Kitchelt, Jan. 12, 1950, Kitchelt papers, box 7 (178), SL.

59. Local interview no. 5.

60. Interview no. 6.

61. Interview with Margaret Hickey, Feb. 14, 1974, conducted by Ann Lever and Susan Hartmann, Hickey papers, UMSL.

62. Interview no. 8.

63. Interview no. 13.

Chapter 7. The "Isolationism" of the Women's Rights Movement

1. Russell L. Curtis, Jr. and Louis A. Zurcher, Jr., "Stable Resources of Protest Movements: The Multi-Organizational Field," *Social Forces*, 52 (September 1973), 52–61.

2. Mayer N. Zald and John D. McCarthy, "Social Movement Industries: Competition and Cooperation Among Movement Organizations," in *Research in Social Movements, Conflict and Change*, ed. Louis Kriesberg, vol. 3 (Greenwich, Conn.: JAI Press, 1980), pp. 1–20.

3. Agnes Wells to Elizabeth Forbes, May 16, 1949, NWP papers, reel 95.

4. Mabel Griswold to Florence Armstrong, Sept. 9, 1949, NWP papers, reel 96. See also Mrs. Charles Robbins to Anita Pollitzer, Feb. 1, 1947, NWP papers, reel 90; and Irene Brooks to Agnes Wells, Aug. 9, 1950, NWP papers, reel 97.

5. Interview no. 2.

6. Alice Paul, "Conversations with Alice Paul: Woman Suffrage and the Equal Rights Amendment," an oral history conducted in 1972 and 1973 by Amelia R. Fry, Regional Oral History Office, University of California, 1976, p. 436.

7. Ethel Ernest Murrell to Ernestine Bellamy, March 26, 1952, NWP papers, reel 98.

8. Virginia S. Freedom to Alice Paul, April 25, 1950, NWP papers, reel 96.

9. Interview no. 2.

10. Speech of Doris Stevens before the School of Politics Women's National Republican Club, Feb. 8, 1949, Stevens papers, SL.

11. Doris Stevens, diary entries, May 6, 1949, Aug. 16, 1950, Oct. 14, 1950, Stevens papers, SL.

12. Doris Stevens, diary entries, Dec. 31, 1949, July 26, 1950, May 21, 1951, Stevens papers, SL.

13. Doris Stevens to Charlotte Boissevain, July 1, 1952, Stevens papers, SL.

14. "Suggested Questions for the Voice of America Interview with Judge Lucy Howorth," Sept. 19, 1951 [?], Somerville-Howorth papers, box 7 (147), SL.

15. Merle W. Cerulla to Burnita Shelton Matthews, Nov. 29, 1952, Matthews papers, SL.

16. Patricia O'Connell to NWP, April 5, 1945, NWP papers, reel 85.

17. Committee on Un-American Activities, U.S. House of Representatives, Report on the Congress of American Women, Oct. 23, 1949, pp. 102–103.

18. Nora Stanton Barney to Bella Dodd, Jan. 17, 1945, NWP papers, reel 84; Nora Stanton Barney to Emanuel Celler, June 5, 1945, NWP papers, reel 86.

19. Nora Stanton Barney to Rose Arnold Powell, April 15, 1956, Powell papers, box 2 (24), SL.

20. Mary C. Kennedy to Agnes Wells, Feb. 21, 1950, NWP papers, reel 96.

21. Jane Norman Smith to Alice Paul, Sept. 29, 1948, NWP papers, reel 94.

22. Lola Stanley to NWP, Sept. 18, 1948, NWP papers, reel 94.

23. Wilma Soss to Florence Kitchelt, March 17, 1950, Kitchelt papers, box 8 (234), SL.

24. Lisa Sergio, *A Measure Filled* (New York: Robert B. Luce, 1972), p. 224.

25. Sarah T. Hughes to Lucy Somerville Howorth and Mary Agnes Brown, July 30, 1946, Somerville-Howorth papers, box 13 (253), SL.

26. Florence Kitchelt to Wilma Soss, March 19, 1950, Kitchelt papers, box 8 (234), SL.

27. Alma Lutz to Caroline Babcock, Oct. 20, 1948, Babcock-Hurlburt papers, box 9 (117), SL.

28. Luella S. Laudin to Rose Arnold Powell, Jan. 29, 1946, Powell papers, box 3 (46), SL.

29. Lizabeth Wiley to Emma Guffey Miller, March 15, 1950, Miller papers, box 3 (38), SL.

30. Marguerite Rawalt, *History of the NFBPWC*, vol. 2 (Washington, D.C.: NFBPWC, 1969), pp. 135, 171, 230–231, 237, 239.

31. Interview no. 12.

32. Mary Murray to Anita Pollitzer, n.d. [December 1946], NWP papers, reel 90.

33. Alice Paul to Elsie Hill, Aug. 9, 1947, NWP papers, reel 91.

34. Jane Norman Smith to Anita Pollitzer, Sept. 24, 1947, NWP papers, reel 92.

35. Interview no. 7.

36. Elizabeth Forbes to Agnes Wells, Nov. 26, 1949, NWP papers, reel 96; Alice Paul to Elizabeth Forbes, Dec. 19, 1949, NWP papers, reel 96; Alice M. Katchadourian to Agnes Wells, Feb. 22, 1950, NWP papers, reel 96.

37. Mabel Griswold to Jennie Scott Griffiths, July 23, 1947, NWP papers, reel 91.

38. Jane Norman Smith to Alice Paul, Oct. 31, 1947, NWP papers, reel 92.

39. HUAC, Report on the Congress of American Women, pp. 1, 3, 21, 23, 99.

40. Mary Markajani to Mabel Griswold, June 10, 1948, NWP papers, reel 93.

41. Excerpt from the text of the Communist Party platform, the *New York Times*, Aug. 7, 1948.

42. See Elizabeth Gurley Flynn, *The Rebel Girl: An Autobiography* (New York: International Publishers, 1973); Peggy Dennis, *The Autobiography of an American Communist: A Personal View of a Political Life, 1925–1975* (Westport, Conn./Berkeley, Calif.: Lawrence Hill & Co./Creative Arts Book Co., 1977); Jessica Mitford, *A Fine Old Conflict* (New York: Vintage Books, 1977); and, especially, Ellen Kay Trimberger, "Women in the Old and New Left: The Evolution of a Politics of Personal Life," *Feminist Studies* 5 (Fall 1979), 432–450; and Peggy Dennis, "Response," *Feminist Studies* 5 (Fall 1979), 451–460.

43. Elizabeth Gurley Flynn, "Life of the Party," *Daily Worker*, March 9, 1950.

44. Alice Paul to Barbara Wilkin, Dec. 13, 1949, NWP papers, reel 96.

45. Interview no. 2.

46. Nina Horton Avery to James B. Frazer, Jr., March 10, 1956, Emma Guffey Miller papers, box 3 (48), SL.

47. Jane Norman Smith to Mildred Palmer, Oct. 12, 1950, NWP papers, reel 97.

48. Emma Guffey Miller to Joseph Clark, March 13, 1962, NWP papers, reel 107.

49. [Frieda Liebman] to Mary E. Owens, Nov. 17, 1959, NWP papers, reel 105.

50. Interview no. 2.

51. Elizabeth Gurley Flynn to Lavinia Dock, June 18, 1944, NWP papers, reel 81.

52. Ella Sherwin to Alice Paul, July 5, 1945, NWP papers, reel 87; Edna Sickmon to Olive Beale, July 26, 1945, NWP papers, reel 87.

53. [Peggy Capewell] to Caroline Babcock, March 10, 1946, Babcock-Hurlburt papers, box 7 (97), SL.

54. See Peter H. Irons, "American Business and the Origins of McCarthyism: The Cold War Crusade of the United States Chamber of Commerce," in *The Specter: Original Essays on the Cold War and the Origins of McCarthyism*, eds. Robert Griffith and Athan Theoharis (New York: New Viewpoints, 1974), pp. 72–89.

55. Caroline Babcock to Ella Maercklein, Jan. 9, 1945, NWP papers, reel 84.

56. Caroline Babcock to Mrs. V. A. Rea, Jan. 12, 1945, NWP papers, reel 84.

57. Alice M. Birdsall to W. J. Eden, Jan. 22, 1945, NWP papers, reel 84.

58. Caroline Babcock to Ella Maercklein, Jan. 9, 1945, NWP papers, reel 84.

59. Florence Kitchelt to Caroline Babcock, March 23, 1950, Babcock-Hurlburt papers, box 9 (109), SL.

60. Alice Paul to Mary Agnes Brewer, April 20, 1958, NWP papers, reel 104.

61. Alice Paul to Clara Wolfe, June 14, 1958, NWP papers, reel 104; [Frieda Liebman] to Helen Borsick, June 16, 1958, NWP papers, reel 104.

62. Memo from Jacob Clayman to Legislative Representatives of IUD Affiliated Unions, June 14, 1960, Esther Peterson papers, SL.

63. Florence Kitchelt to Anita Pollitzer, Feb. 2, 1944, Kitchelt papers, box 8 (211), SL.

64. Caroline Babcock to Mrs. V. A. Rea, Jan. 12, 1945, NWP papers, reel 84.

65. Fannie Ackley to Caroline Babcock, Nov. 6, 1946, NWP papers, reel 90.

66. Alice Paul to Mabel Griswold, Sept. 2, 1954, NWP papers, reel 100.

67. See Jo Freeman, *The Politics of Women's Liberation* (New York: David McKay, 1975), pp. 80–81.

68. President of AFL to Frances P. Bolton, June 11, 1953, NWP papers, reel 99.

69. Report on interview of Agnes Wells and Lucy Rice Winkler with Elisabeth Christman, April 25, 1951, NWP papers, reel 97.

70. On the Women's Bureau, see Judith Anne Sealander, "The Women's Bureau, 1920–1950: Federal Reaction to Female Wage Earning," Diss., Duke University, 1977; and Cynthia E. Harrison, "Prelude to Feminism: Women's Organizations, the Federal Government and the Rise of the Women's Movement, 1942 to 1968," Diss., Columbia University, 1982.

71. Lucille Buchanan to Pauline Newman, Aug. 17, 1944, Frieda Miller papers, box 1 (9), SL; Pauline Newman to Frieda Miller, Feb. 20, 1947, Frieda Miller papers, box 1 (8), SL.

72. Alice Paul to Lucy Rice Winkler, Dec. 15, 1953, NWP papers, reel 99.

73. Alice Paul to Florence Kitchelt, Dec. 15, 1953, Kitchelt papers, box 7 (198), SL; Florence Kitchelt to Alice Leopold, Jan. 23, 1954, Kitchelt papers, box 6 (169), SL; Florence Kitchelt to Alice Leopold, Feb. 4, 1954, Kitchelt papers, box 6 (169), SL.

74. Wilfred G. Helmes to Helen Loy, Oct. 15, 1954, Alice Leopold papers, box 2 (8), SL.

75. Alice Leopold to Florence Kitchelt, Jan. 20, 1955, Kitchelt papers, box 6 (169), SL.

76. Memo from Morag M. Simchak to Esther Peterson, March 4, 1963, Peterson papers, SL; Memo from Jacob Clayman to legislative representatives of IUD Affiliated Unions, June 14, 1960, Peterson papers, SL.

77. Alice M. Birdsall to Anthony Eden, Jan. 22, 1945, NWP papers, reel 84.

78. Ella V. Allen to Alice Paul, July 14, 1956, NWP papers, reel 102.

79. Interview with Pauline Newman, by Barbara Wertheimer, Nov. 1976, Twentieth-Century Trade Union Women: Vehicle for Social Change, Institute of Labor and Industrial Relations, University of Michigan-Wayne State University.

80. Report on the National Conference of the Problems of Working Women, May 2–3, 1953, NWP papers, reel 99.

81. Nancy Gabin, "Female Activism on Gender Issues in the UAW-CIO," paper presented to the North American Labor History Conference, 1985.

82. Lucy Rice Winkler to Mamie Mizen, Nov. 7, 1948, NWP papers, reel 94.

83. Eleanor Taylor Nelson to Alice Paul, Oct. 26, 1958, NWP papers, reel 105.

84. Alma Lutz to Emma Guffey Miller, Jan 26, 1962, NWP papers, reel 107.

85. Interview no. 2.

86. Edna Capewell to Harry Truman, May 20, 1945, NWP papers, reel 86; Edna Capewell to Caroline Babcock, Sept. 14, 1945, Babcock-Hurlburt papers, box 3 (29), SL.

87. Pauline Newman to Frieda Miller, Feb. 28, 1947, Frieda Miller papers, box 1 (8), SL.

88. Minutes to 1951 Biennial Convention, May 25–27, NWP papers, reel 98; Frances Lide, "Woman's Party Asks New Agency in Fight on Women's Bureau," Washington *Star*, May 26, 1951.

89. Emma Guffey Miller to David L. Lawrence, May 17, 1963, NWP papers, reel 108.

90. Interview no. 8.

91. Nora Stanton Barney to Alice Paul, Jan. 4, 1945, NWP papers, reel 84.

92. Emma Guffey Miller to Phillip Murray, Nov. 16, 1945, NWP papers, reel 88.

93. Alice Paul to Nora Stanton Barney, Oct. 17, 1945, NWP papers, reel 88.

94. [Alice Paul] to Mary Kennedy, March 19, 1946, NWP papers, reel 88; Doris Stevens to Katharine Norris, Feb. 5, 1947, Norris papers, box 1 (5), SL.

95. Edna Capewell to Alice Paul, Sept. 21, 1945, NWP papers, reel 87; Edna Capewell to Alice Paul, Nov. 24, 1945, Babcock-Hurlburt papers, box 3 (29), SL; Edna Capewell to Caroline Babcock, May 27, 1946, NWP papers, reel 89; Edna Capewell to Caroline Babcock, March 10, 1946, Babcock-Hurlburt papers, box 7 (97), SL; Edna Capewell to Caroline Babcock, March 18, 1946, Babcock-Hurlburt papers, box 7 (97), SL.

96. Ella Sherwin to Caroline Babcock, March 20, 1946, NWP papers, reel 88; Alice Paul to Mabel Griswold, Nov. 24, 1954, NWP papers, reel 100.

97. Alice Paul to George Meany, Jan. 25, 1956, NWP papers, reel 102.

98. Memo from Mrs. N. Broderick Price to George Meany, Nov. 22, 1957, NWP papers, reel 104.

99. Caroline Babcock to Fannie Ackley, April 25, 1946, NWP papers, reel 89.

100. Mildred Palmer to Fannie Ackley, Feb. 9, 1951, NWP papers, reel 97.

101. Interview no. 1; Anita Pollitzer to Agnes Wells and Mildred Palmer, May 17, 1951, NWP papers, reel 98.

102. Memo from Lucile W. Heming to Anna Lord Strauss, April 8, 1947, LWV papers, box 694, LC.

103. Clara Snell Wolfe to Emma Guffey Miller, Jan. 15, 1946, NWP papers, reel 88.

104. Alice Paul to Martha Willoughby, Nov. 7, 1955, NWP papers, reel 101.

105. Ella M. Sherwin to Florence Kitchelt, Nov. 27, 1944, Kitchelt papers, box 8 (229), SL.

106. Ella M. Sherwin to Florence Kitchelt, Jan. 14, 1945, Kitchelt papers, box 8 (229), SL; Jeannette Marks to Ella Sherwin, Feb. 16, 1945, NWP papers, reel 84.

107. Florence Armstrong to Mary C. Kennedy, Feb. 25, 1945, NWP papers, reel 84.

108. National Chairman to Ella Sherwin, May 2, 1945, NWP papers, reel 86.

109. Rosalind Goodrich Bates to Alice Paul, June 24, 1945, NWP papers, reel 86.

110. Ella M. Sherwin to Caroline Babcock, Oct. 9, 1945, Babcock-Hurlburt papers, box 7 (96), SL.

111. Josephine Casey to Florence Kitchelt, March 3 (1947), Kitchelt papers, box 8 (229), SL.

112. Ella Sherwin to Alice Paul, Feb. 12. 1945, NWP papers, reel 84; Ella Sherwin to Caroline Babcock, Oct. 9, 1945, Babcock-Hurlburt papers, box 7 (96), SL; Ella Sherwin to Alice Paul, Feb. 26, 1945, NWP papers, reel 84; Ella Sherwin

to Helena Hill Weed, April 19, 1946, NWP papers, reel 89; Ella Sherwin to Florence Kitchelt, July 15, 1946, Kitchelt papers, box 8 (229), SL.

113. Elia M. Sherwin to Florence Kitchelt, Nov. 27, 1944, Kitchelt papers, box 8 (220), SL.

114. Ella Sherwin to Florence Kitchelt, Jan. 24, 1945, Kitchelt papers, box 8 (229), SL.

115. Alice Paul to Ella Sherwin, May 31, 1945, NWP papers, reel 86.

116. Fannie Ackley to Mildred Palmer, Feb. 15, 1951, NWP papers, reel 97.

117. Mildred Taylor to Alice Paul, April 15, 1945, NWP papers, reel 85.

118. Ruth Nelson Edelman to Caroline Babcock, June 28, 1945, NWP papers, reel 86.

119. Emma Guffey Miller to Joseph Clark, April 3, 1958, Miller papers, box 3 (19/3), SL.

120. Mary Markajani to Mabel Griswold, June 10, 1948, NWP papers, reel 93; Mary Markajani to Alice Paul, March 10, 1958, NWP papers, reel 104.

121. Telegram from Edna Capewell to Emma Guffey Miller, n.d. [November 1945], NWP papers, reel 88; Edna Capewell to Caroline Babcock, Dec. 18, 1945, Babcock-Hurlburt papers, box 4 (45), SL.

122. "Miss Kellems Writes Unions in Support of Equal Rights," Hartford *Times*, Jan. 5, 1944.

123. Susan M. Hartmann, "Women's Organizations During World War II: The Interaction of Class, Race, and Feminism," in *Woman's Being, Woman's Place*, ed. Mary Kelley (Boston: G. K. Hall, 1979), pp. 313–328.

124. See Nancy Gabin, "Women Workers and the UAW in the Post-World War II Period: 1945–1954," *Labor History* 21 (Winter 1979–80), 5–30; and Ruth Meyerowitz, "Women Unionists and World War II: New Opportunities for Leadership," unpublished paper presented at the Organization of American Historians conference, San Francisco, 1980.

125. Mary Sinclair Crawford to Alice Paul, Aug. 1, 1956, NWP papers, reel 103.

126. See Bettina Aptheker, *Woman's Legacy* (Amherst: University of Massachusetts Press, 1982); Angela Davis, *Women, Race and Class* (New York: Random House, 1981); Ellen Carol DuBois, *Feminism and Suffrage* (Ithaca, N.Y.: Cornell University Press, 1978); Paul Giddings, *When and Where We Enter: The Impact of Black Women on Race and Sex in America* (New York: Bantam, 1984); Aileen Kraditor, *The Ideas of the Woman Suffrage Movement* (New York: Columbia University Press, 1965); and Rosalyn Terborg-Penn, "Discrimination Against Afro-American Women in the Woman's Movement, 1830–1920," in *The Afro-American Woman*, eds. Sharon Harley and Rosalyn Terborg-Penn (Port Washington, N.Y.: Kennikat Press, 1978), pp. 17–27.

127. See Nancy F. Cott, "Feminist Politics in the 1920s: The National Woman's Party," *Journal of American History* 71 (1984), 43–66; Marjory Nelson, "Ladies in the Streets: A Sociological Analysis of the National Woman's Party," Diss., SUNY-Buffalo, 1976. See also Rosalyn Terborg-Penn, "Discontented Black Feminists: Prelude and Postscript to the Passage of the Nineteenth Amendment," in *Decades of Discontent: The Women's Movement, 1920–1940*, eds. Lois Scharf and Joan M. Jensen (Westport, Conn.: Greenwood Press, 1983), pp. 261–278.

128. Mary Church Terrell reported that Alice Paul asked Ida B. Wells, the representative of black Chicago suffragists, not to march with the white Chicago women; see Beverly Washington Jones, "Quest for Equality: The Life of Mary Eliza

Church Terrell, 1863–1954," Diss., University of North Carolina, 1980, p. 99. Amelia R. Fry, in "Alice Paul and the South," an unpublished paper presented at the Southern Historical Association conference, November 1981, reported that she could find no evidence to support this report, since Paul issued invitations to all the black women who inquired about the march.

129. Freda Kirchwey, "Alice Paul Pulls the Strings," *Nation* 112 (March 2, 1921), 332–33.

130. Reported in Thomas C. Pardo, *The National Woman's Party Papers, 1913–1974: A Guide to the Microfilm Edition* (Sanford, N.C.: Microfilming Corporation of America, 1979).

131. Press release, Sept. 2 [no year], Mary Church Terrell papers, reel 13, LC.

132. See Terborg-Penn, "Discontented Black Feminists," pp. 263–264.

133. Mary Church Terrell, "Being a Colored Woman in the U.S.," n.d., Terrell papers, reel 23, LC.

134. Daisy S. George and Dorothy I. Height to Mary Church Terrell, March 1, 1951, Terrell papers, reel 11, LC. See Susan M. Hartmann, *The Home Front and Beyond* (Boston: Twayne Publishers, 1982), p. 148.

135. Mary McLeod Bethune to Burnita Shelton Matthews, May 18, 1950, Matthews papers, SL.

136. Paul, "Conversations with Alice Paul," p. 185.

137. Betsy Reyneau to Caroline Babcock, May 11, 1945, NWP papers, reel 86.

138. Gertrude Crocker to Caroline Babcock, Dec. 20, 1947, Babcock-Hurlburt papers, box 9 (108), SL.

139. Rose Zetzer to Dora Ogle, April 15, 1950, NWP papers, reel 96; Agnes Wells to Elizabeth Forbes, May 2, 1950, NWP papers, reel 96.

140. Lena Madesin Phillips to Lois M. Smith, Jan. 21, 1948, Phillips papers, SL.

141. Lisa Sergio, *A Measure Filled* (New York: Robert B. Luce, 1972), pp. 223–224.

142. Elsa Brita Marcussen to Lena Madesin Phillips, April 27, 1948, Phillips papers, SL.

143. Pauline E. Mandigo to Florence Kitchelt, Aug. 4, 1947, Kitchelt papers, box 2 (29), SL.

144. See Janice Leone, "Degrees of Color: The Integration of the American Association of University Women, 1946–1949," unpublished paper, Ohio State University, 1985.

145. Mary L. Rankin to members of the University Women's Club, Washington Branch AAUW, Oct. 28, 1946, Florence Armstrong papers, box 3 (38), SL.

146. Untitled report, n.d., Mary Church Terrell papers, reel 13, LC.

147. "Liberal AAUW Group Acts to Force End of Racial Bias," Washington *Times-Herald*, May 19, 1948.

148. Mrs. Clarence F. Swift to Helen C. White, Feb. 25, 1947, Mary Church Terrell papers, reel 10, LC.

149. Anna Kelton Wiley to Florence Kitchelt, June 6, 1949, Kitchelt papers, box 9 (245), SL.

150. Florence Kitchelt to Anna Kelton Wiley, June 8, 1949, Kitchelt papers, box 9 (245), SL.

151. Anna Kelton Wiley to Florence Kitchelt, June 9, 1949, Kitchelt papers, box 9 (245), SL.

152. Alma Lutz to Florence Kitchelt, June 11, 1949, Kitchelt papers, box 6

(177), SL; Alma Lutz to Florence Kitchelt, June 22, 1949, Kitchelt papers, box 6 (177), SL.

153. Ruth Lyons to Mary Church Terrell, Aug. 14, 1949, Terrell papers, reel 10, LC.

154. Local interview no. 24.

155. Mary Church Terrell to Althea Hottel, June 8, 1949, Terrell papers, reel 10, LC.

156. Blanche [no last name] to Mary Church Terrell, 18, 1949, Terrell papers, reel 10, LC; Miriam Noll to Mary Church Terrell, Dec. 10, 1950, Terrell papers, reel 11, LC; Helen Duey Hoffman to Mary Church Terrell, April 22, 1948, Terrell papers, reel 10, LC.

157. Paul, "Conversations with Alice Paul," p. 185; Olive Beale to Mary Church Terrell, Oct. 19, 1945, Terrell papers, reel 9, LC.

158. Mabel Vernon, "Speaker for Suffrage and Petitioner for Peace," an oral history conducted in 1972 and 1973 by Amelia R. Fry, Regional Oral History Office, University of California, 1976. pp. 157–158.

159. Evelyn K. Samras to Alice Paul, Feb. 22, 1945, NWP papers, reel 84.

160. Paul, "Conversations with Alice Paul," pp. 185, 473.

161. Alice Paul to Florence Kitchelt, Aug. 17, 1943, Kitchelt papers, box 8 (242), SL.

162. Transcript, "Women's Organizations: Mobilize," Oct. 6, 1950, Somerville-Howorth papers, box 7 (146), SL.

163. Gaeta Wold Boyer to Florence Kitchelt, April 13, 1944, Kitchelt papers, box 7 (196), SL.

164. Gaeta Wold Boyer to Florence Kitchelt, May 11, 1944, Kitchelt papers, box 7 (201), SL.

165. Constance Baker Motley to Florence Kitchelt, Feb. 3, 1948, Kitchelt papers, box 4 (76), SL.

166. Alma Lutz to Florence Kitchelt, Jan. 5, 1948, Kitchelt papers, box 6 (177), SL.

167. Mary Sinclair Crawford to Agnes Wells, July 15, 1945, NWP papers, reel 87; Zelma Huxtable to Caroline Babcock, Dec. 14, 1945, NWP papers, reel 88.

168. Rose Arnold Powell to Anita Pollitzer, Jan. 13, 1949, Powell papers, box 3 (50), SL.

169. Emma E. Newton to Marie Timpona, Sept. 18, 1957, NWP papers, reel 105.

170. Evelyn K. Muzumdar to Karl Hartzell, May 19, 1959, NWP papers, reel 105.

171. Nora Stanton Barney to the editor of the New York *Herald Tribune*, June 18, 1945, NWP papers, reel 86.

172. Telegram from Rebecca Hourwich Reyher to Alice Paul, June 18, 1945, NWP papers, reel 86.

173. Nora Stanton Barney to Thomas E. Dewey, June 25, 1945, NWP papers, reel 86.

174. Eleanor M. Burnet to Nora Stanton Barney, n.d. [June 1945], NWP papers, reel 86.

175. Edith Goode to Katharine Arnett, Feb. 18, 1958, NWP papers, reel 104.

176. Helen Paul to Mary Brandon, July 20, 1956, NWP papers, reel 102.

177. Mary C. Kennedy to Alice Paul, July 17, 1963, NWP papers, reel 108.

178. Telegram from Mildred Robbins to Emma Guffey Miller, Aug. 22, 1963, Miller papers, box 4 (72), SL.

179. Patricia Lochridge, "The Mother Racket," *Woman's Home Companion* 71 (July 1944), 20–21. This description is also based on material in the World War II collection at the Historical Society of Pennsylvania, Philadelphia.

180. Elizabeth Forbes to Florence Kitchelt, Aug. 20 [1943], Kitchelt papers, box 8 (242), SL.

181. Katharine Norris to Nina Horton Avery, April 9, 1946, Norris papers, box 1 (4), SL.

182. Dorothy [Granger] to Katharine Norris, April 17, 1946, Norris papers, box 1 (4), SL.

183. We, the Mothers, Mobilize for America to Alice Paul, Feb. 9, 1947, NWP papers, reel 90.

184. [Mildred Palmer] to Elsie Hill, Oct. 29, 1951, NWP papers, reel 98.

185. Susan B. Anthony II to Rose Arnold Powell, Aug. 6, 1948, Powell papers, box 2 (23), SL.

Chapter 8. Making Connections:
The Resurgence of the Women's Movement

1. Anna Lord Strauss to Mrs. Malcolm L. McBride, Feb. 4, 1947, LWV papers, box 689, LC; Katherine Ellickson, "The President's Commission on the Status of Women," 1976, Ellickson papers, SL. See also Patricia G. Zelman, *Women, Work, and National Policy: The Kennedy-Johnson Years* (Ann Arbor, Mich.: UMI Research Press, 1982); and Cynthia E. Harrison, "Prelude to Feminism: Women's Organizations, the Federal Government and the Rise of the Women's Movement 1942–1968," Diss., Columbia University, 1982. Harrison sees advocacy of a commission on women as one of the two goals of what she aptly names the Women's Bureau coalition.

2. Ellickson, "The PCSW."

3. Local interview no. 17.

4. John F. Kennedy to Emma Guffey Miller, Oct. 1, 1946, NWP papers, reel 90; Alma Lutz to Florence Kitchelt, July 29, 1959, Kitchelt papers, box 7 (178), SL; Marjorie Longwell to Alice Paul, July 16, 1960, NWP papers, reel 106.

5. Emma Guffey Miller to John F. Kennedy, Sept. 3, 1960, NWP papers, reel 106; John F. Kennedy to Emma Guffey Miller, Oct. 7, 1960, Miller papers, box 3 (19/6), SL.

6. John F. Kennedy to Emma Guffey Miller, Sept. 28, 1960, Peterson papers, SL; John F. Kennedy to Emma Guffey Miller, Oct. 7, 1960, Peterson papers, SL; memo from Esther Peterson to Claude Desautels, n.d. [1963], Peterson papers, SL; Esther Peterson, note, May 17, 1975, Peterson papers, SL.

7. Harrison, "Prelude to Feminism," pp. 345–346. See also Zelman, *Women, Work, and National Policy*.

8. Emma Guffey Miller to Alice Paul, Nov. 14, 1960, NWP papers, reel 106.

9. Emma Guffey Miller to John F. Kennedy, Nov. 16, 1960, NWP papers, reel 106; telegram from Emma Guffey Miller to John F. Kennedy, Dec. 11, 1960, NWP papers, reel 106.

10. Interview with Esther Peterson, conducted by Emily Williams, April 26, 1979, for the Franklin D. Roosevelt Library, Peterson papers, SL. See also Zelman, *Women, Work, and National Policy*.

11. Thomas Jeffrey Morain, "The Emergence of the Women's Movement, 1960–1970," Diss., University of Iowa, 1974. Morain based his analysis largely on Judith Hole and Ellen Levine, *Rebirth of Feminism* (New York: Quadrangle/The New York Times Book Co., 1971), pp. 18–28. See also Harrison, "Prelude to Feminism."

12. Esther Peterson, notes, Feb. 28, 1961, Peterson papers, SL.

13. Memo from Esther Peterson to Myer Feldman, May 12, 1961, Peterson papers, SL.

14. Memo from Esther Peterson to Secretary of Labor, June 2, 1961, Peterson papers, SL.

15. Emma Guffey Miller to Ernestine Bellamy, July 24, 1961, NWP papers, reel 107.

16. Arthur J. Goldberg to Grace M. Sloan, n.d. [June 1961], NWP papers, reel 107; Esther Peterson to Grace M. Sloan, June 1, 1961, NWP papers, reel 107.

17. Zelman, *Women, Work, and National Policy*.

18. Ellickson, "The PCSW."

19. Margery Leonard to Emma Guffey Miller, Feb. 16, 1963, NWP papers, reel 108.

20. See, for example, interview no. 8.

21. Emma Guffey Miller to Matilda Fenberg, Dec. 15, 1961, NWP papers, reel 107.

22. Emma Guffey Miller to Alma Lutz, Jan. 23, 1962, NWP papers, reel 107.

23. Mary E. Seebach to Emma Newton, Dec. 24, 1961, NWP papers, reel 107.

24. David L. Lawrence to John F. Kennedy, March 1, 1962, White House Central Subject File, box 206 (FG 737), John F. Kennedy Library, Waltham, Massachusetts (now located in Boston, Massachusetts).

25. Alice Paul to Marjorie Longwell, April 2, 1962, NWP papers, reel 107; Genevieve Blatt to Emma Guffey Miller, March 27, 1962, Miller papers, box 4 (69), SL; memo from Esther Peterson to Claude Desautels, n.d. [1962], Peterson papers, SL.

26. Emma Guffey Miller to John F. Kennedy, n.d. [1963], Miller papers, SL.

27. Alice Paul to Mary Seebach, Dec. 27, 1961, NWP papers, reel 107.

28. Esther Peterson to Byron White, Dec. 29, 1961, Peterson papers, SL.

29. Press release, Office of the White House Press Secretary, Feb. 12, 1962, PCSW papers, box 3, JFKL.

30. Ellickson, "The PCSW."

31. Minutes of First Meeting, Feb. 12–13, 1962, PCSW papers, box 1, JFKL; Transcript of Proceedings, Feb. 12–13, 1962, PCSW papers, box 10, JFKL.

32. Transcript of Proceedings, Committee on Civil and Political Rights, Aug. 24, 1962, PCSW papers, box 14, JFKL.

33. The BPW brief, written by Libby Sacher and Joyce Capps, is included in the Margaret Hickey papers, box 28, UMSL. Holden's draft is included in the Miriam Holden collection, box 1, Princeton University Library, Princeton, New Jersey.

34. Transcript of Proceedings, Committee on Civil and Political Rights, Aug. 24, 1962, PCSW papers, box 14, JFKL.

35. Transcript of Proceedings, Committee on Civil and Political Rights, April 5, 1963, PCSW papers, box 14, JFKL.

36. *American Women: Report of the President's Commission on the Status of Women*, 1963.

37. Marguerite Rawalt to NWP, May 3, 1966, NWP papers, reel 110.

38. Carl Hayden to Hazel Harvey Quaid, Oct. 18, 1963, NWP papers, reel 108.

39. Lyndon Johnson to Emma Guffey Miller, Jan. 26, 1965, Miller papers, box 5 (79), SL.

40. Ruth [Gage-Colby] to Alice Paul, Oct. 29, 1963, NWP papers, reel 108.

41. Emma Guffey Miller to Jane Grant, Nov. 22, 1963, NWP papers, reel 108.

42. Mary Kennedy to Alice Paul, Oct. 21, 1963, NWP papers, reel 108.

43. Jane Grant to Emma Guffey Miller, Oct. 31, 1963, NWP papers, reel 108.

44. Alice Paul to Isabelle M. Allias, Oct. 23, 1963, NWP papers, reel 108.

45. Margaret Hickey, "Talent Search for Womanpower," speech to the American Personnel Guidance Association, March 25, 1964, Margaret Hickey papers, UMSL.

46. Ellickson, "The PCSW." Jo Freeman, in *The Politics of Women's Liberation* (New York: Longman, 1975), emphasizes the role of the PCSW. See also Harrison, "Prelude to Feminism."

47. Mary [no last name] to Frieda Miller, Nov. 19, 1963, Frieda Miller papers, box 1 (6), SL.

48. Mildred B. Dunbar to John F. Kennedy, Dec. 18, 1961, White House Central Subject File, box 206, JFKL; Barbara J. Landfield to Eleanor Roosevelt and Commission members, Feb. 19, 1962, PCSW papers, box 1, JFKL; Judith Beilfuss to John F. Kennedy, March 30, 1962, PCSW papers, box 1, JFKL; Dorothy V. Knibb to PCSW, May 21, 1962, PCSW papers, box 1, JFKL.

49. Memo from Esther Peterson to Myer Feldman, Aug. 5, 1963, White House Staff Files, box 533, JFKL.

50. Virginia R. Allan to Alice Paul, Oct. 30, 1963, NWP papers, reel 108; Alice Paul to Miriam Holden, Nov. 15, 1963, NWP papers, reel 108.

51. Esther Peterson to Carl Hayden, Oct. 10, 1963, Peterson papers, SL.

52. Interview no. 13.

53. Miriam Holden to Alice Paul, Oct. 24, 1962, Holden papers, box 1, Princeton University Library.

54. Emma Guffey Miller to Minnie C. Miles, April 30, 1963, NWP papers, reel 108; Emma Guffey Miller to Governor Scranton, n.d., Miller papers, box 5 (91), SL; Marjorie Tibbs to Emma Guffey Miller, June 1968, Miller papers, box 5 (88), SL.

55. Mary Kennedy to Alice Paul, Dec. 13, 1963, NWP papers, reel 108.

56. Miriam Holden to Alice Paul, Nov. 11, 1963, NWP papers, reel 108.

57. [Emma Guffey Miller] to Mary Kennedy, Jan. 20, 1964, NWP papers, reel 108.

58. Transcript of Proceedings, Committee on Civil and Political Rights, PCSW papers, box 14, JFKL.

59. Ellickson, "The PCSW."

60. Cynthia Wedel to Katherine Ellickson, June 13, 1976, Ellickson papers, SL.

61. Caroline F. Ware to Katherine Ellickson, Nov. 1, 1976, Ellickson papers, SL.

62. Harrison, "Prelude to Feminism," provides an excellent analysis of the

struggle for equal pay, pp. 140–182, 294–337. See also Zelman, *Women, Work, and National Policy*.

63. Lucy Rice Winkler to Alice Paul, Jan. 5, 1945, NWP papers, reel 84.

64. Mildred Palmer, Report on Equal Pay Conference, n.d. [April 1952], NWP papers, reel 98.

65. Memo from Adelia B. Kloak to Frieda Miller, April 1, 1952, Miller papers, box 6 (138), SL.

66. Mabel Griswold to Alice Paul, Oct. 16, 1954, NWP papers, reel 100.

67. Emma Guffey Miller to Alice Paul, June 14, 1963, NWP papers, reel 108.

68. Mary F. Anderson to Alice Paul, June 16, 1963, NWP papers, reel 108.

69. Interview with Esther Peterson, conducted by Emily Williams, April 26, 1979, FDR library, in Peterson papers, SL.

70. Caroline Bird, *Born Female: The High Cost of Keeping Women Down* (New York: David McKay, 1968), chapt. 1; and Hole and Levine, *Rebirth of Feminism*, pp. 30–44, emphasize the attempts of the bill's opponents to ridicule civil rights by including women. Caruthers Gholson Berger, "Equal Pay, Equal Employment Opportunity and Equal Enforcement of the Law for Women," *Valparaiso University Law Review* 5 (1971), 326–373; Freeman, *Politics of Women's Liberation*; Betty Friedan, *It Changed My Life* (New York: Dell, 1977); and Barbara Sinclair Deckard, *The Women's Movement*, 2nd ed. (New York: Harper & Row, 1979), pp. 344–346, present more complex, if very brief, interpretations. The interpretation given here is consistent with the most recent work, in particular Carl M. Brauer, "Women Activists, Southern Conservatives, and the Prohibition of Sex Discrimination in Title VII of the 1964 Civil Rights Act," *Journal of Southern History* 49 (1983), 37–56. See also Harrison, "Prelude to Feminism," pp. 469–483; and Zelman, *Women, Work, and National Policy*, pp. 55–71.

71. Alice Paul to Alice K. Leopold, May 10, 1955, Kitchelt papers, box 6 (169), SL.

72. Helen Paul to Nina Price, July 18, 1956, NWP papers, reel 102.

73. Brauer, "Women Activists," 41–42. Brauer interviewed Butler Franklin and Nina Horton Avery, the authors of the letters.

74. Quoted in Zelman, *Women, Work, and National Policy*, p. 61.

75. Brauer, "Women Activists," 45.

76. Interview no. 8 and interview no. 13.

77. Brauer, "Women Activists," 47.

78. Ibid.

79. Interview no. 8.

80. Emma Guffey Miller to Mary Kennedy, April 15, 1964, NWP papers, reel 109.

81. Memo from Alice Paul. July 10, 1964, NWP papers, reel 109.

82. Marguerite Rawalt to Victoria Gilbert, Aug. 4, 1967, NWP papers, reel 110.

83. Interview no. 8.

84. Interview no. 13.

85. Marjorie Longwell to Alice Paul, July 21, 1964, NWP papers, reel 109.

86. Miriam Holden to Alice Paul, Oct. 16, 1965, Holden papers, box 1, Princeton University Library.

87. "Sex and Nonsense," in *New Republic*; quoted by Alma Lutz to Emma Guffey Miller, Sept. 11, 1965, Miller papers, box 5 (81), SL.

88. Lucinda Klemeyer, " 'Equal Rights' is Battle Cry," Columbus *Dispatch*, Feb. 12, 1967.

89. Mary Kennedy to Alice Paul, June 24, 1964, NWP papers, reel 109.

90. WOMEN (BIPARTISAN) to fellow women, n.d. [1964], NWP papers, reel 109.

91. Helen K. Leslie to Club Presidents, Sept. 21, 1965, Margaret Hickey papers, box 28 (129), UMSL.

92. Interview no. 13.

93. Wilma Soss, *Pocketbook News*, June 27, 1964, Miriam Holden papers, box 1, Princeton University Library.

94. Friedan, *It Changed My Life*, p. 41.

95. Meta Ellis Heller to Alice Paul, March 31, 1963, NWP papers, reel 108.

96. Interview no. 8.

97. Interview no. 6.

98. Zelman, *Women, Work, and National Policy*, pp. 73–88.

99. Betty Friedan, "Up From the Kitchen Floor," *New York Times Magazine*, March 4, 1973, 9; see also Friedan, *It Changed My Life*, pp. 109–123.

100. Emma Guffey Miller to Esther Peterson, April 6, 1966, Peterson papers, SL.

101. Interview with Dorothy Haener, conducted by Lyn Goldfarb, Lydia Kleiner, and Christine Miller, n.d.; interview with Florence Peterson, conducted by Ruth Meyerowitz, July 1976; 20th-Century Trade Union Women: Vehicle for Social Change, Institute of Labor and Industrial Relations, University of Michigan-Wayne State University, microfilmed by the New York Times Oral History Program, 1979.

102. Mary Eastwood to Alice Paul, Oct. 20, 1967, NWP papers, reel 110.

103. Interview no. 11.

104. Interview no. 13.

105. Jo Freeman makes this point in "Resource Mobilization and Strategy: A Model for Analyzing Social Movement Organization Actions," in *The Dynamics of Social Movements*, eds. Mayer N. Zald and John D. McCarthy (Cambridge, Mass.: Winthrop Publishers, 1979), pp. 167–189.

106. Interview no. 6.

107. Interview no. 13.

108. Interview no. 8.

109. Interview no. 12.

110. Interview no. 10.

111. Interview no. 12.

112. Alice Paul, "Conversations with Alice Paul: Woman Suffrage and the Equal Rights Amendment," interviews conducted in 1972 and 1973 by Amelia R. Fry, Regional Oral History Office, University of California, 1976, pp. 530–535; interview no. 11.

113. Communication from Frances Kolb, 1980.

114. Paul, "Conversations with Alice Paul." See also interview no. 8.

115. Lucy Winkler to Caroline Davis, Jan. 10, 1967, NWP papers, reel 110.

116. Inka O'Hanrahan to Alice Paul, March 28, 1967, NWP papers, reel 110; Inka O'Hanrahan to Alice Paul, Sept. 13, 1967, NWP papers, reel 110.

117. Elizabeth Boyer to Alice Paul, Aug. 11, 1967, NWP papers, reel 110.

118. Secretary to Miss Paul to Carol Burris, July 10, 1970, NWP papers, reel 111.

119. Stephanie A. Riepel to NWP, n.d. [April 1970], NWP papers, reel 111.

120. "Overdue Bills," *Newsweek*, March 23, 1970.

121. Paul, "Conversations with Alice Paul."

122. Ernestine Powell to Alice Paul, Jan. 25, 1968, NWP papers, reel 110.

123. Quoted in Katherine Conger Kane, "Hazel Hunkins Hallinan: A Suffrage Survivor Stands Ready for Battle," Washington *Post*, Aug. 21, 1977.

124. Ernestine Powell to Alice Paul, Nov. 20, 1968, NWP papers, reel 111; interview no. 1 and interview no. 2.

125. Miriam Holden to [Mary Glenn] Newell, Feb. 11, 1967, NWP papers, reel 110.

126. "A One-Woman Bridge Spanning the Movement's History," Boston *Globe*, March 3, 1974.

127. Ruth Gage-Colby to Alice Paul, May 5, 1969, NWP papers, reel 111.

128. Jean Witter to Alice Paul, July 27, 1970, NWP papers, reel 111.

129. Jeannette Smyth, "A Day for Celebration," Washington *Post*, March 24, 1972.

130. Interview no. 7.

131. Marjorie Longwell to Alice Paul, March 29, 1972, NWP papers, reel 112.

132. Nancy van Vooren to Alice Paul, June 15, 1970, NWP papers, reel 111.

133. Zelman, *Women, Work, and National Policy*, p. 49.

134. Interview no. 8.

135. Helen Thomas, UPI story, Feb. 13, 1962, NWP papers, reel 107.

136. Memo from Esther Peterson to Pat Moynihan, April 3, 1963, Peterson papers, SL.

137. Anna Lord Strauss to Amy Bush, June 26, 1968, Strauss papers, box 12 (250), SL.

138. Kathryn Paulsen, "The Last Hurrah of the Woman's Party," NWP papers, reel 111.

139. Quoted in Beth Duncan, "You Haven't Come Such a Long Way, Baby," Sunday *Advertiser*, Dec. 12, 1971.

140. Margaret Hickey, speech to the Golden Anniversary Convention of the BPW, July 20, 1969, Hickey papers, box 32 (129), UMSL; interview with Margaret Hickey, conducted by Ann Levert and Susan M. Hartmann, Feb. 14, 1974, UMSL.

141. Judy Klemesrud, "Another Susan B. Anthony Now Speaks Her Mind," the *New York Times*, Sept. 21, 1971.

142. Jeanette Smyth, "Half a Century of Feminism," unidentified newspaper clipping, July 2, 1972.

143. Local interview no. 17.

144. Ruth [Gage-Colby] to Alice Paul, April 4, 1970, NWP papers, reel 111.

145. "Lucy Somerville Howorth Library Fund," Randolph-Macon *Alumnae Bulletin*, Winter 1972.

146. Rena Tederson, "Texas Federal Judge is 'First Lady of Law,' " *Scrantonian*, Jan. 14, 1973.

147. Mary Kennedy to Alice Paul, Feb. 11, 1971, NWP papers, reel 112.

148. Quoted in Paulsen, "Last Hurrah," NWP papers, reel 111.

149. Aldon D. Morris, *The Origins of the Civil Rights Movement: Black Communities Organizing for Change* (New York: Free Press, 1984), pp. 139–173. We are grateful to John McCarthy for pointing out the relevance of Morris's concept for our work.

Chapter 9. Theoretical Overview and Conclusions

1. Paul Wilkinson, *Social Movement* (New York: Praeger, 1971), p. 19.

2. Gary T. Marx and James L. Wood, "Strands of Theory and Research in Collective Behavior," *Annual Review of Sociology* 1 (1975), 363–428.

3. William A. Gamson, *The Strategy of Social Protest* (Homewood, Ill.: Dorsey, 1975), identifies two sets of indicators of success: the acceptance of a social movement by its opponents as a legitimate spokesperson for its interests, and the attainment of advantages by the group's beneficiaries. Success can also be defined more broadly. Social movement scholars such as Ralph Turner and Lewis M. Killian, *Collective Behavior*, 2nd ed. (Englewood Cliffs, N.J.: Prentice-Hall, 1972); Lewis M. Killian, "Social Movements," in *Handbook of Modern Sociology*, ed. Robert E. L. Farris (Chicago: Rand-McNally, 1964), pp. 426–455; and Joseph B. Perry, Jr., and M. D. Pugh, *Collective Behavior* (St. Paul, Minn.: West, 1978), include in their discussions of movement consequences factors such as the acceptance of certain values of the movement among various sectors of society, recognition of the personnel of the movement, continued existence over a long period of time, the acquisition of a large number of adherents, legal change without acceptance of the movement's entire program, residues at the popular culture level, clarification of the collective consciousness, changes in the attitudes and beliefs of individuals, and the development of organizational, leadership, and other specialized skills among participants.

Another way that social movement theorists have approached the subject of movement outcomes is by delineating patterns of social movement decline. Jo Freeman conceptualizes all social movement outcomes in terms of decline because she views social movements as transitory and inherently unstable. Some of the factors responsible for movement decline include resolution of grievances, successful repression by authorities, disappearance of resources, increased cost of participation, destructive internal conflict, inflexibility in goals and ideology, and a lack of success in attaining goals. See Jo Freeman, *Social Movements of the Sixties and Seventies* (New York: Longman, 1983); Perry and Pugh, *Collective Behavior*.

Frederick Miller identifies four patterns of decline: success, co-optation, repression, and failure. See Frederick Miller, "The End of SDS and the Emergence of Weatherman: Demise Through Success," in Freeman, *Social Movements*, pp. 279–297.

In this discussion, we follow David Snyder and William R. Kelly, "Strategies for Investigating Violence and Social Change," in *The Dynamics of Social Movements*, eds. Mayer N. Zald and John D. McCarthy (Cambridge, Mass.: Winthrop Publishers, 1979), pp. 212–237, who suggest that social movements take diverse paths and lead to different outcomes.

4. Turner and Killian, *Collective Behavior*.

5. Leila J. Rupp, "The Women's Community in the National Woman's Party, 1945 to the 1960s," *Signs* 10 (1985), 715–740.

6. Joseph E. Talbot to Florence Kitchelt, Feb. 12, 1945, Kitchelt papers, box 8 (234), SL.

7. Dorothy McAllister to Mrs. Charles Bang, Oct. 8, 1946, LWV of Ohio papers, box 17, Ohio Historical Society, Columbus, Ohio.

8. Jo Freeman, "Resource Mobilization and Strategy: A Model for Analyzing Social Movement Organization Actions," in Zald and McCarthy, *Dynamics*, pp. 167–189.

9. Jo Freeman, *The Politics of Women's Liberation* (New York: David McKay, 1975).

10. See Paula Giddings, *When and Where I Enter: The Impact of Black Women on Race and Sex in America* (New York: Bantam Books, 1984), pp. 340–348.

11. The major formulations include Robert E. Park and Ernest W. Burgess, *Introduction to the Science of Sociology* (Chicago: University of Chicago Press, 1921), pp. 865–952; Herbert Blumer, "Collective Behavior," in *New Outline of Principles of Sociology*, ed. A. M. Lee (New York: Barnes and Noble, 1951), pp. 165–220; Turner and Killian, *Collective Behavior*; Kurt Lang and Gladys Lang, *Collective Dynamics* (New York: Crowell, 1961); and Neil Smelser, *Theory of Collective Behavior* (New York: Free Press, 1962).

12. Later formulations in the collective behavior tradition have rejected some of the tenets of classical theories, especially assumptions about the discontinuity between collective behavior and routine social behavior. See Clark McPhail and David Miller, "The Assembling Process: A Theoretical and Empirical Examination," *American Sociological Review* 38 (December 1973), 721–735; Jack M. Weller and E. L. Quarantelli, "Neglected Characteristics of Collective Behavior," *American Journal of Sociology* 79 (November 1973), 665–685; Ralph H. Turner, "Collective Behavior and Resource Mobilization as Approaches to Social Movements: Issues and Continuity," in *Research in Social Movements, Conflict and Change*, ed. Louis Kriesberg, vol. 4 (Greenwich, Conn.: JAI Press, 1981), pp. 1–24; Benigno E. Aguirre and E. L. Quarantelli, "Methodological, Ideological and Conceptual-Theoretical Criticisms of the Field of Collective Behavior: A Critical Evaluation and Implications for Future Study," *Sociological Focus* 16 (August 1983), 195–216.

13. Major formulations of the resource mobilization perspective include Anthony Oberschall, *Social Conflict and Social Movements* (Englewood Cliffs, N.J.: Prentice-Hall, 1973); John D. McCarthy and Mayer N. Zald, *The Trend of Social Movements in America: Professionalism and Resource Mobilization* (Morristown, N.J.: General Learning Press, 1973); Charles Tilly, Louise Tilly, and Richard Tilly, *The Rebellious Century, 1830–1930* (Cambridge: Harvard University Press, 1975); John D. McCarthy and Mayer N. Zald, "Resource Mobilization and Social Movements: A Partial Theory," *American Journal of Sociology* 82 (1977), 1212–1241; Zald and McCarthy, *Dynamics*; Gamson, *Strategy of Social Protest*; Charles Tilly, *From Mobilization to Revolution* (Reading, Mass.: Addison-Wesley, 1978); Charles Perrow, "The Sixties Observed," in Zald and McCarthy, *Dynamics*, pp. 192–212.

14. Tilly, *From Mobilization to Revolution*; J. Craig Jenkins and Charles Perrow, "Insurgency of the Powerless: Farm Worker Movements, 1946–1972," *American Sociological Review* 42 (1977), 249–258; Anthony Oberschall, "The Decline of the 1960's Social Movements," in *Research in Social Movements, Conflict and Change*, ed. Louis Kriesberg, vol. 1 (Greenwich, Conn.: JAI Press, 1978), pp. 257–289.

15. Abbott L. Ferriss, *Indicators of Trends in the Status of American Women* (New York: Russell Sage Foundation, 1971); see Valerie Kincaid Oppenheimer, "Demographic Influence on Female Employment and the Status of Women," in *Changing Women in a Changing Society*, ed. Joan Huber (Chicago: University of

Chicago Press, 1973), pp. 184–199; Jo Freeman, *Politics*; Joan Huber and Glenna Spitze, *Sex Stratification: Children, Housework, and Jobs* (N.Y.: Academic Press, 1983); and Ethel Klein, *Gender Politics: From Mass Consciousness to Mass Politics* (Cambridge: Harvard University Press, 1984), for discussions of the structural preconditions of the contemporary women's movement.

16. J. Craig Jenkins, "Research Mobilization Theory and the Study of Social Movements," *Annual Review of Sociology* 9 (1983), 527–553.

17. Lyford P. Edwards, *The Natural History of Revolution* (Chicago: University of Chicago Press, 1927); Rex D. Hopper, "The Revolutionary Process," *Social Forces* 18 (1950), 527–553; Blumer, "Social Movements;" Lang and Lang, *Collective Dynamics*; Perry and Pugh, *Collective Behavior*.

18. Mayer Zald and Roberta Ash, "Social Movement Organizations: Growth, Decay, and Change," *Social Forces* 44 (1966), 327–341; Oberschall, "Decline of the 1960's Social Movements"; Charles Perrow, "The Sixties Observed," in Zald and McCarthy, *Dynamics*, pp. 192–211; Snyder and Kelly, "Strategies for Investigating Violence and Social Change"; Zald and McCarthy, *Dynamics*; Joseph R. Gusfield, "Social Movements and Social Change: Perspectives of Linearity and Fluidity," in *Research in Social Movements, Conflict and Change*, ed. Louis Kriesberg, vol. 4 (Greenwich, Conn.: JAI Press), pp. 317–339; Jenkins, "Resource Mobilization Theory"; Miller, "End of SDS."

19. John Lofland, "White-Hot Mobilization: Strategies of a Millenarian Movement," in *Dynamics of Social Movements*, eds. Zald and McCarthy, pp. 157–166; David G. Bromley and Anson D. Shupe, Jr., *"Moonies" in America: Cult, Church, and Crusade* (Beverly Hills, Calif.: Sage Publications, 1979).

20. Aldon Morris, *The Origins of the Civil Rights Movement: Black Communities Organizing for Change* (New York: The Free Press, 1984).

21. The concept of a professional social movement was developed in McCarthy and Zald, *Trend of Social Movements*. Doug McAdam, *Political Process and the Development of Black Insurgency, 1930–1970* (Chicago: University of Chicago Press, 1982); and J. Craig Jenkins and Craig M. Eckert, "Channelling Black Insurgency: Elite Patronage and Professional SMOs in the Development of the Civil Rights Movement," unpublished paper, present research on the role of elite support and professionalization in the rise and decline of the civil rights movement.

22. McAdam, *Political Process*; Morris, *Origins of the Civil Rights Movement*; and Jenkins and Eckert, "Channelling Black Insurgency."

23. Freeman, *Politics*; Joan Cassell, *A Group Called Women: Sisterhood and Symbolism in the Feminist Movement* (New York: David McKay, 1977); Klein, *Gender Politics*; Myra Marx Ferree and Beth B. Hess, *Controversy and Coalition: The New Feminist Movement* (Boston: Twayne, 1985).

24. Olive Banks, *Faces of Feminism: A Study of Feminism as a Social Movement* (Oxford: Martin Robertson, 1981), p. 1.

25. Bert Useem and Mayer N. Zald, "From Pressure Group to Social Movement: Organizational Dilemmas of the Effort to Promote Nuclear Power," *Social Problems* 30 (December 1982), 144–156.

26. Tilly, *From Mobilization to Revolution*; Charles Tilly, "Social Movements and National Politics," Working Paper No. 197 (Ann Arbor: Center for Research on Social Organization, University of Michigan, 1979).

27. Ralph H. Turner, "The Public Perception of Protest," *American Sociological Review* 34 (December 1969), 815–831.

28. Gamson, *Strategy*; Luther P. Gerlach and Virginia H. Hine, *People, Power,*

Change: Movements of Social Transformation (Indianapolis: Bobbs-Merrill, 1970); Frances Fox Piven and R. A. Cloward, *Regulating the Poor* (New York: Pantheon, 1971).

29. Zald and Ash, "Social Movement Organizations."

30. Gerlach and Hine, *People*.

31. Robert J. Ross, "Generational Change and Primary Groups in a Social Movement," in Freeman, *Social Movements*, pp. 177–192.

32. Tilly, *Mobilization*; Freeman, *Social Movements of the Sixties*; Jenkins, "Resource Mobilization Theory"; Morris, *Origins*; Oberschall, *Social Conflict and Social Movements*.

33. Estelle Freedman, "Separatism as Strategy: Female Institution Building and American Feminism, 1870–1930," *Feminist Studies* 5 (1979), 512–529.

34. Myra Marx Ferree and Frederick Miller, "Mobilization and Meaning: Some Social-Psychological Contributions to the Resource Mobilization Perspective on Social Movements," *Sociological Inquiry* 55 (1985), 38–61.

35. John Lofland, *Doomsday Cult* (New York: Irvington Publishers, 1977); Lofland, "White-Hot Mobilization."

36. John Wilson, *Introduction to Social Movements* (New York: Basic Books, 1973).

37. Ralph Turner, "Determinants of Social Movement Strategies," in *Human Nature and Collective Behavior*, ed. Tamotsu Shibutani (Englewood Cliffs, N.J.: Prentice-Hall, 1970).

38. Freeman, "Resource Mobilization."

39. Useem and Zald, "From Pressure Group."

40. Gusfield, "Social Movements and Social Change."

41. William L. O'Neill, *Everyone Was Brave: A History of Feminism in America* (Chicago: Quadrangle Books, 1969), distinguishes between "hard-line feminists" and "social feminists." Myra Marx Ferree and Beth B. Hess, *Controversy and Coalition: The New Feminist Movement* (Boston: Twayne, 1985), label ERA advocates in the post-1920 years "radical."

42. Lewis M. Killian, "Organization, Rationality and Spontaneity in the Civil Rights Movement," *American Sociological Review* 49 (December 1984), 770–783.

43. Verta Taylor, "The Future of Feminism: A Social Movement Analysis," in *Feminist Frontiers: Rethinking Sex, Gender, and Society* (Reading, Mass.: Addison-Wesley, 1983), pp. 434–451; Pamela Johnston Conover and Virginia Gray, *Feminism and the New Right: Conflict Over the American Family* (New York: Praeger, 1983).

APPENDIX

Sample Interview Guide

We prepared a specific set of questions for each interviewee based on her particular area of involvement. Also, some of the interviews were designed to fill in gaps in the archival record. All of the interviewees responded, however, to a number of general questions which are reflected in the following sample interview guide.

Instructions

You have been identified either by another woman we have interviewed or from our archival research as someone who was interested in promoting women's rights in the 1940s, 1950s, and early 1960s. We would like to ask you some questions about your activities on behalf of women in this period.

The questions we will ask focus on two areas. First, we are interested in determining whether or not there was any organized activity taking place on behalf of women's rights in this period and, if so, the nature and extent of this activity. We will be asking you to give us an overview of the key groups and individuals concerned with women's rights, describe what relationships, if any, existed among these groups, discuss your perceptions of the most important issues that concerned women's organizations in this period, and characterize the societal response to these issues.

Second, we want to get a picture of the nature and extent of your own personal involvement in women's rights work in the period. We will be asking you to describe the kinds of activities you participated in, the factors that affected your getting involved in these activities, the groups and individuals with whom you worked, and the impact that these activities had on your own life.

Questions:

Section One: The first series of questions.

1. During the period from 1945 to the mid–1960s, did you work with any *organized* groups, either formal or informal, concerned with women's issues?

Probe: Which group?
How did you participate?
On what issues did it work?
What tactics did it use?
Describe the size and nature of membership
How were participants recruited?
Where did power reside in the organization?
How was it supported financially?
With which groups did it work most closely?

(If more than one group was mentioned, ask these questions for each group and explore differences and similarities between the groups, the extent of overlapping membership, and the groups' relationships to each other.)

2. We want to ask you specifically about a few women's organizations that existed in this period. In each case, could you describe what you know about them (their size, composition, primary activities and issues, relationships to other groups, etc.)?

The National Woman's Party
The League of Women Voters
The Women's Bureau of the Department of Labor
The National Federation of Business & Professional Women's Clubs
The General Federation of Women's Clubs
The National Association of Colored Women
The American Association of University Women
The National Association of Women Deans and Counsellors
The National Association of Women Lawyers
Women's service organizations (Altrusa, Soroptomist, Zonta)

3. Were there any other groups or networks of women actively involved in improving women's status in this period? Describe their composition, activities, and their relationship to these other groups.

4. Were there any important areas of cooperation or disagreement among women's groups in this period? Over what? Why do you think these issues were so important? (*Probe*: Issues, tactics, leadership, recruitment strategies.)

5. How would you characterize the general societal response to groups and individuals interested in promoting women's rights in this period?

Section Two: The second set of questions pertains to your own more personal involvement and perceptions in this period.

1. Can you describe when and how you first became involved in working for women's rights? (*Probe*: Any involvement in the suffrage movement, specific events, people, work experiences, reading matter, family and/ or coworkers.)

2. Did you define yourself as a feminist in this period? If so, when did you

first use that term to characterize your views? If not, did you ever begin using that term to describe yourself, and when?

3. What did feminism mean in the 1940s, 1950s, and 1960s? (*Probe*: Goals, strategies, beliefs.) Have your views changed in any way since the emergence of the later women's movement?

4. Was being a feminist difficult then? More difficult than now? Why? Did people react negatively to the term? Who?

5. What factors do you think made it possible for you to devote your efforts to women's rights in this period? (*Probe*: Support of spouse, family, friends, and others.)

6. What rewards, if any, did you receive for participating in women's rights work? (*Probe*: Material incentives and benefits of participation.)

7. Has being involved in women's rights work affected your personal life in important ways?

8. Looking back, what activities have been most important to you in working for women's rights?

9. Were you involved in activities that led to the founding of the contemporary women's movement that emerged in the mid–1960s?

10. Are you currently involved in any way in efforts to improve the status of women? If so, describe.

11. Finally, as someone with an historical perspective, how do you view the contemporary women's movement? How does it compare to the struggle for women's rights in the 1940s, 1950s, and 1960s? (*Probe*: Goals, strategies, composition, beliefs, societal attitude, etc.)

12. Obtain current demographic information, if it has not already been discussed in the interview.

    a. Age
    b. Marital history
    c. History of parental obligations
    d. Occupational history
    e. Education
    f. Residence

13. Obtain the names and current addresses of any other women who were involved in women's rights work in the post–1945 period.

Bibliography

Manuscript Collections

American Association of University Women, Ohio Division, papers, Ohio Historical Society, Columbus, Ohio.

Florence A. Armstrong papers, Arthur and Elizabeth Schlesinger Library on the History of Women in America, Radcliffe College, Cambridge, Massachusetts.

Caroline Lexow Babcock-Olive M. Hurlburt papers, Schlesinger Library, Cambridge, Mass.

Mary Ritter Beard papers, Schlesinger Library, Cambridge, Mass.

Mary Dewson papers, Schlesinger Library, Cambridge, Mass.

Margaret A. Hickey papers, Western Historical Manuscript Collection, University of Missouri—St. Louis, St. Louis, Missouri.

Miriam Y. Holden collection, Rare Books and Manuscripts, Princeton University Library, Princeton, New Jersey.

Lucy Somerville Howorth papers, Schlesinger Library, Cambridge, Mass.

Sarah T. Hughes papers, Schlesinger Library, Cambridge, Mass.

Florence L. C. Kitchelt papers, Schlesinger Library, Cambridge, Mass.

League of Women Voters papers, Library of Congress, Washington, D.C.

League of Women Voters of Ohio papers, Ohio Historical Society, Columbus, Ohio.

Alice K. Leopold papers, Schlesinger Library, Cambridge, Mass.

Alma Lutz papers, Schlesinger Library, Cambridge, Mass.

Burnita Shelton Matthews papers, Schlesinger Library, Cambridge, Mass.

Emma Guffey Miller papers, Schlesinger Library, Cambridge, Mass.

Frieda S. Miller papers, Schlesinger Library, Cambridge, Mass.

National Woman's Party Papers, 1913–1974, microfilmed and distributed by the Microfilming Corporation of America, 1979.

National Women's Trade Union League papers, Library of Congress, Washington, D.C.

Katharine Augusta Norris papers, Schlesinger Library, Cambridge, Mass.

Esther Peterson papers, Schlesinger Library, Cambridge, Mass.

Lena Madesin Phillips papers, Schlesinger Library, Cambridge, Mass.

E. S. Pollock papers, personal files, Cincinnati, Ohio.

Rose Arnold Powell papers, Schlesinger Library, Cambridge, Mass.

President's Commission on the Status of Women papers, John F. Kennedy Library, Boston, Mass. (formerly in Waltham, Mass.).

Doris Stevens papers, Schlesinger Library, Cambridge, Mass.

Anna Lord Strauss papers, Schlesinger Library, Cambridge, Mass.

Mary Church Terrell papers, Library of Congress, Washington, D.C.

Agnes Wells papers, Schlesinger Library, Cambridge, Mass.

Helen Hunt West papers, Schlesinger Library, Cambridge, Mass.

White House Central Subject Files and White House Staff Files, John F. Kennedy Library, Boston, Mass. (formerly in Waltham, Mass.).

Anna Kelton Wiley papers, Schlesinger Library, Cambridge, Mass.

Chârl Ormond Williams papers, Schlesinger Library, Cambridge, Mass.

Women's Joint Congressional Committee papers, Library of Congress, Washington, D.C.

Interviews

Caruthers Berger, conducted by Verta Taylor and Leila J. Rupp, May 15, 1982, Dallas, Texas.

Elizabeth Chittick, conducted by Verta Taylor, Dec. 16, 1981, Washington, D.C.

Catherine East, conducted by Verta Taylor and Leila J. Rupp, Oct. 15, 1979, Washington, D.C.

Mary Eastwood, conducted by Verta Taylor and Leila J. Rupp, Nov. 26, 1982, Washington, D.C.

Ruth Gage-Colby (with Marian Parks), conducted by Verta Taylor and Leila J. Rupp, April 9, 1983, Daytona Beach, Florida.

Dorothy Haener, conducted by Lyn Goldfarb, Lydia Kleiner, and Christine Miller, Twentieth-Century Trade Union Women: Vehicle for Social Change, Institute of Labor and Industrial Relations, University of Michigan-Wayne State University.

Margaret A. Hickey, conducted by Ann Lever and Susan M. Hartmann, Feb. 14, 1974, Western Historical Manuscript Collection, University of Missouri—St. Louis.

Margaret Hickey, conducted by Verta Taylor and Leila J. Rupp, July 25, 1982, Santa Fe, New Mexico.

Millie Jeffrey, conducted by Ruth Meyerowitz, Twentieth-Century Trade Union Women, 1976.

Monika Kehoe, conducted by Verta Taylor and Leila J. Rupp, June 22, 1982, San Francisco, California.

Florence Luscomb, conducted by Brigid O'Farrell, Twentieth-Century Trade Union Women, 1976.

Burnita Shelton Matthews, "Pathfinder in the Legal Aspects of Women," conducted by Amelia R. Fry in 1973, Suffragists Oral History Project, Regional Oral History Office, University of California, 1975.

Pauli Murray, conducted by Verta Taylor and Leila J. Rupp, June 16, 1983, Baltimore, Maryland.

Pauline Newman, conducted by Barbara Wertheimer, Twentieth-Century Trade Union Women, 1976.

Alice Paul, "Conversations with Alice Paul: Woman Suffrage and the Equal Rights Amendment," conducted in 1972 and 1973 by Amelia R. Fry, Suffragists Oral History Project, Regional Oral History Office, University of California, 1976.

Esther Peterson, conducted by Martha Ross, Twentieth-Century Trade Union Women, 1977.

Esther Peterson, conducted by Emily Williams, April 26, 1979, for Franklin D. Roosevelt Library, in the Peterson papers, Schlesinger Library, Cambridge, Mass.

Florence Peterson, conducted by Ruth Meyerowitz, Twentieth-Century Trade Union Women, 1976.

E. S. Pollock, conducted by Verta Taylor and Leila J. Rupp, Dec. 10, 1979, Cincinnati, Ohio.

Ernestine Powell, conducted by Verta Taylor and Leila J. Rupp, Sept. 28 and Oct. 5, 1979, Columbus, Ohio.

Jeannette Rankin, "Activist for World Peace, Women's Rights, and Democratic Government," conducted by Malca Chall and Hannah Josephson in 1972, Suffragists Oral History Project, Regional Oral History Office, University of California, 1974.

Marguerite Rawalt, conducted by Verta Taylor and Leila J. Rupp, Oct. 15, 1979, Washington, D.C.

Helen Schleman, conducted by Mary Rhodes, Aug. 31, 1980, West Lafayette, Indiana.

Mabel Vernon, "Speaker for Suffrage and Petitioner for Peace," conducted in 1972 and 1973 by Amelia R. Fry, Suffragists Oral History Project, Regional Oral History Office, University of California, 1976.

Local interviews, conducted by Mary Irene Moffitt, July–September 1982, Columbus, Ohio:

Alice Arnett	Katherine Moore
Elizabeth Brownell	Helen Mulholland
Dorothy Cornelius	Mildred Munday
Lucille Curtis	Amalie Nelson
Eileen Evans	Marie Pfeiffer
Phyllis Greene	Mary Alice Price
Frances Harding	Jean Reilly
Martha Heintz	Dorothy Reynolds
Adenelle Hescott	Liz Rounds
Ethel Johnson	Peg Rosenfield
Kay Kauffman	Babette Sirak
Amy Lazarus	Anita Ward
Mary Lazarus	Myrrh Warken
Mary Ellen Ludlum	Juanita Webster
Mrs. Robert McClarren	Charlotte Witkind
Agnes Merritt	Florence Worrell
Mary Miller	

Periodicals

American Magazine	*Look*
American Mercury	*McCall's*
Collier's	*New York Times Magazine*
Coronet	*Newsweek*
Equal Rights	*Reader's Digest*
Good Housekeeping	*Saturday Evening Post*
Independent Woman	*Time*
Ladies' Home Journal	*Woman's Home Companion*
Life	

Other Published Primary Sources

Bernays, Doris Fleischman. *A Wife Is Many Women*. New York: Crown Publishers, 1955.

Committee on Un-American Activities, U.S. House of Representatives. *Report on the Congress of American Women*. Oct. 23, 1949.

Edwards, India. *Pulling No Punches: Memoir of a Woman in Politics*. New York: Putnam's, 1977.

Green, Arnold W. and Eleanor Melnick. "What Has Happened to the Feminist Movement?" In *Studies in Leadership: Leadership and Democratic Action*. Ed. Alvin W. Gouldner. New York: Russell and Russell, 1950, pp. 277–302.

Herschberger, Ruth. *Adam's Rib*. New York: Pellegrini & Cudahy, 1948.

Kass, Babette and Rose C. Feld. *The Economic Strength of Business and Professional Women*. New York: NFBPWC, 1954.

Kinsey, Alfred C. et al. *Sexual Behavior in the Human Female*. Philadelphia: W. B. Saunders, 1953.

Komarovsky, Mirra. *Women in the Modern World*. Boston: Little, Brown, 1953.

Lundberg, Ferdinand and Marynia F. Farnham. *Modern Woman: The Lost Sex*. New York: Harper and Brothers, 1947.

Mead, Margaret and Frances Bagley Kaplan. *American Women: Report of the President's Commission on the Status of Women and Other Publications of the Commission*. New York: Scribner's, 1965.

Mesta, Perle. *Perle: My Story*. New York: McGraw-Hill, 1960.

Meyer, Agnes E. *Out of These Roots: The Autobiography of an American Woman*. Boston: Little, Brown, 1953.

Sergio, Lisa. *A Measure Filled: The Life of Lena Madesin Phillips Drawn From Her Autobiography*. New York: Robert B. Luce, 1972.

Strecker, Edward A. *Their Mothers' Sons: The Psychiatrist Examines an American Problem*. Philadelphia: Lippincott, 1946.

White, Lynn, Jr. *Educating Our Daughters*. New York: Harper, 1950.

Wylie, Philip. *Generation of Vipers*. New York: Rinehart & Co., 1942.

Social Movement Literature

Aguirre, Benigno E. and E. L. Quarantelli. "Methodological, Ideological, and Conceptual-Theoretical Criticisms of the Field of Collective Behavior: A Critical Evaluation and Implications for Future Study." *Sociological Focus* 16 (1983), 195–216.

Aveni, Adrian F. "Organizational Linkages and Resource Mobilization: The Significance of Linkage Strength and Breadth." *Sociological Quarterly* 19 (1978), 185–202.

Banks, Olive. *Faces of Feminism: A Study of Feminism as a Social Movement.* Oxford: Martin Robertson, 1981.

Blumer, Herbert. "Social Movements." In *Principles of Sociology.* Ed. Alfred McClung Lee. New York: Barnes and Noble, 1951, pp. 199–220.

Bromley, David G. and Anson D. Shupe, Jr. *"Moonies" in America: Cult, Church, and Crusade.* Beverley Hills, Calif.: Sage Publications, 1979.

Bromley, David G. and Anson D. Shupe, Jr. "Repression and the Decline of Social Movements: The Case of the New Religions." In *Social Movements of the Sixties and Seventies.* Ed. Jo Freeman. New York: Longman, 1983, pp. 335–347.

Cassell, Joan. *A Group Called Women: Sisterhood and Symbolism in the Feminist Movement.* New York: David McKay, 1977.

Conover, Pamela Johnston and Virginia Gray. *Feminism and the New Right: Conflict over the American Family.* New York: Praeger, 1983.

Coser, Lewis A. *The Functions of Social Conflict.* Glencoe, Ill.: Free Press, 1956.

Curtis, Russell L., Jr. and Louis A. Zurcher, Jr. "Stable Resources of Protest Movements: The Multi-Organizational Field." *Social Forces* 52 (1973), 52–61.

Edwards, Lyford D. *The Natural History of Revolution.* Chicago: University of Chicago Press, 1927.

Ferree, Myra Marx and Beth B. Hess. *Controversy and Coalition: The New Feminist Movement.* Boston: Twayne, 1985.

Ferree, Myra Marx and Frederick Miller. "Mobilization and Meaning: Some Social-Psychological Contributions to the Resource Mobilization Perspective on Social Movements," *Sociological Inquiry* 55 (1985), 38–61.

Feuer, Lewis S. "The Conflict of Generations." In *Collective Behavior*, 2nd ed. Eds. Ralph H. Turner and Lewis M. Killian. Englewood Cliffs, N.J.: Prentice-Hall, 1972, pp. 372–379.

Freeman, Jo. "Crises and Conflicts in Social Movement Organizations," *Chrysalis* 5 (1977), 43–51.

Freeman, Jo. "On the Origins of Social Movements." In *Social Movements of the Sixties and Seventies.* Ed. Jo Freeman. New York: Longman, 1983, pp. 8–30.

Freeman, Jo. *The Politics of Women's Liberation.* New York: David McKay, 1975.

Freeman, Jo. "Resource Mobilization and Strategy: A Model for Analyzing Social Movement Organization Actions." In *The Dynamics of Social Movements.* Eds. Mayer N. Zald and John D. McCarthy. Cambridge, Mass.: Winthrop Publishers, 1979, pp. 167–189.

Freeman, Jo. *Social Movements of the Sixties and Seventies.* New York: Longman, 1983.

Gamson, William A. *The Strategy of Social Protest.* Homewood, Ill.: Dorsey, 1975.

Gerlach, Luther P. and Virginia H. Hine. *People, Power, Change: Movements of Social Transformation*. Indianapolis: Bobbs-Merrill, 1970.

Granovetter, Mark S. "The Strength of Weak Ties," *American Journal of Sociology* 78 (1973), 1360–1380.

Gross, Alan E. et al. "The Men's Movement: Personal Versus Political." In *Social Movements of the Sixties and Seventies*. Ed. Jo Freeman. New York: Longman, 1983, pp. 71–81.

Gusfield, Joseph R. "Social Movements and Social Change: Perspectives of Linearity and Fluidity." In *Research in Social Movements, Conflict and Change*. Ed. Louis Kriesberg. Greenwich, Conn.: JAI Press, 1981, vol. 4, pp. 317–339.

Hopper, Rex D. "The Revolutionary Process, *Social Forces* 28 (1950), 270–279.

Huber, Joan. "Ambiguities in Identity Transformation: From Sugar and Spice to Professor." In *Feminist Frontiers: Rethinking Sex, Gender, and Society*. Eds. Laurel Richardson and Verta Taylor. Reading, Mass.: Addison-Wesley, 1983, pp. 330–336.

Huber, Joan and Glenna Spitze. *Sex Stratification: Children, Housework, and Jobs*. New York: Academic Press, 1983.

Huber, Joan. "Toward a Socio-Technological Theory of the Women's Movement," *Social Forces* 23 (1976).

Jenkins, J. Craig and Craig M. Eckert. "Chanelling Black Insurgency: Elite Patronage and Professional SMOs in the Development of the Civil Rights Movement." Paper presented at the ASA meeting, 1986.

Jenkins, J. Craig and Charles Perrow. "Insurgency of the Powerless: Farm Worker Movements, 1946–1972," *American Sociological Review* 42 (1977), 249–258.

Jenkins, J. Craig. "Resource Mobilization Theory and the Study of Social Movements," *Annual Review of Sociology* 9 (1983), 527–553.

Kanter, Rosabeth Moss. *Commitment and Community: Communes and Utopias in Sociological Perspective*. Cambridge, Mass.: Harvard University Press, 1972.

Killian, Lewis M. "Organization, Rationality and Spontaneity in the Civil Rights Movement," *American Sociological Review* 49 (1984), 770–783.

Killian, Lewis M. "Social Movements." In *Handbook of Modern Sociology*. Ed. Robert E. L. Farris. Chicago: Rand-McNally, 1964, pp. 426–455.

Lang, Kurt and Gladys Engel Lang. *Collective Dynamics*. New York: Thomas Y. Crowell, 1961.

Lofland, John. *Doomsday Cult*. New York: Irvington Publishers, 1977.

Lofland, John. "White-Hot Mobilization: Strategies of a Millenarian Movement." In *The Dynamics of Social Movements*. Eds. Mayer N. Zald and John D. McCarthy. Cambridge, Mass.: Winthrop Publishers, 1979, pp. 157–166.

McAdam, Doug. *Political Process and the Development of Black Insurgency, 1930–1970*. Chicago: University of Chicago Press, 1982.

McCarthy, John D. and Mayer N. Zald. "Resource Mobilization and Social Movements: A Partial Theory," *American Journal of Sociology* 82 (1977), 1212–1241.

McCarthy, John D. and Mayer N. Zald. *The Trend of Social Movements in America: Professionalization and Resource Mobilization*. Morristown, N.J.: General Learning Press, 1973.

Marx, Gary T. and James L. Wood. "Strands of Theory and Research in Collective Behavior," *Annual Review of Sociology* 1 (1975), 363–428.

Miller, Frederick. "The End of SDS and the Emergence of Weatherman: Demise Through Success." In *Social Movements of the Sixties and Seventies*. Ed. Jo Freeman. New York: Longman, 1983, pp. 279–297.

Molotch, Harvey. "Media and Movements." In *The Dynamics of Social Movements*. Eds. Mayer N. Zald and John D. McCarthy. Cambridge, Mass.: Winthrop Publishers, 1979, pp. 71–93.

Morris, Aldon. *The Origins of the Civil Rights Movement: Black Communities Organizing for Change*. New York: The Free Press, 1984.

Oberschall, Anthony. "The Decline of the 1960's Social Movements." In *Research in Social Movements, Conflict and Change*. Ed. Louis Kriesberg. Greenwich, Conn.: JAI Press, 1978, vol. 1, pp. 257–289.

Oberschall, Anthony. *Social Conflict and Social Movements*. Englewood Cliffs, N.J.: Prentice-Hall, 1973.

Olson, M. *The Logic of Collective Action*. New York: Schocken, 1968.

Park, Robert E. and Ernest W. Burgess. *An Introduction to the Science of Sociology*. Greenwich, Conn.: JAI Press, 1978.

Perrow, Charles. "The Sixties Observed." In *The Dynamics of Social Movements*. Eds. Mayer N. Zald and John D. McCarthy. Cambridge, Mass.: Winthrop Publishers, 1979, pp. 192–211.

Perry, Joseph B. and M. D. Pugh. *Collective Behavior*. St. Paul, Minn.: West, 1978.

Piven, Frances Fox and R. A. Cloward. *Regulating the Poor*. New York: Pantheon, 1971.

Rosenthal, Naomi et al. "Social Movements and Network Analysis: A Case Study of Nineteenth-Century Women's Reform in New York State," *American Journal of Sociology* 90 (1985), 1022–1954.

Ross, Robert J. "Generational Change and Primary Groups in a Social Movement." In *Social Movements of the Sixties and Seventies*. Ed. Jo Freeman. New York: Longman, 1983, pp. 177–192.

Snyder, David and William R. Kelly. "Strategies for Investigating Violence and Social Change." In *The Dynamics of Social Movements*. Eds. Mayer N. Zald and John D. McCarthy. Cambridge, Mass.: Winthrop Publishers, 1979, pp. 212–237.

Stark, Rodney and William Sims Bainbridge. "Networks of Faith: Interpersonal Bonds and Recruitment to Cults and Sects," *American Journal of Sociology* 85 (1980), 1376–1395.

Stoper, Emily. "The Student Nonviolent Coordinating Committee: Rise and Fall of a Redemptive Organization." In *Social Movements of the Sixties and Seventies*. Ed. Jo Freeman. New York: Longman, 1983, pp. 320–334.

Taylor, Verta. "The Continuity of the American Women's Movement: An Elite-Sustained Stage." Paper presented at the American Sociological Association conference, New York, August 1986.

Taylor, Verta. "The Future of Feminism in the 1980s: A Social Movement Analysis." In *Feminist Frontiers: Rethinking Sex, Gender, and Society*. Eds. Laurel Richardson and Verta Taylor. Reading, Mass.: Addison-Wesley, 1983, pp. 434–451.

Tilly, Charles. *From Mobilization to Revolution*. Reading, Mass.: Addison-Wesley, 1978.

Tilly, Charles et al. *The Rebellious Century, 1830–1930*. Cambridge, Mass.: Harvard University Press, 1975.

Tilly, Charles. "Social Movements and National Politics." Working Paper No. 197. Ann Arbor: Center for Research on Social Organization, University of Michigan, 1979.

Turner, Ralph H. and Lewis M. Killian. *Collective Behavior*, 2nd ed. Englewood Cliffs, N.J.: Prentice-Hall, 1972.

Turner, Ralph H. "Collective Behavior and Resource Mobilization as Approaches to Social Movements: Issues and Continuity." In *Research in Social Movements, Conflict and Change*. Ed. Louis Kriesberg. Greenwich, Conn.: JAI Press, 1981, vol. 4, pp. 1–24.

Turner, Ralph. "Determinants of Social Movement Strategies." In *Human Nature and Collective Behavior*. Ed. Tamotsu Shibutani. Englewood Cliffs, N.J.: Prentice-Hall, 1970.

Turner, Ralph H. "The Public Perception of Protest," *American Sociological Review* 34 (1969), 815–831.

Useem, Bert and Mayer N. Zald. "From Pressure Group to Social Movement: Organizational Dilemmas of the Effort to Promote Nuclear Power," *Social Problems* 30 (1982), 144–156.

Wilkinson, Paul. *Social Movement*. New York: Praeger, 1971.

Wilson, John. *Introduction to Social Movements*. New York: Basic Books, 1973.

Wilson, John Q. *Political Organizations*. New York: Basic Books, 1973.

Zald, Mayer N. and John D. McCarthy. *The Dynamics of Social Movements*. Cambridge, Mass.: Winthrop Publishers, 1979.

Zald, Mayer N. and John D. McCarthy. "Social Movement Industries: Competition and Cooperation Among Movement Organizations." In *Research in Social Movements, Conflict and Change*. Ed. Louis Kriesberg. Greenwich, Conn.: JAI Press, 1980, vol. 3, pp. 1–20.

Zald, Mayer and Roberta Ash. "Social Movement Organizations: Growth, Decay and Change," *Social Forces* 44 (1966), 327–341.

Historical Literature

Aptheker, Bettina. *Woman's Legacy*. Amherst: University of Massachusetts Press, 1982.

Banks, Olive. *Faces of Feminism: A Study of Feminism as a Social Movement*. Oxford: Martin Robertson, 1981.

Berger, Caruthers Gholson. "Equal Pay, Equal Employment Opportunity and Equal Enforcement of the Law for Women," *Valparaiso University Law Review* 5 (1971), 326–373.

Becker, Susan D. *The Origins of the Equal Rights Amendment: American Feminism Between the Wars*. Westport, Conn.: Greenwood Press, 1981.

Bird, Caroline. *Born Female: The High Cost of Keeping Women Down*. New York: David McKay, 1968.

Brauer, Carl M. "Women Activists, Southern Conservatives, and the Prohibition of Sex Discrimination in Title VII of the 1964 Civil Rights Act," *Journal of Southern History* 49 (1983), 37–56.

Bremner, Robert H. "Families, Children, and the State." In *Reshaping America:*

Society and Institutions, 1945–1960. Eds. Robert H. Bremner and Gary Reichard. Columbus: Ohio State University Press, 1982, pp. 3–32.

Bullough, Vern and Bonnie Bullough. "Lesbianism in the 1920's and 1930's: A Newfound Study," *Signs* 2 (1977), 895–904.

Campbell, D'Ann. "Wives, Workers and Womanhood: America During World War II." Diss., University of North Carolina, 1979.

Campbell, D'Ann. *Women at War with America: Private Lives in a Patriotic Era.* Cambridge, Mass.: Harvard University Press, 1984.

Carter, Paul A. *Another Part of the Fifties* . New York: Columbia University Press, 1983.

Chafe, William H. *The American Woman: Her Changing Social, Economic, and Political Roles, 1920–1970* . New York: Oxford University Press, 1972.

Cook, Blanche Wiesen. "Female Support Networks and Political Activism: Lillian Wald, Crystal Eastman, Emma Goldman," *Chrysalis* 3 (1977), 43–61.

Cook, Blanche Wiesen. "The Historical Denial of Lesbianism," *Radical History Review* 20 (1979), 60–65.

Coover, Edwin Russell. "Status and Role Change Among Women in the U.S., 1940–1970." Diss., University of Minnesota, 1973.

Cott, Nancy F. *The Bonds of Womanhood*. New Haven, Conn.: Yale University Press, 1977.

Cott, Nancy F. "Feminist Politics in the 1920s: The National Woman's Party," *Journal of American History* 71 (1984), 43–68.

Cuniberti, Betty. "Young, Old Feminists Split," *Los Angeles Times*, Oct. 6, 1981.

Daly, Mary. *Gyn/Ecology: The Metaethics of Radical Feminism*. Boston: Beacon Press, 1978.

Davis, Angela. *Women, Race and Class*. New York: Random House, 1981.

Davis, Madeline, Liz Kennedy, and Avra Michelson. "Aspects of the Buffalo Lesbian Community in the Fifties." Unpublished paper presented at the National Women's Studies Association conference, Bloomington, Indiana, 1980.

Deckard, Barbara Sinclair. *The Women's Movement*. 2nd ed. New York: Harper & Row, 1979.

D'Emilio, John. "The Homosexual Menace: The Politics of Sexuality in Cold War America." Unpublished paper presented at the Organization of American Historians conference, Philadelphia, 1982.

D'Emilio, John. *Sexual Politics, Sexual Communities: The Making of a Homosexual Minority in the U.S., 1940–1970*. Chicago: University of Chicago Press, 1983.

Dennis, Peggy. *The Autobiography of an American Communist: A Personal View of a Political Life, 1925–1975*. Westport, Conn./Berkeley, Calif.: Lawrence Hill & Co., 1977.

Dennis, Peggy. "Response" [to Ellen Kay Trimberger], *Feminist Studies* 5 (1979), 451–460.

Douvan, Elizabeth. "Differing Views on Marriage—1957 to 1976," University of Michigan Center for Continuing Education of Women *Newsletter* 12 (1979), 1–2.

DuBois, Ellen Carol. *Feminism and Suffrage: The Emergence of an Independent Women's Movement in America, 1848–1869*. Ithaca, N.Y.: Cornell University Press, 1978.

Erskine, Hazel. "The Polls: Women's Role," *Public Opinion Quarterly* 35 (1971), 282–287.

Evans, Sara. *Personal Politics: The Roots of Women's Liberation in the Civil Rights Movement and the New Left*. New York: Knopf, 1979.

Ewen, Stuart. *Captains of Consciousness: Advertising and the Social Roots of the Consumer Culture*. New York: McGraw-Hill, 1976.

Faber, Doris. *The Life of Lorena Hickok: E.R.'s Friend*. New York: William Morrow, 1980.

Faderman, Lillian. *Surpassing the Love of Men*. New York: William Morrow, 1981.

Ferree, Myra Marx and Beth B. Hess. *Controversy and Coalition: The New Feminist Movement*. Boston: Twayne, 1985.

Ferriss, Abbot L. *Indicators of Trends in the Status of American Women*. New York: Russell Sage, 1971.

Flexner, Eleanor. *Century of Struggle*. Rev. ed. Cambridge, Mass.: Belknap Press, 1975.

Flynn, Elizabeth Gurley. *The Rebel Girl: An Autobiography*. New York: International Publishers, 1973.

Freedman, Estelle. "Separatism as Strategy: Female Institution Building and American Feminism, 1870–1930," *Feminist Studies* 5 (1979), 512–529.

Freeman, Jo. *The Politics of Women's Liberation*. New York: David McKay, 1975.

Friedan, Betty. *It Changed My Life*. New York: Dell, 1977.

Friedan, Betty. *The Feminine Mystique*. New York: Dell, 1963.

Friedan, Betty. "Up From the Kitchen Floor," *New York Times Magazine*, March 4, 1973, p. 9.

Fry, Amelia R. "Alice Paul and the South." Unpublished paper presented at the Southern Historical Association conference, 1981.

Fry, Amelia R. "The Divine Discontent: Alice Paul and Militancy in the Suffrage Campaign." Unpublished paper presented at the Berkshire Conference of Women Historians, 1981.

Fry, Amelia R. "Suffragist Alice Paul's Memoirs: The Pros and Cons of Oral History," *Frontiers* 2 (1977), 65–69.

Gabin, Nancy. "Female Activism on Gender Issues in the UAW-CIO." Unpublished paper presented at the North American Labor History Conference, Detroit, 1985.

Gabin, Nancy. " 'They Have Placed a Penalty on Womanhood:' The Protest Actions of Women Auto Workers in Detroit-Area UAW Locals, 1945–1947," *Feminist Studies* 8 (1982), 373–398.

Gabin, Nancy. "Women Workers and the UAW in the Post-World War II Period: 1945–1954," *Labor History* 21 (1979–80), 5–30.

Gans, Herbert J. *The Levittowners*. New York: Random House, 1967.

Geidel, Peter. "The National Woman's Party and the Origins of the Equal Rights Amendment, 1920–1923," *Historian* 42 (1980), 557–582.

Geise, L. Ann. "The Female Role in Middle Class Women's Magazines from 1955 to 1976: A Content Analysis of Nonfiction Selections," *Sex Roles* 5 (1979), 51–62.

Giddings, Paula. *When and Where I Enter: The Impact of Black Women on Race and Sex in America*. New York: Bantam, 1984.

Goldfarb, Lyn. *Separated and Unequal: Discrimination Against Women Workers After World War II*. Washington, D.C.: Union for Radical Political Economics Political Education Project, n.d.

Goldman, Eric. *The Crucial Decade—And After: America, 1945–1960*. New York: Knopf, 1960.

Gorman, Phyllis. "An Analysis of the Social Movement Organization the Daughters of Bilitis, 1955–1963." M.A. Thesis, Ohio State University, 1985.

Griffith, Robert. "Harry S. Truman and the Burden of Modernity," *Reviews in American History* 9 (1981), 295–306.

Griffith, Robert and Athan Theoharis. *The Specter: Original Essays on the Cold War and the Origins of McCarthyism*. New York: New Viewpoints, 1974.

Hamby, Alonzo L. *Beyond the New Deal: Harry S. Truman and American Liberalism*. New York: Columbia University Press, 1973.

Harrington, Michael. *The Other America*. New York: Macmillan, 1962.

Harrison, Cynthia E. "A 'New Frontier' for Women: The Public Policy of the Kennedy Administration," *Journal of American History* 67 (1980), 630–646.

Harrison, Cynthia E. "Prelude to Feminism: Women's Organizations, the Federal Government and the Rise of the Women's Movement, 1942 to 1968." Diss., Columbia University, 1982.

Hart, Jeffrey. *When the Going Was Good! American Life in the Fifties*. New York: Crown Publishers, 1982.

Hartmann, Susan M. *The Home Front and Beyond: American Women in the 1940's*. Boston: Twayne, 1982.

Hartmann, Susan M. "Women's Organizations During World War II: The Interaction of Class, Race and Feminism." In *Woman's Being, Woman's Place*. Ed. Mary Kelley. Boston: G. K. Hall, 1979, pp. 313–328.

Hole, Judith and Ellen Levine. *Rebirth of Feminism*. New York: Quadrangle/The New York Times Book Co., 1971.

Huber, Joan and Glenna Spitze. *Sex Stratification: Children, Housework, and Jobs*. New York: Academic Press, 1983.

Irons, Peter H. "American Business and the Origins of McCarthyism: The Cold War Crusade of the United States Chamber of Commerce." In *The Specter: Original Essays on the Cold War and the Origins of McCarthyism*. Eds. Robert Griffith and Alan Theoharis. New York: New Viewpoints, 1974.

Irwin, Inez Haynes. *The Story of Alice Paul and the National Woman's Party*. 1921; reprinted, Fairfax, Va.: Denlinger's Publishers, 1977.

Jezer, Marty. *The Dark Ages: Life in the U.S., 1945–1960*. Boston: South End Press, 1982.

Jones, Beverly Washington. "Quest for Equality: The Life of Mary Eliza Church Terrell, 1863–1954." Diss., University of North Carolina, 1980.

Kaledin, Eugenia. *Mothers and More: American Women in the 1950's*. Boston: Twayne, 1984.

Kellogg, Peter J. "Civil Rights Consciousness in the 1940's," *Historian* 42 (1979), 18–41.

Kinsey, Alfred C. et al. *Sexual Behavior in the Human Female*. Philadelphia: W. B. Saunders Co., 1953.

Klein, Ethel. *Gender Politics: From Consciousness to Mass Politics*. Cambridge, Mass.: Harvard University Press, 1984.

Klein, Ethel. "A Social Learning Perspective on Political Mobilization: Why the Women's Movement Happened When It Did." Diss., University of Michigan, 1979.

Komarovsky, Mirra. "Women Then and Now: A Journey of Detachment and Engagement," *Women's Studies Quarterly* 10 (1982), 5–9.

Kraditor, Aileen. *The Ideas of the Woman Suffrage Movement, 1890–1920*. New York: Columbia University Press, 1965.

Lane, Ann J. *Mary Ritter Beard: A Sourcebook*. New York: Schocken Books, 1977.

Lash, Joseph P. *Eleanor: The Years Alone*. New York: New American Library, 1972.

Lemons, J. Stanley. *The Woman Citizen: Social Feminism in the 1920's*. Urbana: University of Illinois Press, 1973.

Leone, Janice. "Degrees of Color: The Integration of the American Association of University Women, 1946–1949." Unpublished paper, Ohio State University, 1985.

Lewis, Sasha Gregory. *Sunday's Women: A Report on Lesbian Life Today*. Boston: Beacon Press, 1979.

Lochridge, Patricia. "The Mother Racket," *Woman's Home Companion* 71 (July 1944), 20–21.

McGowan, Barbara. "Postwar Attitudes Toward Women and Work." In *New Research on Women and Sex Roles*. Ed. Dorothy McGuigan. Ann Arbor: University of Michigan Center for Continuing Education of Women, 1976.

Meyerowitz, Ruth. "Women Unionists and World War II: New Opportunities for Leadership." Unpublished paper presented at the Organization of American Historians conference, San Francisco, 1980.

Miller, Douglas T. and Marion Nowak. *The Fifties: The Way We Really Were*. Garden City, N.Y.: Doubleday, 1977.

Miller, Robert Stevens, Jr. "Sex Discrimination and Title VII of the Civil Rights Act of 1964," *Minnesota Law Review* 51 (1967), 877–897.

Mitford, Jessica. *A Fine Old Conflict*. New York: Vintage Books, 1977.

Morain, Thomas Jeffrey. "The Emergence of the Women's Movement, 1960–1970." Diss., University of Iowa, 1974.

Nelson, Marjory. "Ladies in the Streets: A Sociological Analysis of the National Woman's Party, 1910–1930." Diss., SUNY-Buffalo, 1976.

Ondercin, David G. "The Compleat Woman: The Equal Rights Amendment and Perceptions of Womanhood, 1920–1972." Diss., University of Minnesota, 1973.

O'Neill, William. *Everyone Was Brave*. Chicago: Quadrangle Press, 1971.

Oppenheimer, Valerie Kincade. "Demographic Influence on Female Employment and the Status of Women." In *Changing Women in a Changing Society*. Ed. Joan Huber. Chicago: University of Chicago Press, 1973, pp. 184–199.

Oppenheimer, Valerie Kincade. *The Female Labor Force in the U.S.: Demographic and Economic Factors Governing its Growth and Changing Composition*. Population Monograph Series, no. 5. Berkeley: Institute of International Studies, University of California, 1970.

Pardo, Thomas C. *The National Woman's Party Papers, 1913–1974: A Guide to the Microfilm Edition*. Sanford, N.C.: Microfilming Corporation of America, 1979.

Patterson, Cynthia M. "New Directions in the Political History of Women: A Case Study of the National Woman's Party's Campaign for the Equal Rights Amendment, 1920–1927," *Women's Studies International Forum* 5 (1982), 585–597.

Polenberg, Richard. *One Nation Divisible: Class, Race, and Ethnicity in the U.S. Since 1938.* New York: Penguin Books, 1980.

Randall, Susan L. "A Legislative History of the ERA, 1923–1960." Diss., University of Utah, 1979.

Rawalt, Marguerite. *History of the National Federation of Business and Professional Women's Clubs.* Vol. 2. Washington, D.C.: NFBPWC, 1969.

Reinartz, Kay F. "The Paper Doll: Images of American Women in Popular Song." In *Women: A Feminist Perspective.* Ed. Jo Freeman. Palo Alto, Calif.: Mayfield Publishing Co., 1975, pp. 293–308.

Rich, Adrienne. " 'Disloyal to Civilization:' Feminism, Racism, and Gynephobia," *Chrysalis* 7 (1979), 9–27.

Rossi, Alice. *The Feminist Papers.* New York: Bantam, 1973.

Rupp, Leila J. " 'Imagine My Surprise:' Women's Relationships in Historical Perspective," *Frontiers* 5 (1980), 61–70.

Rupp, Leila J. "The Survival of American Feminism: The Women's Movement in the Postwar Period." In *Reshaping America: Society and Institutions, 1945–1960.* Eds. Robert H. Bremner and Gary W. Reichard. Columbus: Ohio State University Press, 1982, pp. 33–65.

Rupp, Leila J. "The Women's Community in the National Woman's Party, 1945 to the 1960s," *Signs* 10 (1985), 715–740.

Rupp, Leila J. and Verta Taylor. "The Women's Movement Since 1960: Structure, Strategies, and New Directions." In *American Choices: Social Dilemmas and Public Policy Since 1960.* Eds. Robert H. Bremner, Gary W. Reichard, Richard J. Hopkins. Columbus: Ohio State University Press, 1986, pp. 75–104.

Sahli, Nancy. "Smashing: Women's Relationships Before the Fall," *Chrysalis* 8 (1979), 17–27.

Scharf, Lois. *To Work and To Wed: Female Employment, Feminism and the Great Depression.* Westport, Conn.: Greenwood Press, 1980.

Scharf, Lois and Joan M. Jensen. *Decades of Discontent: The Women's Movement, 1920–1940.* Westport, Conn.: Greenwood Press, 1983.

Sealander, Judith Anne. "The Women's Bureau, 1920–1950: Federal Reaction to Female Wage Earning." Diss., Duke University, 1977.

Sicherman, Barbara and Carol Hurd Green. *Notable American Women: The Modern Period.* Cambridge, Mass.: Belknap Press, 1980.

Sitkoff, Harvard. "American Blacks in World War II: Rethinking the Militancy-Watershed Hypothesis." In *The Home Front and War in the Twentieth Century: The American Experience in Perspective.* Ed. James Titus. Washington, D.C.: U.S. Government Printing Office, 1984, pp. 147–155.

Sitkoff, Harvard. *The Struggle for Black Equality, 1954–1980.* New York: Hill and Wang, 1981.

Skold, Karen Beck. "The Job He Left Behind: American Women in the Shipyards During World War II." In *Women, War, and Revolution.* Eds. Carol R. Berkin and Clara M. Lovett. New York: Holmes and Meier, 1980, pp. 55–75.

Smith-Rosenberg, Carroll. "The Female World of Love and Ritual: Relations Between Women in Nineteenth-Century America," *Signs* 1 (1975), 1–29.

Sochen, June. *Movers and Shakers: American Women Thinkers and Activists, 1900–1970.* New York: Quadrangle/The New York Times Book Co., 1973.

Spender, Dale. *There's Always Been a Women's Movement This Century.* London. Pandora Press, 1983.

Stevens, Doris. *Jailed for Freedom.* 1920; reprinted New York: Schocken Books, 1976.

Sudler, Elizabeth. "Reclaiming Our Heritage: A Trip to the NWP Convention," *Congressional Union* 1 (1981), 4.

Taylor, Verta. "The Continuity of the American Women's Movement: An Elite-Sustained Stage." Paper presented at the American Sociological Association conference, New York, August 1986.

Taylor, Verta and Leila J. Rupp. "Researching the Women's Movement: We Make Our Own History, But Not Just As We Please." Unpublished paper, 1985.

Terborg-Penn, Rosalyn. "Discontented Black Feminists: Prelude and Postscript to the Passage of the Nineteenth Amendment." In *Decades of Discontent: The Women's Movement, 1920–1940.* Eds. Lois Scharf and Joan M. Jensen. Westport, Conn.: Greenwood Press, 1983, pp. 261–278.

Terborg-Penn, Rosalyn. "Discrimination Against Afro-American Women in the Woman's Movement, 1830–1920." In *The Afro-American Woman.* Eds. Sharon Harley and Rosalyn Terborg-Penn. Port Washington, N.Y.: Kennikat Press, 1978, pp. 17–27.

Tobias, Sheila and Lisa Anderson. *What Really Happened to Rosie the Riveter.* MSS Modular Publication, Module 9, 1974.

Trimberger, Ellen Kay. "Women in the Old and New Left: The Evolution of a Politics of Personal Life," *Feminist Studies* 5 (1979), 432–450.

Waite, Linda J. "Working Wives, 1940–1960," *American Sociological Review* 41 (1976), 65–80.

Ware, Susan. *Beyond Suffrage: Women in the New Deal.* Cambridge, Mass.: Harvard University Press, 1981.

Weibel, Kathryn. *Mirror, Mirror: Images of Women Reflected in Popular Culture.* Garden City, N.Y.: Doubleday, 1977.

Wells, Anna Mary. *Miss Marks and Miss Woolley.* Boston: Houghton Mifflin, 1978.

Welsch, Janice. "Actress Archetypes in the 1950's: Doris Day, Marilyn Monroe, Elizabeth Taylor, Audrey Hepburn." In *Women and the Cinema.* Eds. Karyn Kay and Gerald Peary. New York: Dutton, 1977, pp. 99–111.

Welter, Barbara. "The Cult of True Womanhood, 1820–1860," *American Quarterly* 18 (1966), 151–174.

Willis, Jean L. "Alice Paul: The Quintessential Feminist." In *Feminist Theorists.* Ed. Dale Spencer. New York: Pantheon Books, 1983.

Zelman, Patricia G. *Women, Work, and National Policy: The Kennedy-Johnson Years.* Ann Arbor: UMI Research Press, 1982.

Index

282INDEX